ALLEN COUNTY PUBLIC LIBRARY

D0742221

Sensors and Transducers

Sensors and Transducers

Third edition

Ian R. Sinclair

OXFORD AUCKLAND BOSTON JOHANNESBURG MELBOURNE NEW DELHI

Newnes
An imprint of Butterworth-Heinemann
Linacre House, Jordan Hill, Oxford OX2 8DP
225 Wildwood Avenue, Woburn, MA 01801-2041
A division of Reed Educational and Professional Publishing Ltd

A member of the Reed Elsevier plc group

First published by BSP Professional Books 1988
Reprinted by Butterworth-Heinemann 1991
Second edition published by Butterworth-Heinemann 1992
Third edition 2001

British Library Cataloguing in Publication Data
A catalogue record for this book is available from the British Library

ISBN 0 7506 4932 1

Typeset by David Gregson Associates, Beccles, Suffolk
Printed and bound in Great Britain by Biddles Ltd, *www.biddles.co.uk*

Contents

Preface to Third Edition

This third edition of *Sensors and Transducers* has been thoroughly revised to take account of the ever-increasing role of these components and of improvements in design. New tables of properties and illustrations have also been added. The topic of switches and switching actions has also been added because so many types of sensor are intended ultimately to provide a switching action.

Ian Sinclair

Preface to First Edition

The purpose of this book is to explain and illustrate the use of sensors and transducers associated with electronic circuits. The steady spread of electronic circuits into all aspects of life, but particularly into all aspects of control technology, has greatly increased the importance of sensors which can detect, as electrical signals, changes in various physical quantities. In addition, the conversion by transducers of physical quantities into electronic signals and vice versa has become an important part of electronics.

Because of this, the range of possible sensors and transducers is by now very large, and most textbooks that are concerned with the interfaces between electronic circuits and other devices tend to deal only with a few types of sensors for specific purposes. In this book, you will find described a very large range of devices, some used industrially, some domestically, some employed in teaching to illustrate effects, some used only in research laboratories. The important point is that the reader will find reference to a very wide range of devices, much more than it would be possible to present in a more specialized text.

In addition, I have assumed that the physical principles of each sensor or transducer will not necessarily be familiar. To be useful, a book of this kind should be accessible to a wide range of users, and since the correct use of sensors and transducers often depends critically on an understanding of the physical principles involved, these principles have been explained in as much depth as is needed. I have made the reasonable assumption that electrical principles will not be required to be explained in such depth as the principles of, for example, relative humidity. In order for the book to be as serviceable as possible to as many readers as possible, the use of mathematics has been avoided unless absolutely essential to the understanding of a device. I have taken here as my guide the remark by Lord Kelvin that if he needed to use mathematics to explain something it was probably because he didn't really understand it. The text should prove useful to anyone who encounters sensors and transducers, whether from the point of view of specification, design, servicing, or education.

I am most grateful to RS Components for much useful and well-organized information, and to Bernard Watson, of BSP Professional Books, for advice and encouragement.

Ian Sinclair
April 1988

Introduction

A *sensor* is a device that detects or measures a physical quantity, and in this book the types of sensors that we are concerned with are the types whose output is electrical. The opposite device is an *actuator*, which converts a signal (usually electrical) to some action, usually mechanical. A *transducer* is a device that converts energy from one form into another, and here we are concerned only with the transducers in which one form of energy is electrical. Actuators and sensors are therefore forms of transducers, and in this book we shall deal with actuators under the heading of transducers.

The differences between sensors and transducers are often very slight. A sensor performs a transducing action, and the transducer must necessarily sense some physical quantity. The difference lies in the efficiency of energy conversion. The purpose of a sensor is to detect and measure, and whether its efficiency is 5% or 0.1% is almost immaterial, provided the figure is known. A transducer, by contrast, is intended to convert energy, and its efficiency is important, though in some cases it may not be high. Linearity of response, defined by plotting the output against the input, is likely to be important for a sensor, but of much less significance for a transducer. By contrast, efficiency of conversion is important for a transducer but not for a sensor. The basic principles that apply to one, however, must apply to the other, so that the descriptions that appear in this book will apply equally to sensors and to transducers.

- Switches appear in this book both as transducers/sensors in their own right, since any electrical switch is a mechanical–electrical transducer, and also because switch action is such an important part of the action of many types of sensors and transducers.

Classification of sensors is conventionally by the conversion principle, the quantity being measured, the technology used, or the application. The organization of this book is, in general, by the physical quantity that is sensed or converted. This is not a perfect form of organization, but no form is, because there are many 'one-off' devices that sense or convert for some unique purpose, and these have to be gathered together in an 'assortment' chapter. Nevertheless, by grouping devices according to the sensed quantity, it is much easier for the reader to find the information that is needed, and that is the guiding principle for this book. In addition, some of the devices that are dealt with early in the book are those which form part of other sensing or transducing systems that appear later. This avoids having to repeat a description, or refer forward for a description.

Among the types of energy that can be sensed are those classed as radiant, mechanical, gravitational, electrical, thermal, and magnetic. If we consider the large number of principles that can be used in the design of sensors and transducers, some 350 to date, it is obvious that not all are of equal importance. By limiting the scope of this book to sensors and transducers with electrical/electronic inputs or outputs of the six forms listed above, we can reduce this number to a more manageable level.

Several points should be noted at this stage, to avoid much tedious repetition in the main body of the book. One is that a fair number of physical effects are sensed or measured, but have no requirement for transducers – we do not, for example, generate electricity from earthquake shocks though we certainly want to sense them. A second point is that the output from a sensor, including the output from electronic circuits connected to the sensor, needs to be proportional in some way to the effect that is being sensed, or at least to bear some simple mathematical relationship to the quantity. This means that if the output is to be used for measurements, then some form of calibration can be carried out. It also implies that the equation that connects the electrical output with the input that is being sensed contains various constants such as mass, length, resistance and so on. If any of these quantities is varied at any time, then recalibration of the equipment will be necessary.

Sensors can be classed as active or passive. An active or self-generating sensor is one that can generate a signal without the need for any external power supply. Examples include photovoltaic cells, thermocouples and piezoelectric devices. The more common passive sensors need an external source of energy, which for the devices featured in this book will be electrical. These operate by modulating the voltage or current of a supply. Another class of passive sensors, sometimes called modifiers, use the same type of energy at the output as at the input. Typical of these types is a diaphragm used to convert the pressure or velocity oscillations of sound waves into movements of a solid sheet.

Another point that we need to be clear about is the meaning of *resolution* as applied to a sensor. The resolution of a sensor measures its ability to detect a change in the sensed quantity, and is usually quoted in terms of the smallest change that can be detected. In some cases, resolution is virtually infinite, meaning that a small change in the sensed quantity will cause a small change in the electrical output, and these changes can be detected to the limits of our measuring capabilities. For other sensors, particularly when digital methods are used, there is a definite limit to the size of change that can be either detected or converted.

It is important to note that very few sensing methods provide a digital output directly, and most digital outputs are obtained by converting from analogue quantities. This implies that the limits of resolution are determined by the analogue to digital conversion circuits rather than by the sensor itself. Where a choice of sensing methods exists, a method that causes a change of frequency of an oscillator is to be preferred. This is because frequency is a quantity that lends itself very easily to digital handling methods with no need for other analogue to digital conversion methods.

The sensing of any quantity is liable to error, and the errors can be static or dynamic. A static error is the type of error that is caused by reading problems, such as the parallax of a needle on a meter scale, which causes the apparent reading to vary according to the position of the observer's eye. Another error of this type is the interpolation error, which arises when a needle is positioned between two marks on a scale, and the user has to make a guess as to the amount signified by this position. The amount of an interpolation error is least when the scale is linear. One distinct advantage of digital readouts is that neither parallax nor interpolation errors exist, though this should not be taken to mean that errors corresponding to interpolation errors are not present. For example, if a digital display operates to three places of decimals, the user has no way of knowing if a reading should be 1.2255 because this will be shown as 1.225, and a slight increase in the measured quantity will change the reading to 1.226.

The other form of error is dynamic, and a typical error of this type is a difference between the quantity as it really is and the amount that is measured, caused by the loading of the measuring instrument itself. A familiar example of this is the false voltage reading measured across a high-resistance potential divider with a voltmeter whose input resistance is not high enough. All forms of sensors are liable to dynamic errors if they are used only for sensing, and to both dynamic and static errors if they are used for measurement.

Since the development of microprocessors, a new breed of sensors has been developed, termed *intelligent* or *smart* sensors. This type of system uses a miniature sensor that is integrated on a single chip with a processor. Strictly speaking, this is a monolithic integrated sensor to distinguish it from the hybrid type in which the sensor and the processor are fabricated on the same substrate but not on the same chip. This book is concerned mainly with sensor and transducer principles rather than with the details of signal processing. The advantages of such integration methods include:

- Improved signal-to-noise ratio
- improved linearity and frequency response
- improved reliability.

Finally, two measurable quantities can be quoted in connection with any sensor or transducer. These are responsivity and detectivity, and although the names are not necessarily used by the manufacturer of any given device, the figures are normally quoted in one form or another. The responsivity is:

$$\frac{\text{output signal}}{\text{input signal}}$$

which will be a measure of transducing efficiency if the two signals are in comparable units (both in watts, for example), but which is normally expressed with very different units for the two signals. The detectivity is defined as:

$$\frac{\text{S/N of output signal}}{\text{size of output signal}}$$

where S/N has its usual electrical meaning of signal to noise ratio. This latter definition can be reworked as:

$$\frac{\text{responsivity}}{\text{output noise signal}}$$

if this makes it easier to measure.

Chapter 1

Strain and pressure

1.1 Mechanical strain

The words stress and strain are often confused in everyday life, and a clear definition is essential at this point. Strain is the result of stress, and is defined as the fractional change of the dimensions of an object. By fractional change, I mean that the change of dimension is divided by the original dimension, so that in terms of length, for example, the strain is the change of length divided by the original length. This is a quantity that is a pure number, one length divided by another, having no physical dimensions. Strain can be defined for area or for volume measurements in a similar way as change divided by original quantity. For example, area strain is change of area divided by original area, and volume strain is change of volume divided by original volume.

A stress, by contrast, is a force divided by an area. As applied to a wire or a bar in tension or compression, for example, the *tensile* (pulling) stress is the applied force divided by the area over which it is applied, which will be the area of cross section of the wire or bar. For materials such as liquids or gases which can be compressed uniformly in all dimensions, the *bulk* stress is the force per unit area, which is identical to the pressure applied, and the strain is the change of volume divided by the original volume. The most common strain transducers are for tensile mechanical strain. The measurement of strain allows the amount of stress to be calculated through a knowledge of the elastic modulus. The definition of any type of *elastic modulus* is stress/strain (which has the units of stress, since strain has no physical units), and the most commonly used elastic moduli are the linear Young's modulus, the shear (twisting) modulus, and the (pressure) bulk modulus.

For small amounts of strain, the strain is proportional to stress, and an elastic modulus is a quantity that expresses the ratio stress/strain in the

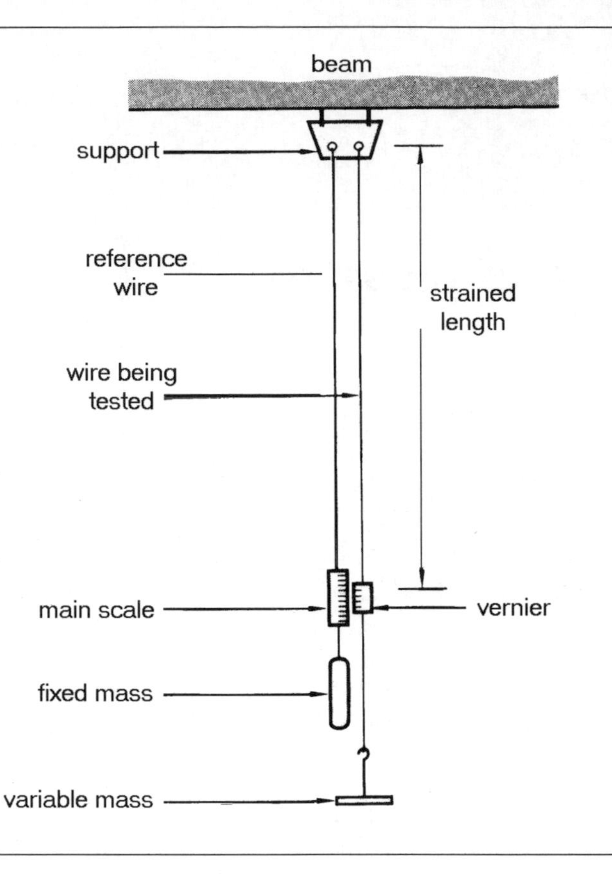

Figure 1.1 The classic method of measuring tensile stress and strain for a wire.

elastic region, i.e. the portion of the stress–strain graph that is linear. For example, Young's modulus is the ratio tensile stress/tensile strain, typically measured for a material in the form of a wire (Figure 1.1). The classic form of measurement, still used in school demonstrations, uses a long pair of wires, one loaded, the other carrying a vernier scale.

Sensing tensile strain involves the measurement of very small changes of length of a sample. This is complicated by the effect of changes of temperature, which produce expansion or contraction. For the changes of around 0–30°C that we encounter in atmospheric temperature, the expansion or contraction of length will be about the same size as the changes caused by large amounts of stress. Any system for sensing and measuring strain must therefore be designed in such a way that temperature effects can be compensated for. The principles used to sense linear or area strain are piezoresistive and piezoelectric.

The commonest form of strain measurement uses resistive strain gauges. A resistive strain gauge consists of a conducting material in the form of a

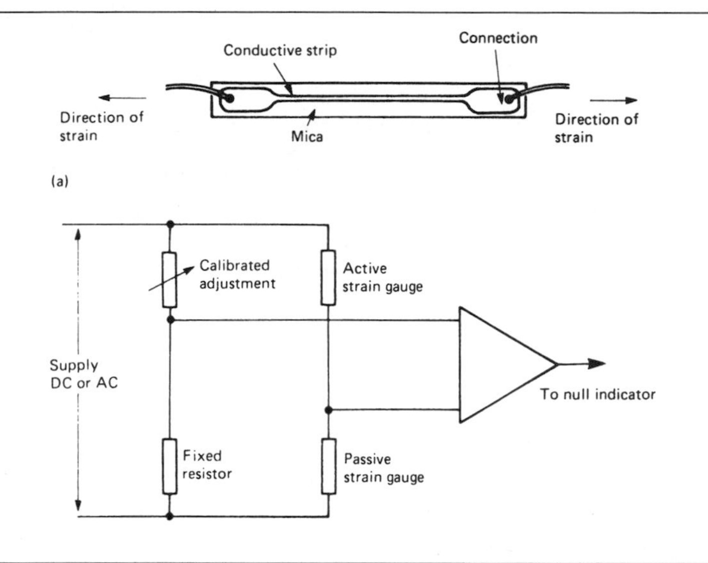

Figure 1.2 Strain gauge use. (a) Physical form of a strain gauge. (b) A bridge circuit for strain gauge use. By using an active (strained) and a passive (unstrained) gauge in one arm of the bridge, temperature effects can be compensated if both gauges are identically affected by temperature. The two gauges are usually side by side but with only one fastened to the strained surface.

thin wire or strip which is attached firmly to the material in which strain is to be detected. This material might be the wall of a building, a turbine blade, part of a bridge, anything in which excessive stress could signal impending trouble. The fastening of the resistive material is usually by means of epoxy resins (such as 'Araldite'), since these materials are extremely strong and are electrical insulators. The strain gauge strip will then be connected as part of a resistance bridge circuit (Figure 1.2). This is an example of the piezoresistive principle, because the change of resistance is due to the deformation of the crystal structure of the material used for sensing.

The effects of temperature can be minimized by using another identical unstrained strain gauge in the bridge as a comparison. This is necessary not only because the material under investigation will change dimensions as a result of temperature changes, but because the resistance of the strain gauge element itself will vary. By using two identical gauges, one unstrained, in the bridge circuit, these changes can be balanced against each other, leaving only the change that is due to strain. The sensitivity of this type of gauge, often called the *piezoresistive* gauge, is measured in terms of the *gauge factor*. This is defined as the fractional change of resistance divided by the change of strain, and is typically about 2 for a metal wire gauge and about 100 for a semiconductor type.

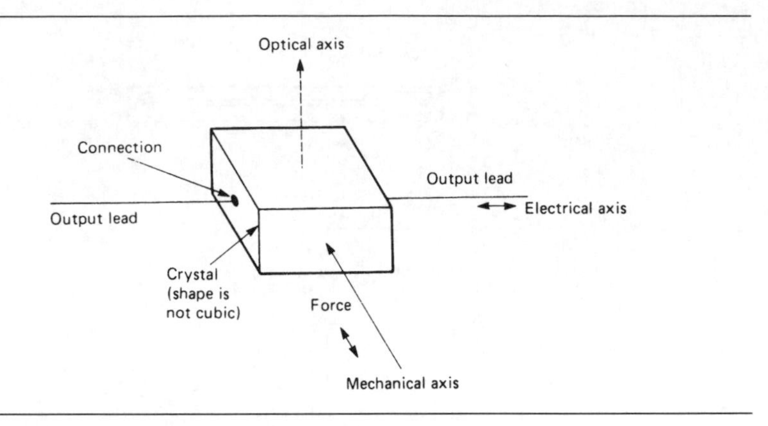

Figure 1.3 Piezoelectric crystal principles. The crystal shape is not cubic, but the directions of the effects are most easily shown on a cube. The maximum electric effect is obtained across faces whose directions are at right angles to the faces on which the force is applied. The third axis is called the optical axis because light passing through the crystal in this direction will be most strongly affected by polarization (see Chapter 3).

The change of resistance of a gauge constructed using conventional wire elements (typically thin Nichrome wire) will be very small, as the gauge factor figures above indicate. Since the resistance of a wire is proportional to its length, the fractional change of resistance will be equal to the fractional change of length, so that changes of less than 0.1% need to be detected. Since the resistance of the wire element is small, i.e., of the order of an ohm or less, the actual change of resistance is likely to be very small compared to the resistance of connections in the circuit, and this can make measurements very uncertain when small strains have to be measured.

The use of a semiconductor strip in place of a metal wire makes measurement much easier, because the resistance of such a strip can be considerably greater, and so the changes in resistance can be correspondingly greater. Except for applications in which the temperature of the element is high (for example, gas-turbine blades), the semiconductor type of strain gauge is preferred. Fastening is as for the metal type, and the semiconductor material is surface passivated – protected from atmospheric contamination by a layer of oxidation on the surface. This latter point can be important, because if the atmosphere around the gauge element removes the oxide layer, then the readings of the gauge will be affected by chemical factors as well as by strain, and measurements will no longer be reliable.

Piezoelectric strain gauges are useful where the strain is of short duration, or rapidly changing in value. A piezoelectric material is a crystal whose ions move in an asymmetrical way when the crystal is strained, so that an EMF is generated between two faces of the crystal (Figure 1.3). The EMF can be very large, of the order of several kV for a heavily strained crystal,

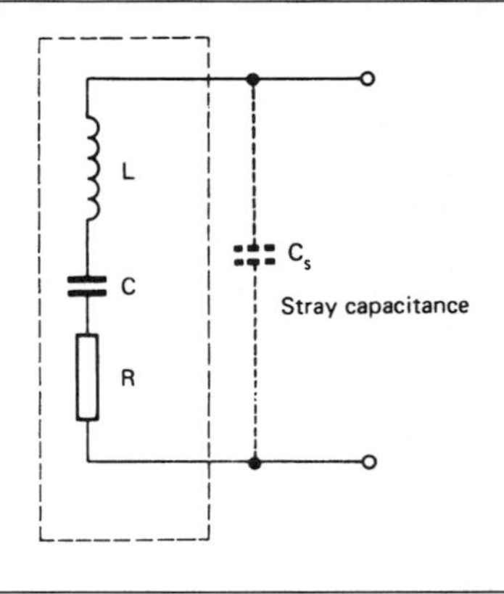

Figure 1.4 The equivalent circuit of a crystal. This corresponds to a series resonant circuit with very high inductance, low capacitance and almost negligible resistance.

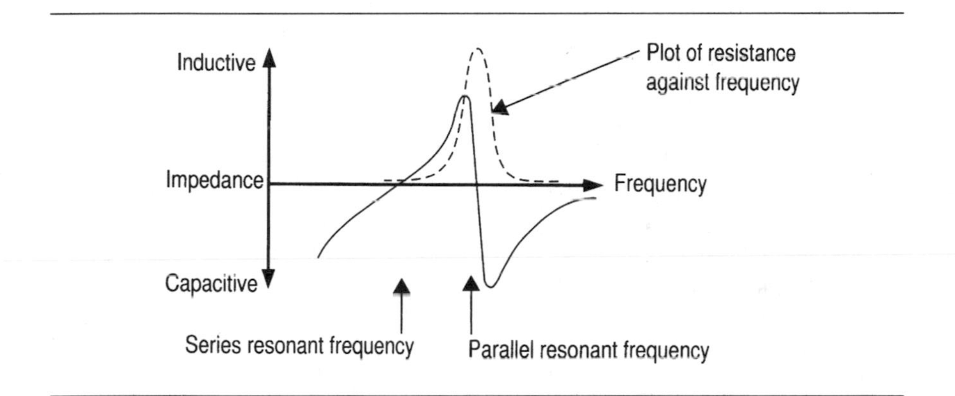

Figure 1.5 The electrical characteristics of a typical quartz crystal.

so that the gauge can be sensitive, but the output impedance is very high and usually capacitive. Figure 1.4 illustrates the electrical equivalent circuit, and Figure 1.5 shows the response around the main resonant frequencies for a quartz crystal. The output of a piezoelectric strain gauge is not DC, so this type of gauge is not useful for detecting slow changes, and its main application is for acceleration sensing (see Chapter 2).

Two major problems of strain gauge elements of any type are *hysteresis* and *creep*. Hysteresis means that a graph of resistance change plotted against length change does not follow the same path of decreasing stress as for increasing stress (Figure 1.6). Unless the gauge is over-stretched, this effect should be small, of the order of 0.025% of normal readings at the

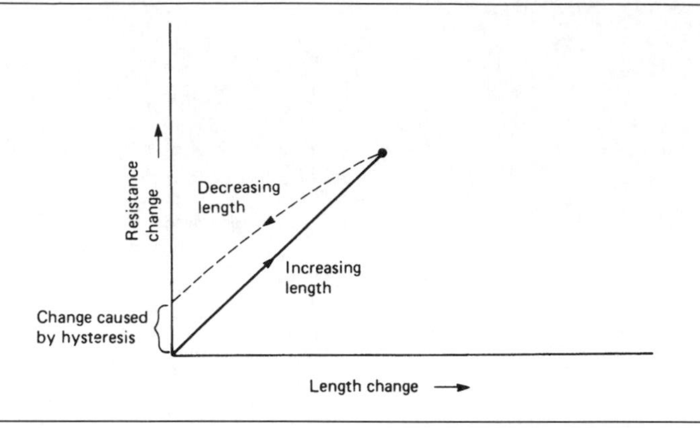

Figure 1.6 The hysteresis effect on a strain gauge, greatly exaggerated. The graph is linear for increasing strain, but does not take the same path when the strain is decreasing. This results in the gauge having permanently changed resistance when the strain is removed.

most. Overstretching of a strain gauge will cause a large increase in hysteresis, and, if excessive, will cause the gauge to show a permanent change of length, making it useless until it can be recalibrated. The other problem, creep, refers to a gradual change in the length of the gauge element which does not correspond to any change of strain in the material that is being measured. This also should be very small, of the order of 0.025% of normal readings. Both hysteresis and creep are non-linear effects which can never be eliminated but which can be reduced by careful choice of the strain gauge element material. Both hysteresis and creep increase noticeably as the operating temperature of the gauge is raised.

LOAD CELLS

Load cells are used in electronic weighing systems. A load cell is a force transducer that converts force or weight into an electrical signal. Basically, the load cell uses a set of strain gauges, usually four connected as a Wheatstone-bridge circuit. The output of the bridge circuit is a voltage that is proportional to the force on the load cell. This output can be processed directly, or digitized for processing.

1.2 Interferometry

Laser interferometry is another method of strain measurement that presents considerable advantages, not least in sensitivity. Though the principles of the method are quite ancient, its practical use had to wait until suitable lasers and associated equipment had been developed, along with practicable electronic methods of reading the results. Before we can look at

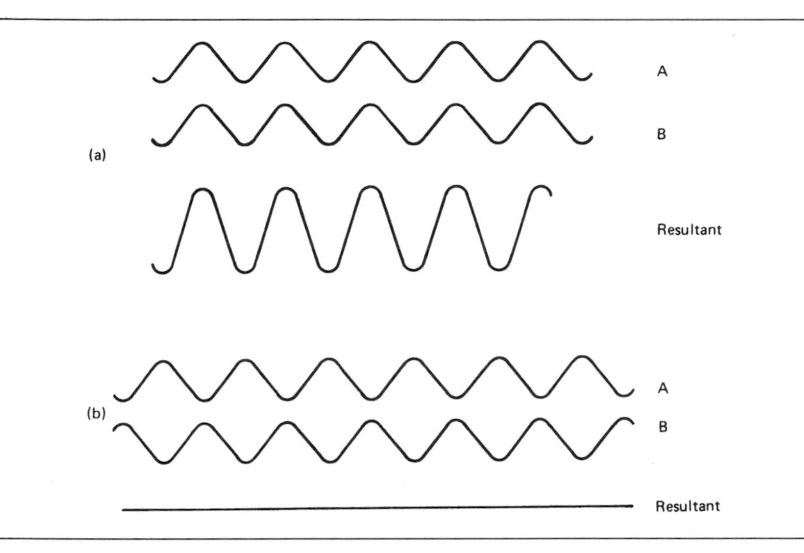

Figure 1.7 Wave interference. When waves meet and are in phase (a), the amplitudes add so that the resultant wave has a larger amplitude. If the waves are in antiphase (b), then the resultant is zero or a wave of small amplitude.

what is involved in a laser interferometer strain gauge, we need to understand the basis of wave interference and why it is so difficult to achieve with light.

All waves exhibit interference (Figure 1.7). When two waves meet and are in phase (peaks of the same sign coinciding), then the result is a wave of greater amplitude, a reinforced wave. This is called constructive interference. If the waves are in opposite phase when they meet, then the sum of the two waves is zero, or a very small amplitude of wave, and this is destructive interference. The change from constructive to destructive interference therefore occurs for a change of phase of one wave relative to another of half a cycle. If the waves are emitted from two sources, then a movement of one source by a distance equal to half a wavelength will be enough to change the interference from constructive to destructive or vice versa. If the waves that are used have a short wavelength, then the distance of half a wavelength can be very short, making this an extremely sensitive measurement of change of distance.

The wavelength of red light is about 700 nm, i.e., 10^{-7} m or 10^{-4} mm, so that a shift of half this distance between two red light sources could be expected to cause the change between fully constructive and fully destructive interference – in practice we could detect a considerably smaller change than this maximum amount.

This method would have been used much earlier if it were not for the problem of coherence. Interference is possible only if the waves that are interfering are continuous over a sufficiently long period. Conventional

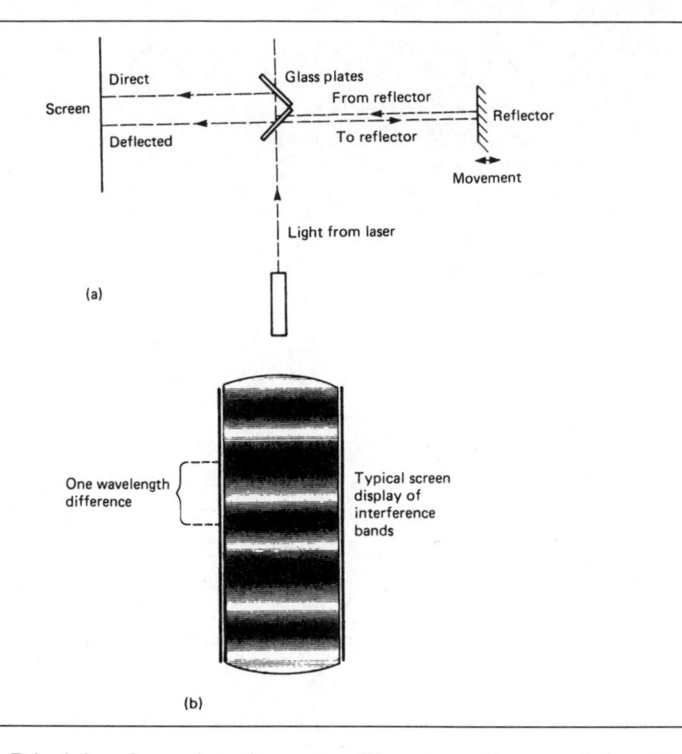

Figure 1.8 Principles of wave interferometry. The set-up of laser and glass plates is shown in (a). The glass plates will pass some light and reflect some, so that both the reflector and the screen will receive some light from the laser beam. In addition, the light reflected from the reflector will also strike the screen, causing an interference pattern (b). For a movement of half of one wavelength of the reflector, the pattern will move a distance equal to the distance between bands on the screen.

light generators, however, do not emit waves continuously. In a light source such as a filament bulb or a fluorescent tube, each atom emits a pulse of light radiation, losing energy in the process, and then stops emitting until it has regained energy. The light is therefore the sum of all the pulses from the individual atoms, rather than a continuous wave. This makes it impossible to obtain any interference effects between two separate normal sources of light, and the only way that light interference can normally be demonstrated using such sources is by using light that has passed through a pinhole to interfere with its own reflection, with a very small light path difference.

The laser has completely changed all this. The laser gives a beam in which all the atoms that contribute light are oscillating in synchronization; this type of light beam is called *coherent*. Coherent light can exhibit interference effects very easily, and has a further advantage of being very easy to obtain in accurately parallel beams from a laser. The interferometer makes use of both of these properties as illustrated in Figure 1.8.

Light from a small laser is passed to a set of semi-reflecting glass plates and some of the light is reflected onto a screen. The rest of the light is aimed at a reflector, so that the reflected beam will return to the glass plates and also be reflected to the screen. Now this creates an interference pattern between the light that has been reflected from the outward beam and the light that has been reflected from the returning beam. If the distant reflector moves by one quarter of a wavelength of light, the light path of the beam to and from the reflector will change by half a wavelength, and the interference will change between constructive and destructive. Since this is a light beam, this implies that the illumination on the screen will change between bright and dark. A photocell can measure this change, and by connecting the photocell through an amplifier to a digital counter, the number of quarter wavelengths of movement of the distant reflector can be measured electronically.

The interferometer is often much too sensitive for many purposes. For example, the effect of changing temperatures is not easy to compensate for, though this can be done by using elaborate light paths in which the two interfering beams have travelled equal distances, one in line with the stress and the other in a path at right angles. An advantage of this method is that no physical connection is made between the points whose distance is being measured; there is no wire or semiconductor strip joining the points; the main body of the interferometer is in one place and the reflector in another. The distance between the main part of the device and the reflector is not fixed, the only restraint being that the distance must not exceed the *coherence distance* for the laser. This is the average distance over which the light remains coherent, and is usually at least several metres for a laser source.

1.3 Fibre optic methods

Developments in the manufacture and use of optical fibres have led to these devices being used in the measurement of distance changes. The optical fibre (Figure 1.9) is composed of glass layers and has a lower refractive index for the outer layer than for the inner. This has the effect of trapping a light beam inside the fibre because of the total internal reflection effect (Figure 1.10). When a light ray passes straight down a fibre, the number of internal reflections will be small, but if the fibre is bent, then the number of reflections will be considerably increased, and this leads to an increase in the distance travelled by the light, causing a change in the time needed, and hence to a change in the phase.

This change of phase can be used to detect small movements by using the type of arrangement shown in Figure 1.11. The two jaws will, as they move together, force the optical fibre to take up a corrugated shape in which the light beam in the fibre will be reflected many times. The extra

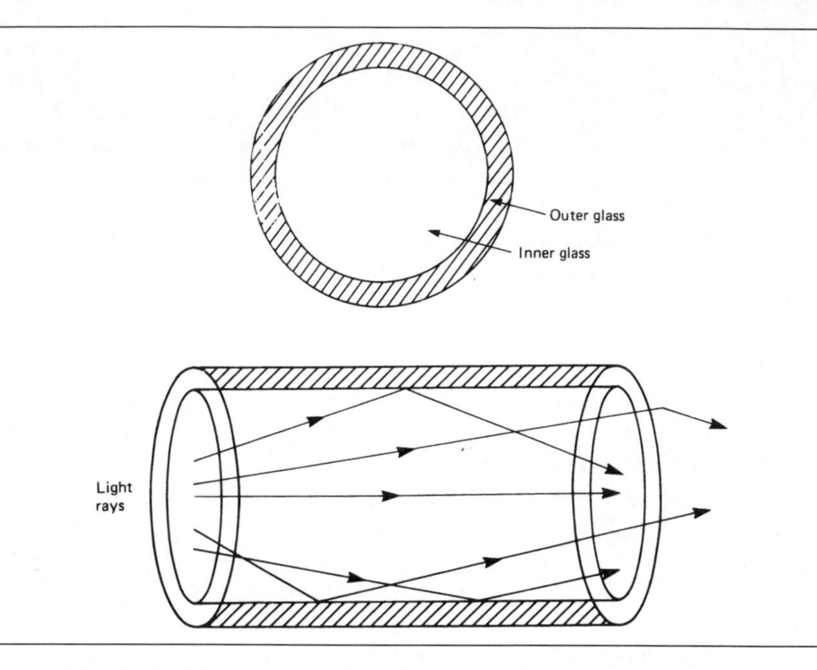

Figure 1.9 Optical fibre construction. The optical fibre is not a single material but a coaxial arrangement of transparent glass or (less usefully) plastics. The materials are different and refract light to different extents (refractivity) so that any light ray striking the junction between the materials is reflected back and so trapped inside the fibre.

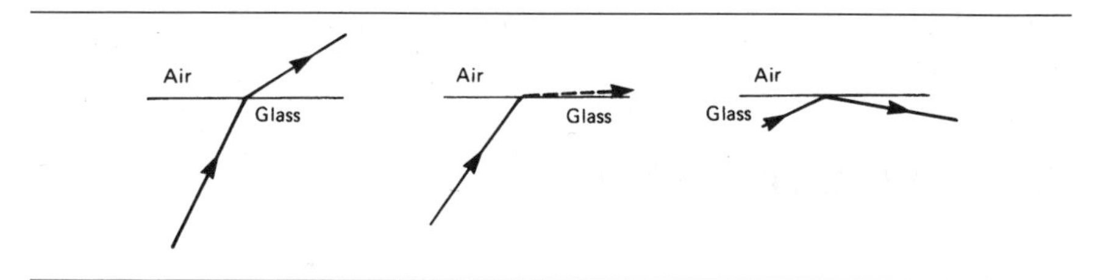

Figure 1.10 Total internal reflection. When a ray of light passes from an optically dense (highly refractive) material into a less dense material, its path is refracted away from the original direction (a) and more in line with the surface. At some angle (b), the refracted beam will travel parallel to the surface, and at glancing angles (c), the beam is completely reflected. The use of two types of glass in an optical fibre ensures that the surface is always between the same two materials, and the outer glass is less refractive than the inner so as to ensure reflection.

distance travelled by the beam will cause a delay that can be detected by interferometry, using a second beam from an unchanged fibre. The sensor must be calibrated over its whole range, because there is no simple relationship between the amount of movement and the amount by which the light is delayed.

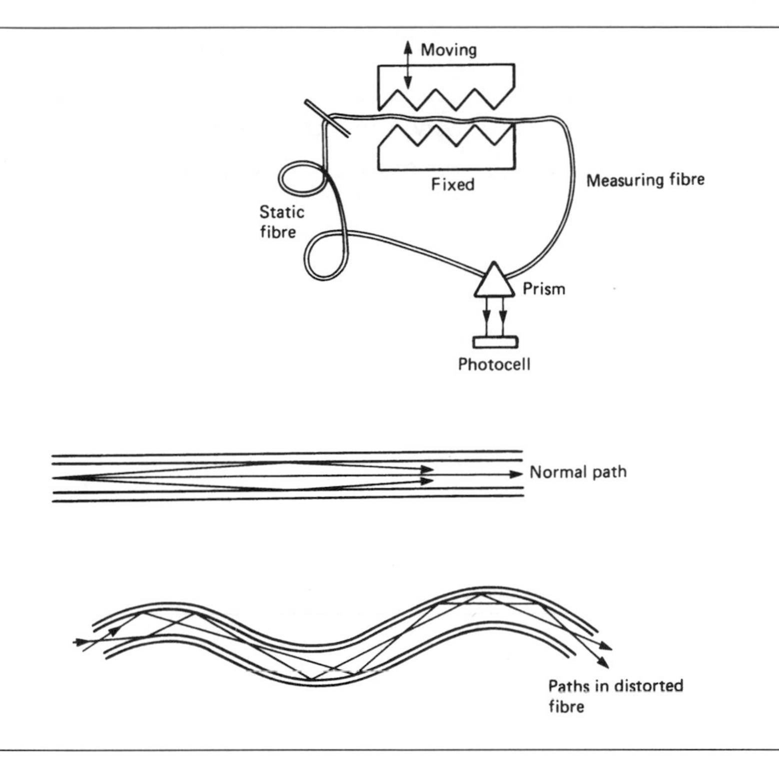

Figure 1.11 Using optical fibres to detect small distance changes. The movement of the jam distorts one fibre, forcing the light paths to take many more reflections and thus increasing the length of the total light path. An interference pattern can be obtained by comparing this to light from a fibre that is not distorted, and the movement of the pattern corresponds to the distortion of one fibre. The sensitivity is not so great as that of direct interferometry, and the use of fibres makes the method more generally useful, particularly in dark liquids or other surroundings where light beams could not normally penetrate.

1.4 Pressure gauges

Pressure in a liquid or a gas is defined as the force acting per unit area of surface. This has the same units as mechanical stress, and for a solid material, the force/area quantity is always termed stress rather than pressure. For a solid, the amount of stress would be calculated, either from knowledge of force and area of cross-section, or from the amount of strain. Where the stress is exerted on a wire or girder, the direct calculation of stress may be possible, but since strain can be measured by electronic methods, it is usually easier to make use of the relationship shown in Table 1.1.

Young's modulus is a quantity that is known for each material, or which can be measured for a sample of material. The stress is stated in units of

Table 1.1 Stress, strain and the elastic constants of Young's modulus and the bulk modulus.

Stress = strain × Young's modulus (for tensile stress)

Example: If measured strain is 0.001 and the Young's modulus for the material is 20×10^{10} N/m^2 then stress is: $20 \times 10^{10} \times 0.001 = 20 \times 10^{7}$ n/m^2
For bulk stress use:

Stress = strain × bulk modulus

$$\text{with volume stress} = \frac{\text{change of volume}}{\text{original volume}}$$

N/m^2 (newton per square metre), and is normally a large quantity. When pressure in a liquid or gas is quoted, the units of N/m^2 can also be termed pascals (Pa). Since the pascal or N/m^2 is a small unit, it is more usual to work with kilo-pascals (kPa), equal to 1000 Pa. For example, the 'normal' pressure of the atmosphere is 101.3 kPa.

The measurement of pressure in liquids and gases covers two distinct ranges. Pressure in liquids usually implies pressures greater than atmospheric pressure, and the methods that are used to measure pressures of this type are similar for both liquids and gases. For gases, however, it may be necessary also to measure pressures lower than atmospheric pressure, in some cases very much lower than atmospheric pressure. Such measurements are more specialized and employ quite different methods. We shall look first at the higher range of pressures in both gases and liquids.

The pressure sensors for atmospheric pressure or higher can make use of both indirect and direct effects. The indirect effects rely on the action of the pressure to cause displacement of a diaphragm, a piston or other device, so that an electronic measurement or sensing of the displacement will bear some relationship to the pressure. The best-known principle is that of the aneroid barometer, illustrated in Figure 1.12. The diaphragm is acted on by the pressure that is to be measured on one side, and a constant (usually lower) pressure on the other side. In the domestic version of the barometer, the movement of the diaphragm is sensed by a system of levers which provide a pointer display of pressure.

For electronic measurement, the diaphragm can act on any displacement transducer and one well-suited type is the capacitive type, illustrated in Figure 1.13. The diaphragm is insulated from the fixed backplate, and the capacitance between the diaphragm and the backplate forms part of the resonant circuit of an oscillator. Reducing the spacing between the diaphragm and the backplate will increase the capacitance, in accordance with the formula shown in Figure 1.13(b), and so reduce the resonant

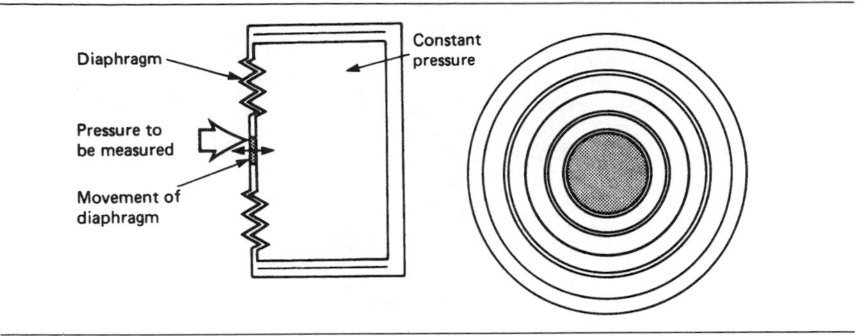

Figure 1.12 The aneroid barometer principle. The domestic barometer uses an aneroid capsule with a low pressure inside the sealed capsule. Changes of external pressure cause the diaphragm to move, and in the domestic barometer these movements are amplified by a set of levers.

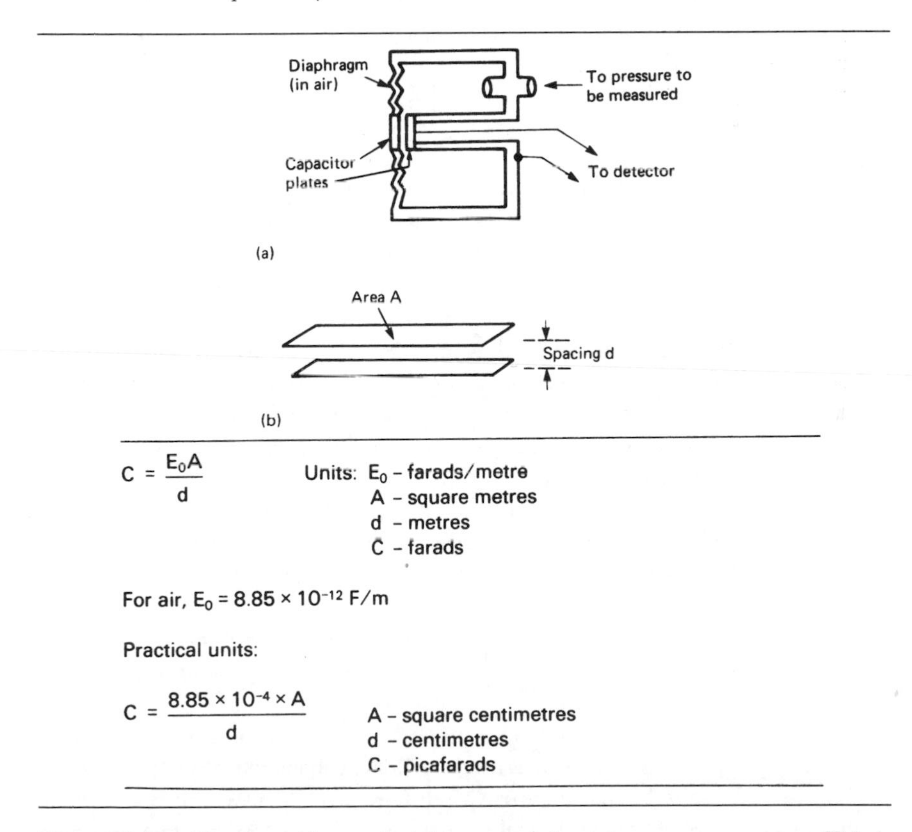

Figure 1.13 The aneroid capsule (a) arranged for pressure measurement. This is an inside-out arrangement as compared to the domestic barometer. The pressure to be measured is applied inside the capsule, with atmospheric air or some constant pressure applied outside. The movement of the diaphragm alters the capacitance between the diaphragm and a fixed plate, and this change of capacitance can be sensed electronically. The formula relating capacitance to spacing is shown in (b).

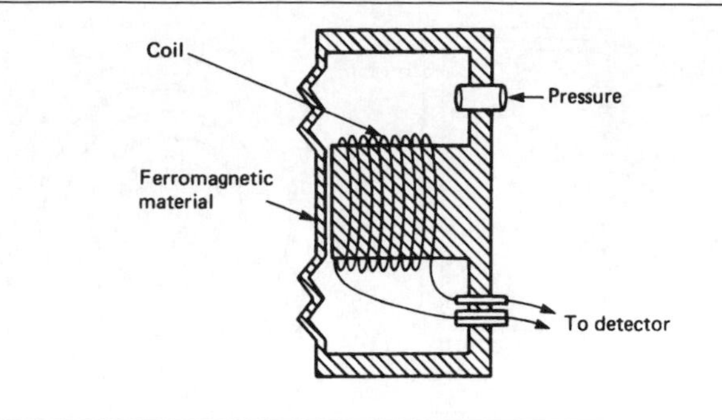

Figure 1.14 Using a variable reluctance type of sensing system. The movement of the diaphragm causes considerable changes in the reluctance of the magnetic path, and so in the inductance of the coil.

frequency of the oscillator. This provides a very sensitive detection system, and one which is fairly easy to calibrate.

Although the thin metal corrugated diaphragm makes the device suitable only for detecting pressures of about atmospheric pressure, the use of a thicker diaphragm, even a thick steel plate, can permit the method to be used with very much higher pressures. For such pressure levels, the sensor can be made in the form of a small plug that can be screwed or welded into a container. The smaller the cross-section of the plug the better when high pressures are to be sensed, since the absolute amount of force is the product of the pressure and the area of cross-section. The materials used for the pressure-sensing plate or diaphragm will also have to be chosen to suit the gas or liquid whose pressure is to be measured. For most purposes, stainless steel is suitable, but some very corrosive liquids or gases will require the use of more inert metals, even to the extent of using platinum or palladium.

When a ferromagnetic diaphragm can be used, one very convenient sensing effect is variable reluctance, as illustrated in principle in Figure 1.14. The variable-reluctance type of pressure gauge is normally used for fairly large pressure differences, and obviously cannot be used where diaphragms of more inert material are required. The method can also be used for gases, and for a range of pressures either higher or lower than atmospheric pressure.

The aneroid barometer capsule is just one version of a manometer that uses the effect of pressure on elastic materials. Another very common form is the coiled flattened tube, as illustrated in Figure 1.15, which responds to a change of pressure inside the tube (or outside it) by coiling or uncoiling. This type of sensor can be manufactured for various ranges of pressure simply by using different materials and thicknesses of tubing, so that this method can be used for both small and large pressure changes. The main drawback as far as electronics is concerned is the conversion from the

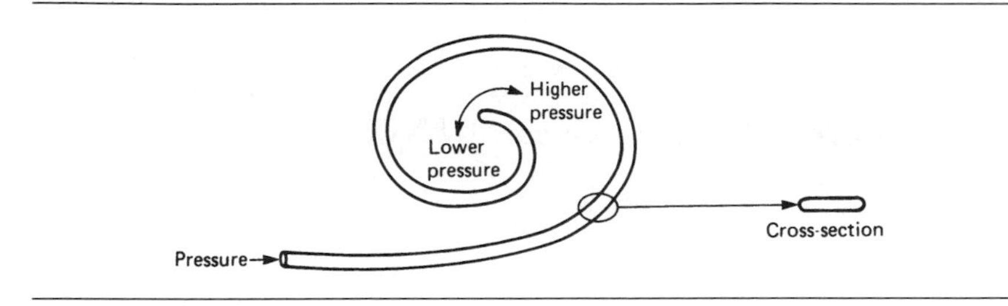

Figure 1.15 The flattened-tube form of a pressure sensor.

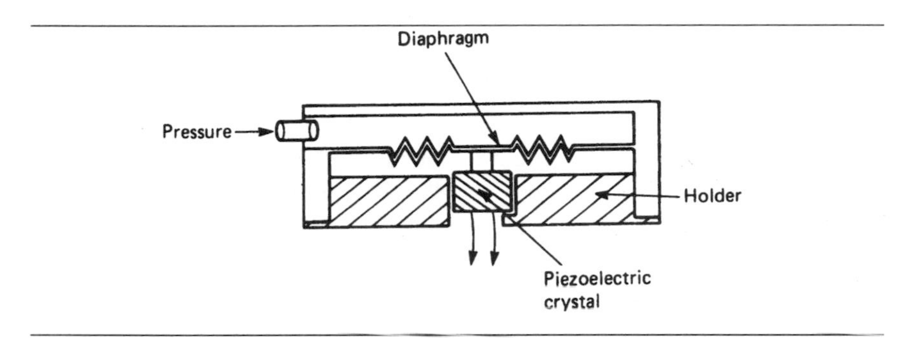

Figure 1.16 Using a piezoelectric crystal detector coupled to a diaphragm for sensing pressure changes.

coiling/uncoiling of the tube into electronic signals, and one common solution is to couple the manometer to a potentiometer.

Another transducing method uses a piezoelectric crystal, usually of barium titanate, to sense either displacement of a diaphragm connected to a crystal, or pressure directly on the crystal itself. As explained earlier, this is applicable more to short duration changes than to steady quantities. For a very few gases, it may be possible to expose the piezoelectric crystal to the gas directly, so that the piezoelectric voltage is proportional to the pressure (change) on the crystal. For measurements on liquids and on corrosive gases, it is better to use indirect pressure, with a plate exposed to the pressure which transmits it to the crystal, as in Figure 1.16. This type of sensor has the advantage of being totally passive, with no need for a power supply to an oscillator and no complications of frequency measurement. Only a high input impedance voltmeter or operational amplifier is needed as an indicator, and if the sensor is used for switching purposes, the output from the crystal can be applied directly to a FET op-amp.

Piezoresistive, piezoelectric, and capacitive pressure gauges can be fabricated very conveniently using semiconductor techniques. Figure 1.17 illustrates the principle of a piezoresistive pressure gauge constructed on a silicon base by oxidizing the silicon (to form an insulator) and then deposit-

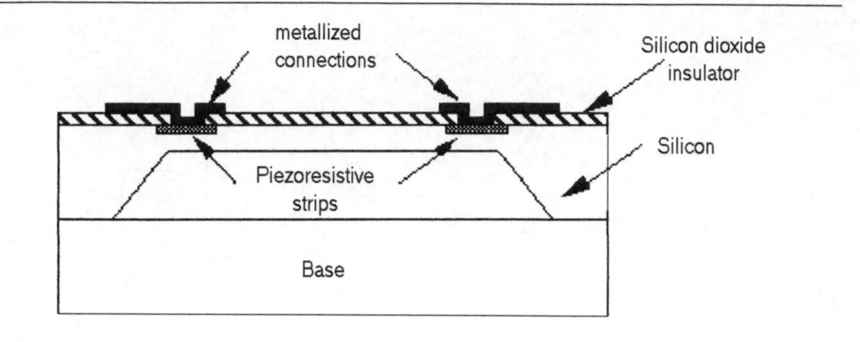

Figure 1.17 A piezoresistive semiconductor pressure gauge element.

ing the piezoresistive elements and the metal connections. Piezoelectric and capacitive pressure-sensing units can be created using the same methods.

1.5 Low gas pressures

The measurement of low gas pressures is a much more specialized subject. Pressures that are only slightly lower than the atmospheric pressure of around 100 kPa can be sensed with the same types of devices as have been described for high pressures. These methods become quite useless, however, when the pressures that need to be measured are very low, in the range usually described as 'vacuum'. Pressure sensors and transducers for this range are more often known as vacuum gauges, and many are still calibrated in the older units of millimetres of mercury of pressure. The conversion is that 1 mm of mercury is equal to 133.3 Pa. The high-vacuum region is generally taken to mean pressures of 10^{-3} mm, of the order of 0.1 Pa, although methods for measuring vacuum pressures generally work in the region from about 1 mm (133.3 Pa) down. Of some 20 methods used for vacuum measurement, the most important are the *Pirani gauge* for the pressures in the region 1 mm to 10^{-3} mm (about 133 Pa to 0.13 Pa), and the *ion gauge* for significantly lower pressures down to about 10^{-9} mm, or 1.3×10^{-7} Pa. A selection of measuring methods is illustrated in Table 1.2.

- All vacuum gauge heads need recalibration when a head is replaced.

The Pirani gauge, named after its inventor, uses the principle that the thermal conductivity of gases decreases in proportion to applied pressure for a wide range of low pressures. The gauge (Figure 1.18) uses a hot wire element, and another wire as sensor. The temperature of the sensor wire is deduced from its resistance, and it is made part of a resistance measuring

Table 1.2 Vacuum gauge types and approximate pressure limits.

Gauge type	Pressure range (Pa)
Diaphragm	10^{5} to 10^{-2}
Manometer	10^{5} to 10^{-3}
Pressure balance	1 to 10^{5}
Radioactive ionization gauge	10^{-2} to 10^{5}
Compression gauge	10^{-6} to 10^{3}
Viscosity gauge	10^{-6} to 10^{3}
Pirani gauge	10^{-3} to 10^{4}
Thermomolecular gauge	10^{-7} to 10^{-1}
Penning gauge	10^{-7} to 10^{-1}
Cold-cathode magnetron gauge	10^{-8} to 10^{-2}
Hot-cathode ionization gauge	10^{-5} to 1
High-pressure ionization gauge	10^{-4} to 10
Hot cathode gauge	10^{-7} to 10^{-2}
Modulator gauge	10^{-8} to 10^{-2}
Suppressor gauge	10^{-9} to 10^{-2}
Extractor gauge	10^{-10} to 10^{-2}
Bent beam gauge	10^{-11} to 10^{-2}
Hot-cathode magnetron gauge	10^{-11} to 10^{-2}

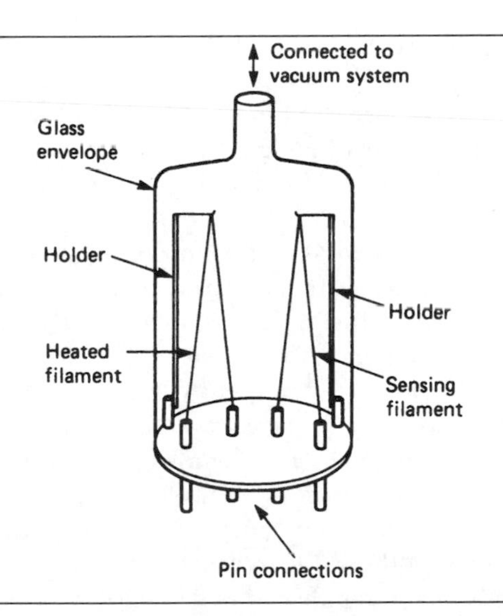

Figure 1.18 The Pirani gauge. One filament is heated, and the other is used as a sensor of temperature by measuring its resistance. As the pressure in the air surrounding the filaments is decreased, the amount of heat conducted between the filaments drops, and the change in resistance of the cold filament is proportional to the change in pressure.

bridge circuit identical to that used for resistive strain gauges. As the gas pressure around the wires is lowered, less heat will be conducted through the gas, and so the temperature of the sensor wire will drop, since the amount of heat transmitted by convection is negligible (because of the arrangement of the wires) and the amount radiated is also very small because of the comparatively low temperature of the 'hot' wire. Commercially available Pirani gauges, such as those from Leybold, are robust, easy to use, fairly accurate, and are not damaged if switched on at normal air pressures. They can be obtained calibrated for various pressure ranges, each with a range (high/low) of around 10^4.

1.6 Ionization gauges

For very low pressure, or high vacuum, measurement, some form of ionization gauge is invariably used. There are many gauges of this type, but the principles are much the same and the differences are easily understood when the principles are grasped. The ionization gauge operates by using a stream of electrons to ionize a sample of the remaining gas in the space in which the pressure is being measured. The positive gas ions are then attracted to a negatively charged electrode, and the amount of current carried by these ions is measured. Since the number of ions per unit volume depends on the number of atoms per unit volume, and this latter figure depends on pressure, the reading of ion current should be reasonably proportional to gas pressure. The proportionality is fairly constant for a fixed geometry of the gauge (Figure 1.19) and for a constant level of electron emission. The range of the gauge is to about 10^{-7} mm (0.013 Pa), which is about the pressure used in pumping transmitting radio valves and specialized cathode ray tubes.

The most serious problem in using an ionization gauge is that it requires electron emission into a space that is not a perfect vacuum. The type of electron emitter that is used in the hot-cathode or *Bayard–Alpert* gauge is invariably a tungsten filament. If this is heated at any time when the gas pressure is too high (above 10^{-3} mm, 133 Pa), then the filament will be adversely affected. If, as is usual, the gas whose pressure is being reduced is air, the operation of the filament at these pressures will result in oxidation, which will impair electron emission or result in the total burnout of the filament. If hot-cathode ionization gauges are used, as they nearly always are, in conjunction with other gauges, usually Pirani gauges, then it should be possible to interlock the supplies so that the ionization gauge cannot be turned on until the pressure as indicated by the other gauge, is sufficiently low. If this can be done, then the ionization gauge can have a long and useful life. A spare gauge head should always be held in stock, however, in case of filament damage, because tungsten filaments are delicate, particularly when at full working temperature. Each gauge head will

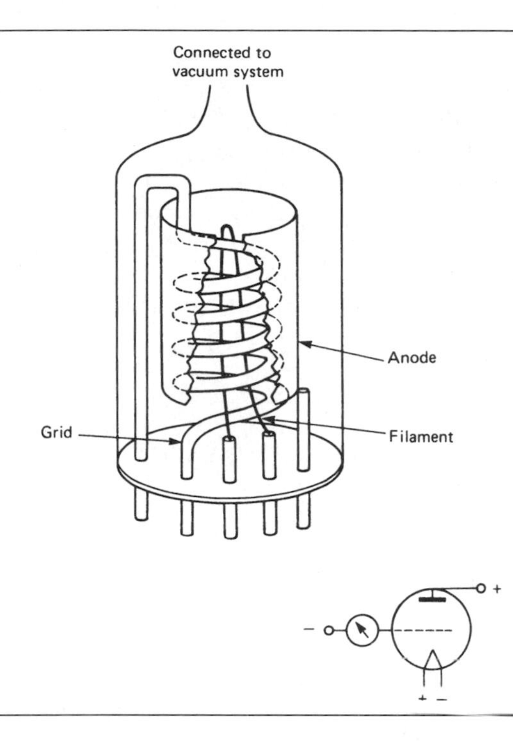

Figure 1.19 The simplest form of an ionization gauge. The grid is a loosely wound spiral of wire surrounding the filament, and exerts little control on the electron stream. With a constant high current of electrons to the anode, positive ions from the remaining gas are attracted to the grid and the resulting grid current is measured and taken as proportional to gas pressure.

need to be calibrated if precise measurements of low pressure are required.

A common variation on the ionization method is the *Penning gauge*, which uses electron emission from a point (a cold-cathode emitter). This avoids cathode damage from oxidation and from fluorine, and the same advantage is claimed for ionization gauges that use thoria-coated iridium (ThOIr) cathodes. A tungsten filament is not poisoned by halogen gases, and is preferred for applications that involve fluorine, chlorine or iodine gases.

Other variants on the ionization gauge arise because a simple electron beam in a confined space is not necessarily a very efficient means of ionizing the residual gas in that space, because only the atoms in the path of the beam can be affected. If the electron beam is taken through a longer path, more atoms can be bombarded, and more ions generated from a given volume of gas, and so the sensitivity of the device is greatly increased. The usual scheme is to use a magnetic field to convert the normal straight path of the electron beam into a spiral path that can be of a much greater total length. This is the *magnetron* principle, used in the magnetron tube to generate microwave frequencies by spinning electrons into a circular path that just touches a metal cavity, so that the cavity resonates and so modulates the electron beam.

The much greater sensitivity that can be obtained in this way is bought at the price of having another parameter, the magnetic field flux density, that will have to be controlled in order to ensure that correct calibration is maintained. The magnetic field is usually applied by means of a permanent magnet, so that day-to-day calibration is good, but since all permanent magnets lose field strength over a long period, the calibration should be checked annually. Gauges of this type can be used down to very low pressures, of the order of 10^{-11} Pa.

- On the other end of the pressure range, a radioactive material can be used as a source of ionization, and this allows measurements up to much higher ranges of pressure, typically up to 10^5 Pa.

1.7 Transducer use

The devices that have been described are predominantly used as sensors, because with a few exceptions, their efficiency of conversion is very low and to achieve transducer use requires the electrical signals to be amplified. The piezoelectric device used for pressure sensing is also a useful transducer, and can be used in either direction. Transducer use of piezoelectric crystals is mainly confined to the conversion between pressure waves in a liquid or gas and electrical AC signals, and this use is described in detail in Chapter 5. The conversion of energy from an electrical form into stress can be achieved by the magnetically cored solenoid, as illustrated in Figure 1.20. A current flowing in the coil creates a magnetic field, and the core will move so as to make the magnetic flux path as short as possible. The amount of force can be large, so that stress can be exerted (causing strain) on a solid material. If the core of the solenoid is mechanically connected to a diaphragm, then the force exerted by the core can be used to apply pressure to a gas or a liquid. In general, though, there are few applications for electronic transducers for strain or pressure and the predominant use of devices in this class is as sensors.

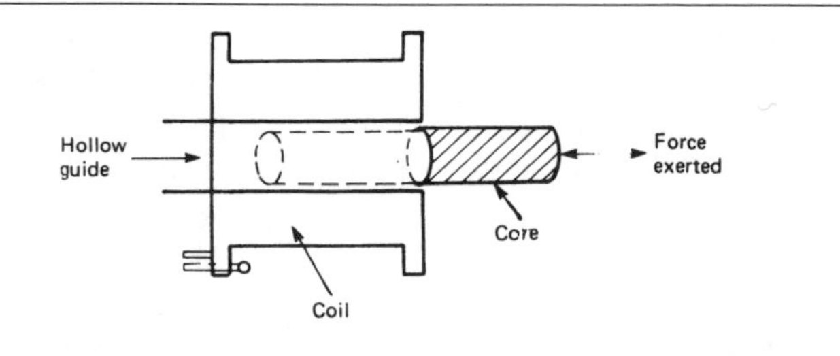

Figure 1.20 The solenoid, which is a current-to-mechanical stress transducer.

Chapter 2

Position, direction, distance and motion

2.1 Position

Position, as applied in measurement, invariably means position relative to some point that may be the Earth's north pole, the starting point of the motion of an object, or any other convenient reference point. Methods of determining position make use of distance and direction (angle) information, so that a position can be specified either by using rectangular (Cartesian) co-ordinates (Figure 2.1) or by polar co-ordinates (Figure 2.2). Position on flat surfaces, or even on the surface of the Earth, can be specified using two dimensions, but for air navigational purposes three-dimensional co-ordinates are required. For industrial purposes, positions are usually confined within a small space (for example, the position of a robot tug) and it may be possible to specify position with a single number, such as the distance travelled along a rail.

In this chapter we shall look at the methods that are used to measure direction and distance so that position can be established either for large- or small-scale ranges of movement. There are two types of distance sensing: the sensing of distance to some fixed point, and the sensing of distance moved, which are different both in principle and in the methods that have to be used. The methods that are applied for small-scale sensing of position appear at first glance to be very different, but they are in fact very similar in principle.

Since position is related to distance (the difference between two positions), velocity (rate of change of position) and acceleration (rate of change of velocity), we shall look at sensors for these quantities also. Rotational movement is also included because it is very often the only movement in a system and requires rather different methods. In addition,

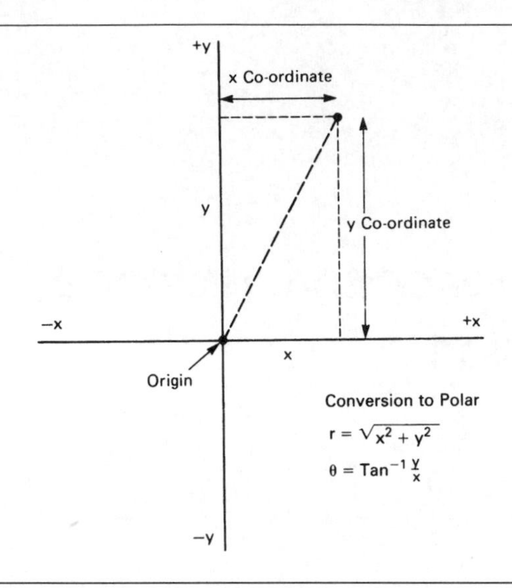

Figure 2.1 The Cartesian co-ordinate system. This uses measurements in two directions at right angles to each other as reference axes, and the position of a point is plotted by finding its distance from each axis. For a three-dimensional location, three axes, labelled x, y and z, can be used. The figure also shows the conversion of two-dimensional Cartesian co-ordinates to polar form.

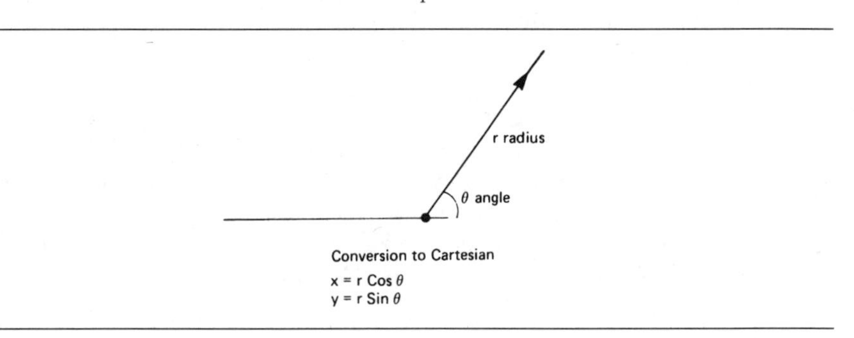

Figure 2.2 Polar co-ordinates make use of a fixed point and direction. The distance from the fixed point, and the angle between this line and the fixed direction, are used to establish a two-dimensional position. For a three-dimensional location, an additional angle is used. The figure also shows conversion of two-dimensional polar co-ordinates to Cartesian.

of course, the rotation of a wheel is often a useful measurement of linear distance moved.

2.2 Direction

The sensing of direction on the Earth's surface can be achieved by observing

Table 2.1 A true celestial navigation method.

- For each of several identified stars, measure the altitude of a star and the Greenwich time.
- Calculate the position of the star at the time of your observation, using the Almanac.
- From this position calculation, calculate for each star you have observed what altitude and azimuth (direction) you should have observed.
- Compare each measured altitude with each calculated altitude to give a figure of *offset.*
- Plot each offset on a chart as a line of position.
- Find your true position as the point where several lines of position cross.

and measuring the apparent direction of distant stars, by using the Earth's magnetic field, by making use of the properties of gyroscopes, or by radio methods, the most modern of which are satellite direction-finders.

Starting with the most ancient method, observation of stars, otherwise known as Celestial navigation, depends on making precise angle measurements. The basic (two-dimensional) requirements are a time measurement and tables of data. For example, a *sextant* can be used to measure the angle of a known star above the horizon, a precise clock (a *chronometer*) that can be read to the nearest second (one second error corresponds to about $\frac{1}{4}$ nautical mile in distance) is used to keep Greenwich mean time, and a copy of a databook such as the 'Nautical Almanac' will allow you to find your position from these readings.

The simplest form of celestial navigation is the observation of local noon. The sextant is used to measure the angle of the sun above the horizon at local noon, and the Almanac will find the latitude corresponding to this angle value. By referring to the chronometer you can find the difference between local noon and Greenwich noon, and so find, using the Almanac, the longitude. The latitude and longitude figures establish your position.

Navigation by the local noon method is simple, but it is not necessarily always available, and although it has been the mainstay of navigation methods in the past, it was superseded several centuries ago by true celestial navigation, which relies on making a number of observations on known stars. The advantage of using stars is that you do not have to wait for a time corresponding to local noon. The process is summarized in Table 2.1.

The traditional compass uses the effect of the Earth's magnetic field on a small magnetized needle that is freely suspended so that the needle points along the line of the field, in the direction of magnetic north and south. The qualifying word 'magnetic' is important here. The magnetic north pole of the Earth does not coincide with the geographical north pole, nor is it a fixed point. Any direction that is found by use of a magnetic form of

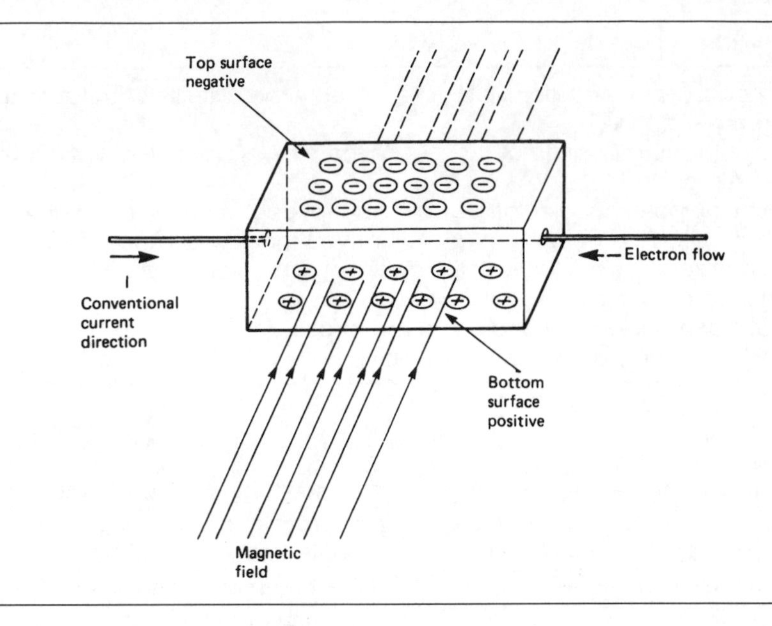

Figure 2.3 The Hall effect. Hall showed that the force of a magnetic field on a current carrier was exerted on the carriers, and would cause deflection. The deflection leads to a difference in voltage across the material, which is very small for a metal because of the high speeds of the carriers, but much larger for a semiconductor.

compass must therefore be corrected for true north if high accuracy is required. The size and direction of this correction can be obtained from tables of magnetic constants (the *magnetic elements*) that are published for the use of navigators. The drift speed and direction of the magnetic north pole can be predicted to some extent, and the predictions are close enough to be useful in fairly precise navigation in large areas on the Earth's surface.

For electronic sensing of direction from the Earth's magnetic field, it is possible to use a magnetic needle fastened to the shaft of a servo-generator, but this type of mechanical transducer is rarely used now that Hall-effect sensors are available. The Hall effect is an example of the action of a magnetic effect on moving charged particles, such as electrons or holes, and it was the way in which hole movement in metals and semiconductors was first proved. The principle is a comparatively simple one, but for most materials, detecting the effect requires very precise measurements.

The principle is illustrated in Figure 2.3. If we imagine a slab of material carrying current from left to right, this current, if it were carried entirely by electrons, would consist of a flow of electrons from right to left. Now for a current and a magnetic field in the directions shown, the force on the conductor will be upwards, and this force is exerted on the particles that carry the current, the electrons. There should therefore be more electrons on the top surface than on the bottom surface, causing a voltage difference, the Hall voltage, between the top and bottom of the slab. Since the electrons are negatively charged, the top of the slab is negative and the bottom positive. If the main carriers are holes, the voltage direction is reversed.

The Hall voltage is very small in good conductors, because the particles move so rapidly that there is not enough time to deflect a substantial number in this way unless a very large magnetic field is used. In semiconductor materials, however, the particles move more slowly, and the Hall voltages can be quite substantial, enough to produce an easily measurable voltage for relatively small magnetic fields such as the horizontal component of the Earth's field. Small slabs of semiconductor are used for the measurement of magnetic fields in Hall-effect fluxmeters and in electronic compasses. A constant current is passed through the slab, and the voltage between the faces is set to zero in the absence of a magnetic field. With a field present, the voltage is proportional to the size of the field, but the practical difficulty is in determining direction.

The direction of maximum field strength is in a line drawn between the magnetic north and south poles, but because the Earth is (reasonably exactly) a sphere, such a line, except at the equator, is usually directed into the Earth's surface, and the angle to the horizontal is known as the *angle of dip* (Figure 2.4). The conventional magnetic compass needle gets around this problem by being pivoted and held so that it can move only in a horizontal plane, and this is also the solution for the Hall-effect detector.

A precision electronic compass uses a servomotor to rotate the Hall slab under the control of a discriminator circuit which will halt the servomotor in the direction of maximum field strength with one face of the Hall slab positive. By using an analogue to digital converter for angular rotation, the direction can be read out in degrees, minutes and seconds. The advantages of this system are that the effects of bearing friction that plague a conventional compass are eliminated, and the reading is not dependent on a human estimate of where a needle is placed relative to a scale. Many conventional needle compasses are immersed in spirit, and the refractivity of the liquid causes estimates of needle position to be very imprecise, unless the scale is backed by a mirror in order that parallax can be avoided by placing the eye so that the needle and its reflection coincide.

The global nature of the Earth's magnetic field makes it particularly convenient for sensing direction, but the irregular variations in the field cause problems, and other methods are needed for more precise direction-finding, particularly over small regions. Magnetic compasses served the Navy well in the days of wooden ships, and when iron (later, steel) construction replaced wood, magnetic compasses could still be used provided that the deviation between true magnetic north and apparent north (distorted by the magnetic material in the ship) could be calculated and allowed for, using deviation tables. By the early part of the 20th century, it was found

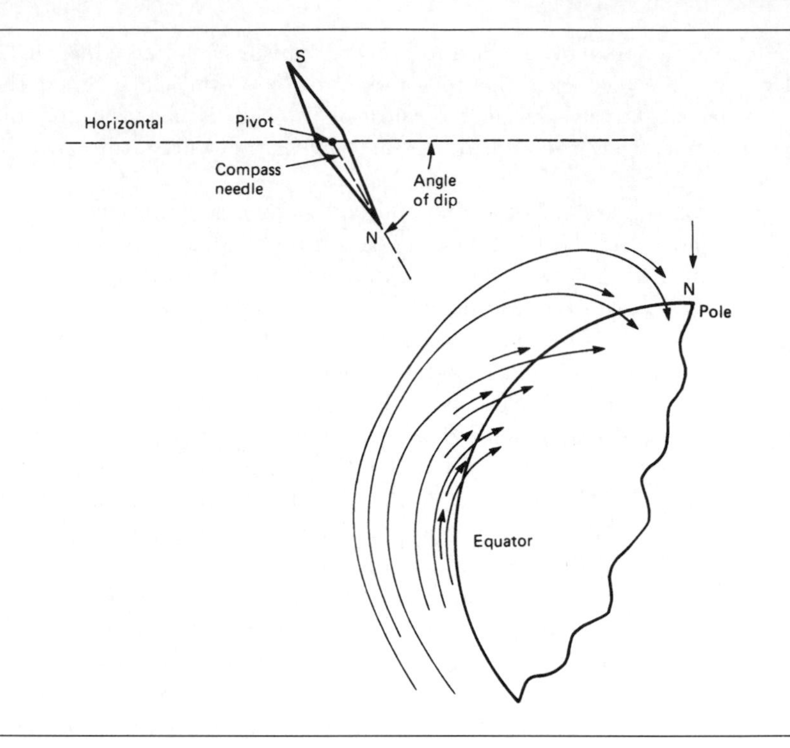

Figure 2.4 The angle of dip shows the actual direction of the Earth's field, which in the northern hemisphere is always into the surface of the Earth.

that the magnetization of a warship could be affected by firing guns or by steering the same course for a long period, and that deviation tables could not be relied upon to correct for these alterations. Submarines provided even greater difficulties because of their use of electric motors, and also because the interior is almost completely shielded by ferrous metal from the Earth's field.

This led in 1910 to the development of the Anschütz gyrocompass. The principle is that a spinning flywheel has directional inertia, meaning that it resists any attempt to alter the direction of its axis. If the flywheel is suspended so that the framework around it can move in any direction without exerting a force on the flywheel, then if the axis of the flywheel has been set in a known position, such as true north, this direction will be maintained for as long as the flywheel spins.

The early Anschütz models were disturbed by the rolling motion of a ship, and a modified model appeared in 1912. This compass model was superseded, in 1913, by the Sperry type of gyrocompass. Full acceptance of gyrocompasses did not occur until errors caused by the ships' movement could be eliminated. Suspension frameworks were developed from the old-fashioned gimbals that were used for ships' compasses, and the wartime

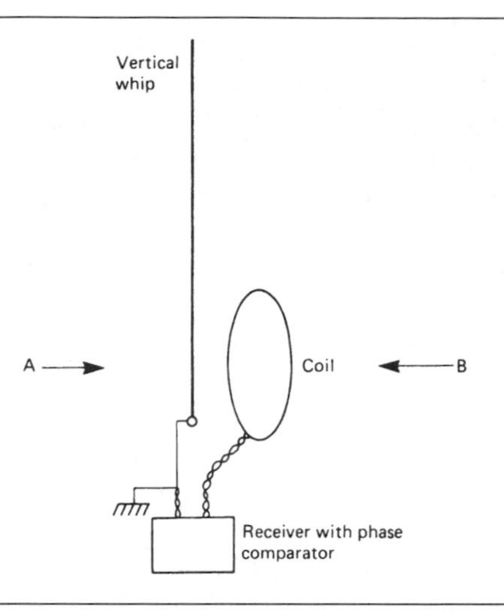

Figure 2.5 The radio direction-finder principle. The output from the vertical aerial is obtained from the electrostatic field of the wave, and does not depend on direction. The magnetic portion of the wave will induce signals in a coil, but the phase of these signals depends on the direction of the transmitter. By combining the signals from the two aerials, and turning the coil, the direction of the transmitter can be found as the direction of maximum signal.

gyrocompasses maintained the rotation of the spinning wheel by means of compressed air jets.

Gyrocompass design was considerably improved for use in air navigation in World War II. The gyrocompass has no inherent electrical output, however, and it is not a simple matter to obtain an electrical output without placing any loading on the gyro wheel. Laser gyroscopes making use of rotating light beams have been developed, but are extremely specialized and beyond the scope of this book. In addition, gyroscopes are not used to any extent in small-scale direction finding for industrial applications.

Radio has been used for navigational purposes for a long time, in the form of radio beacons that are used in much the same way as light beacons were used in the past. The classical method of using a radio beacon is illustrated in Figure 2.5 and consists of a receiver that can accept inputs from two aerials, one a circular coil that can be rotated and the other a vertical whip. The signal from the coil aerial is at maximum when the axis of the coil is in line with the transmitter, and the phase of this maximum signal will be either in phase with the signal from the vertical whip aerial or in antiphase, depending on whether the beacon transmitter is ahead or astern of the coil. By using a phase-sensitive receiver that indicates when the phases are identical, the position of maximum signal ahead can be found, and this will be the direction of the radio beacon.

The form of radio direction-finding that dated from the early part of the 20th century was considerably improved by Watson-Watt, who also invented radar. The original Watson-Watt system used multiple-channel reception with two dipoles, arranged to sense directions at right angles to each other and a single whip aerial connected to separate receivers. A later improvement used a single channel, and modern methods make use of digital signal processing to establish direction much more precisely.

Satellite direction-finding is an extension of these older systems and depends on the supply of *geostationary* satellites. A geostationary satellite is one whose angular rotation is identical to that of the Earth, so that as the Earth rotates the satellite is always in the same position relative to the surface of the planet. The navigation satellites are equipped with transponders that will re-radiate a coded received signal. At the surface, a vessel can send out a suitably coded signal and measure the time needed for the response. By signalling to two satellites in different positions, the position on the Earth's surface can be established very precisely – the precision depends on the frequency that is used, and this is generally in the millimetre range.

2.3 Distance measurement – large scale

The predominant method of measuring distance to a target point on a large scale is based on wave reflection of the type used in radar or sonar. The principle is that a pulse of a few waves is sent out from a transmitter, reflected back from some distant object and detected by a receiver when it returns. Since the speed of the waves is known, the distance of the reflector can be calculated from the time that elapses between sending and receiving. This time can be very short, of the order of microseconds or less, so that the duration of the wave pulse must also be very short, a small fraction of the time that is to be measured. Both radar and sonar rely heavily on electronic methods for generating the waveforms and measuring the times, and although we generally associate radar with comparatively long distances, we should remember that radar intruder alarms are available whose range is measured in metres rather than in kilometres. Figure 2.6 shows a block diagram of a radar system for distance measurement, such as would form the basis of an aircraft altimeter. A sonar system for water depth would take the same general form, but with different transducers (see Chapter 5). The important difference is in wave speeds; 3×10^8 m/s for radio waves in air, but only 1.5×10^3 m/s for sound waves in sea-water.

Where radar or sonar is used to provide target movement indications, the time measurements will be used to provide a display on a cathode ray tube, but for altimeters or depth indications, the time can be digitally measured and the figure for distance displayed. Before the use of radar alti-

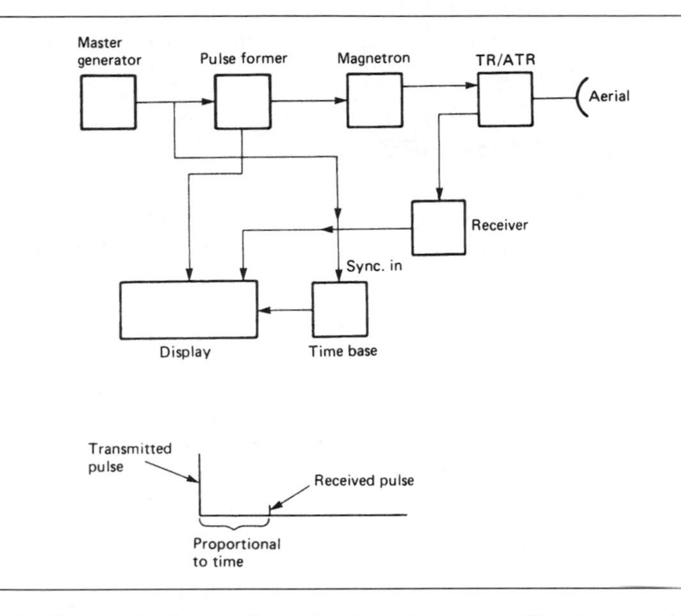

Figure 2.6 The block diagram for a simple radar system. The time required for a pulse of microwave signal to travel to the target and back is displayed in the form of a distance on a cathode ray tube. The transmitter and receiver share the same aerial, using a TR/ATR (transmit/anti-transmit) stage to short-circuit the receiver while the transmitted pulse is present.

meters, the only method available was barometric, measuring the air pressure by an aneroid capsule and using the approximate figure of 3800 Pa change of pressure per kilometre of altitude. The air pressure, however, alters with other factors such as humidity, wind-speed and temperature, so that pressure altimeters are notoriously unreliable. Even if such an altimeter were to give a precise reading, the height that it measures will either be height above sea-level or the height relative to the altitude of the place in which the altimeter was set, rather than true height. It is, in fact, remarkable that air travel ever became a reality with such a crude method of height measurement.

Position measurement on a smaller scale (e.g. factory floor scale) can make use of simpler methods, particularly if the movement is confined in some way, such as by rails or by the popular method of making a robot trolley follow buried wires or painted lines. For confined motions on rails or over wires, the distance from a starting point may be the only measurement that is needed, but it is more likely that the movement is two-dimensional. Over small areas of a few square metres, an artificially generated magnetic field can be used along with magnetic sensors of the types already described. Radio beacon methods, using very low power transmitters, are also useful, and ultrasonic beacons can be used; although problems arise if there are strong reflections from hard surfaces. For a full discussion of the methods as distinct from the sensors, the reader should consult a text on robotics.

2.4 Distance travelled

The sensing of distance travelled, as distinct from distance from a fixed reference point, can make use of a variety of sensors. In this case, we shall start with the sensors for short distance movements, because for motion over large distances the distance travelled will generally be calculated by comparing position measurements rather than directly. Sensors for small distances can make use of resistive, capacitive or inductive transducers in addition to the use of interferometers (see Chapter 1) and the millimetre-wave radar methods that have been covered earlier. The methods that are described here are all applicable to distances in the range of a few millimetres to a few centimetres. Beyond this range the use of radar methods becomes much more attractive.

A simple system of distance sensing is the use of a linear (in the mechanical sense) potentiometer (Figure 2.7). The moving object is connected to the slider of the potentiometer, so that each position along the axis will correspond to a different output from the slider contact – either AC or DC can be used since only amplitude needs to be measured. The output can be displayed on a meter, converted to digital signals to operate a counter, or used in conjunction with voltage level sensing circuits to trigger some action when the object reaches some set position. The main objections to this potentiometric method are: that the range of movement is limited by the size of potentiometers that are available (although purpose-built potentiometers can be used), and that the friction of the potentiometer is an obstacle to the movement. The precision that can be obtained depends on how linear (in the electrical sense) the winding can be made, and 0.1% should be obtainable with reasonable ease.

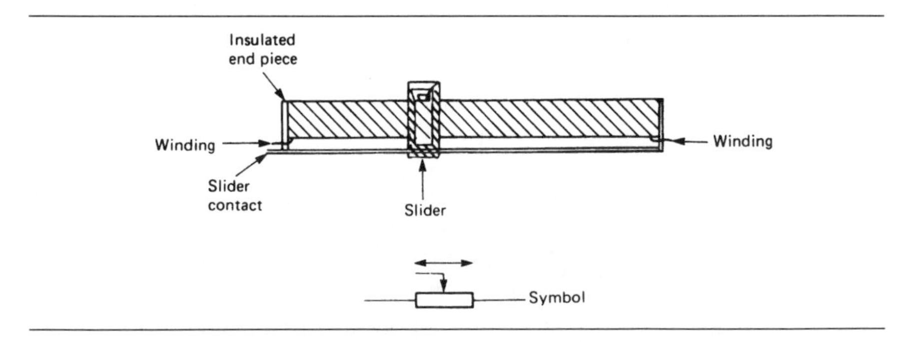

Figure 2.7 A sensor for linear displacement in the form of a linear potentiometer. The advantage of this type of sensor is that the output can be a steady DC or AC voltage that changes when the displacement changes.

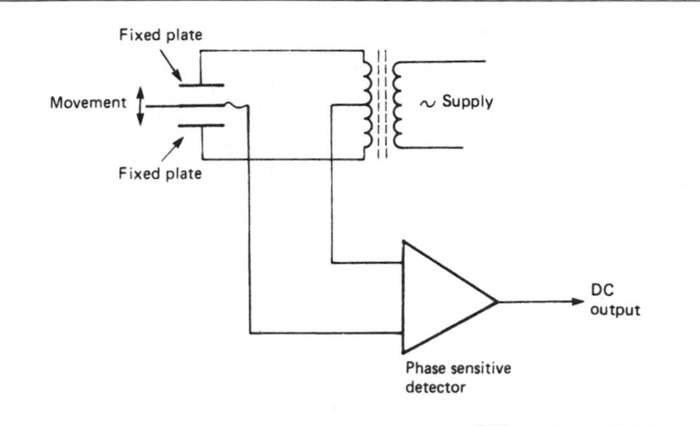

Figure 2.8 The capacitor plate sensor in one of its forms. A change in the position of the moving plate will cause the voltage between this plate and the centre tap of the transformer to change phase, and this phase change can be converted into a DC output from the phase-sensitive detector.

An alternative that is sometimes more attractive, but often less practical, is the use of a capacitive sensor. This can take the form of a metal plate located on the moving object and moving between two fixed plates that are electrically isolated from it. The type of circuit arrangement is illustrated in Figure 2.8, showing that the fixed plates are connected to a transformer winding so that AC signals in opposite phase can be applied. The signal at the moveable plate will then have a phase and amplitude that depends on its position, and this signal can be processed by a phase-sensitive detector to give a DC voltage that is proportional to the distance from one fixed plate. Because the capacitance between plates is inversely proportional to plate spacing, this method is practicable only for very short distances, and is at its most useful for distances of a millimetre or less.

An alternative physical arrangement of the plates is shown in Figure 2.9, in which the spacing of the fixed plates relative to the moving plate is small and constant, but the movement of the moving plate alters the area that is common to the moving plate and a fixed plate. This method has the advantage that an insulator can be used between the moving plate and the fixed plates, and that the measurable distances can be greater, since the sensitivity depends on the plate areas rather than on variable spacing.

The most commonly used methods for sensing distance travelled on the small scale, however, depend on induction. The basic principle of induction methods is illustrated in Figure 2.10, in which two fixed coils enclose a moving ferromagnetic core. If one coil is supplied with an AC signal, then the amplitude and phase of a signal from the second coil depends on the position of the ferromagnetic core relative to the coils. The amplitude of signal, plotted against distance from one coil, varies as shown

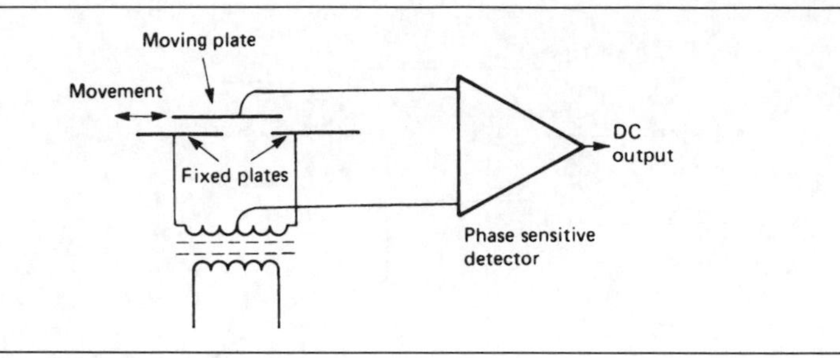

Figure 2.9 Another form of the three-plate capacitor displacement detector. In this case, the spacing between the plates is fixed, but the relative area of centre plate covering the other two can vary. The range of movement for this type can be considerably greater than for the previous type.

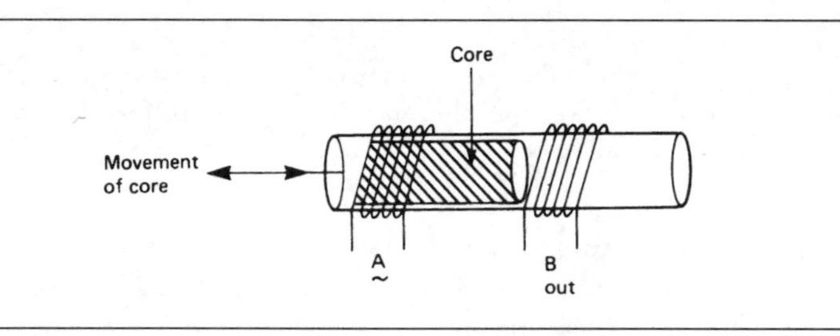

Figure 2.10 The most basic inductive displacement sensor. An AC voltage is applied to one coil, and the position of the core determines how much will be picked up by the other coil.

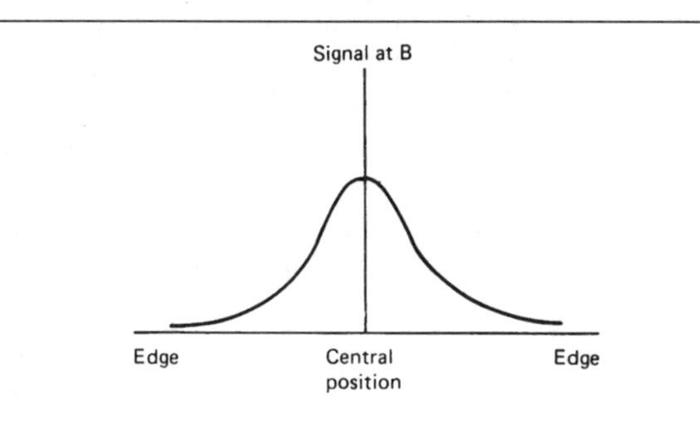

Figure 2.11 A graph of output voltage plotted against core position for the arrangement of Figure 2.10.

in Figure 2.11, and the disadvantage of this simple arrangement is that a given amplitude other than the maximum can correspond to more than one distance. In addition, the shape of the graph means that even if the range is restricted the output is never linearly proportional to the distance.

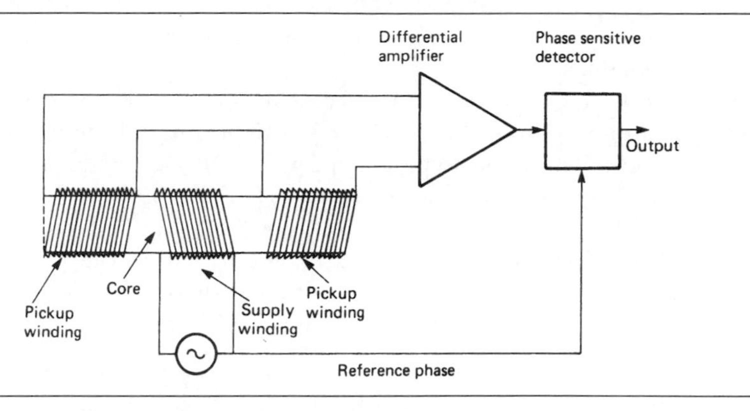

Figure 2.12 The linear variable differential transformer, or LVDT. Movement of the core alters the voltage levels and phases of the voltages across the outer coils, and these voltages can be converted into DC by the phase-sensitive detector.

A development of the simple inductive sensor is the *linear variable differential transformer* (LVDT), which is now the most commonly used sensor for distance in the range of millimetres to centimetres. The principle is illustrated by the circuit diagram shown in Figure 2.12. The device consists basically of three fixed coils, one of which is connected to an AC supply. The other two coils are connected to a phase-sensitive detector, and as a core of ferromagnetic material moves in the coil axis, the output from the detector will be proportional to the distance of the core from one end of the coils. As the name suggests, the output from the phase-sensitive detector will be fairly linearly proportional to distance, and there are considerable advantages as compared to other types of distance sensors, as follows.

1. Virtually zero friction, since the core need not be in contact with the coils, and so no wear.
2. Linear output.
3. Very high resolution, depending mainly on the detector.
4. Good electrical isolation between the core and the coils.
5. A large output signal from the coils so that the phase-sensitive detector needs little or no amplification.
6. No risk of damage if the core movement is excessive.
7. Strong construction that is resistant to shock and vibration.

Commercially available LVDTs can be used with either AC or DC supplies; the DC types use a miniature built-in oscillator inverter to provide the AC for the coils. The performance of the AC types is always superior, but in some systems where, because of lack of space, no separate AC supply can be provided, the DC type must be used. The AC supply is at a high frequency and for the AC type of LVDT it is supplied by a separate inverter system, and is available cased, on a plug-in card, or as a hybrid thick-film module.

Short-stroke LVDTs have a useful linear range of movement of a few millimetres only, from $+1$ mm to ± 5 mm depending on design and construction. The long-stroke type can provide a displacement range of as large as ± 62 mm. Typically, a frequency of 5 kHz is used for the AC drive. A miniature LVDT with ± 3 mm stroke, type SM3, has body dimensions of only 9.52 mm diameter × 35 mm length, not including the plunger arm. The longer stroke versions are larger, with a body diameter of 25 mm and body length (excluding arm) of 150 mm for a ± 25 mm stroke.

- Because of the seven advantages listed above, the LVDT has superseded most other types of distance sensors. For very small distances, the strain gauge (see Chapter 1) can be used; laser interferometers are applicable when very precise changes must be sensed, and radar methods are used for longer distances.

One peculiar advantage of the laser interferometer is that its output can readily be converted to digital form, since it is based on the counting of wavepeaks. Most sensors give analogue outputs, and where a sensor is described as having a digital output, this usually implies that an analogue to digital conversion has been carried out. The interferometer is one of the few sensors which is capable of providing a genuinely digital output. Another is the linear digital encoder, and since digital methods are of ever-increasing importance in measurement and control, a description of this device in detail is appropriate here.

The linear digital encoder gives an output which is a binary number that can be proportional to the distance of the encoder relative to a fixed point. The encoder can, of course, be fixed with an object moving relative to it. The simplest form of encoder of this type is optical, and the principle is illustrated in Figure 2.13. A glass strip is printed with a pattern of the form shown, using large blocks rather than the more practical lines, alternately opaque and transparent. At one edge, the pattern is comparatively fine, and the size of each unit determines the resolution of the device – the smallest change of distance that can be sensed is the width of one block or bar. The next strip along contains blocks or bars of double the width of the first, again alternately opaque and transparent. The next strip in turn has bars that are twice the width of its predecessor and so on. The number of strips determines the number of binary digits in the output and the number of increments of position that can be detected. For example, if the pattern consists of four strips, then the maximum number of increments of position is $2^4 = 16$ which is not likely to be of practical use except for specialized purposes. For eight strips, however, the number of resolvable positions is $2^8 = 256$, and for 16 strips, the number is 65536.

The optical linear encoder is read by using a sensor for each strip or track. The usual scheme is to use a strip of photocells so that light from a source

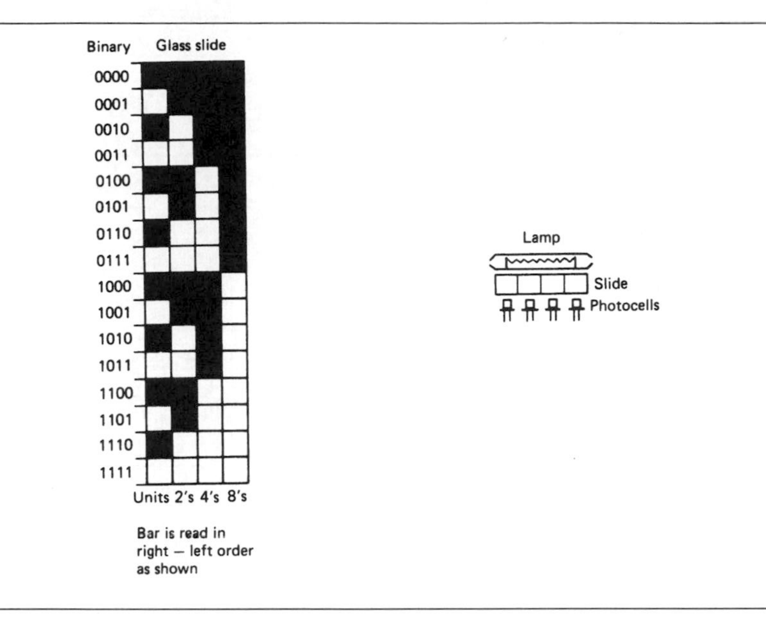

Figure 2.13 The binary optical encoder. Moving the glass strip between a light source and a set of photocells will result in outputs from each photocell that will make up a binary number. The main problem of using 8-4-2-1 binary code with this system is that incorrect readings can be obtained when the position of the slide is such that sections overlap the photocells. The resolution is equal to the width of the units mark.

on the other side of the encoder strip will pass through the transparent sections to the appropriate photocell. The output of each photocell will therefore, after amplification, be a 1 or a 0 signal, and the set of photocells provides a binary number if the cells are read in order of the broadest bars towards the finest. If an eight bit number is used, with eight strips and eight photocells, then the number range is 0 to 256, and this determines the resolution as being 1/256 of the total length of the encoder. For example, if the encoder is 10 cm long, then the resolution is 10/256 cm, or 0.39 mm. The device can be used with the light and photocells at rest and the encoder slide moving (the normal use), or with the encoder slide at rest and the light source and/or photocells moving. The photocells must be placed almost in contact with the encoder slide, and with suitable light shielding to ensure that stray light does not cause false responses.

One problem that arises with optical linear (or angular) encoders is the suitability of the coding system. The usual binary number system is often termed 8-4-2-1 code, in recognition of the fact that each digit place represents a doubling or halving of the value of the adjacent place. This can mean that some changes of number can involve changes in more than one of the digits. For example, the change from 7 to 8 in binary terms is the

Table 2.2 The Grey scale, which is widely used for optical encoders and for industrial purposes. Unlike 8-4-2-1 binary, the Grey code alters by only one digit for each unit change, so that errors due to positioning a slide, causing a section to affect more than one photocell, are negligible.

Denary	8-4-2-1 binary	Grey code	Denary	8-4-2-1 binary	Grey code
0	0000	0000	8	1000	1100
1	0001	0001	9	1001	1101
2	0010	0011	10	1010	1111
3	0011	0010	11	1011	1110
4	0100	0110	12	1100	1010
5	0101	0111	13	1101	1011
6	0110	0101	14	1110	1001
7	0111	0100	15	1111	1000

change from 0111 to 1000, in which each of four digits has changed simply to indicate a value change of just one unit.

For this reason, another code, the Grey code (illustrated in Table 2.2), is used to a much greater extent for devices such as optical encoders. In a Grey code scale, a change of one unit will affect only one digit of the binary number, so that the chances of error caused by misalignment of the sectors of the coding glass, or by half-way positions between sectors, are greatly reduced. ICs exist which will convert Grey code into normal 8-4-2-1 binary, so that arithmetic can be carried out on the numbers if needed. For some applications, conversion may not be necessary.

The linear digital encoder systems provide a set of binary digital signals directly, but since this requires a slide which is very precisely printed and which will use one detector for each binary digit, less direct conversion methods are often used. A slide using only a single set of equally spaced markings can be used in conjunction with a two-phase pick-up system to provide pulse signals whose phasing can be used to indicate direction and whose count indicates distance moved. A system of this type has only a single output to a counter (with a second connection if direction is to be sensed), and the counter will then convert into binary code. This type of system is considered more fully under the heading of *Rotation* later in this chapter.

Another optical method that is particularly useful for small displacements, particularly vibration amplitudes, is the optical grating method whose principles are illustrated in Figure 2.14. Each grating consists of a glass or plastic plate on which fine lines are engraved. One grating is fixed and the other is part of the object whose movement is to be detected. As the object moves, the amount of light that passes through both gratings will be altered – ideally in the pattern of a sine wave for a movement that is of an amplitude equal to the distance between grating lines. This altera-

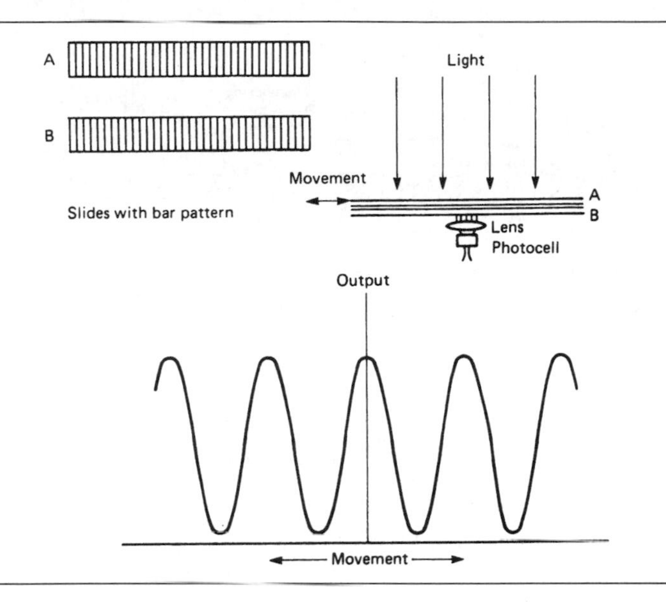

Figure 2.14 The moiré fringe detector. The two slides each have identical thin line patterns printed on them. When one slide moves relative to the other, the amount of light passing through varies considerably depending on the relative positions of lines and spaces in the two slides. A photocell can detect this output and count peaks to find a total displacement.

tion of the light amplitude, termed moiré fringes, can be detected by a photocell and used to provide an electrical output. Large movements are read in terms of the number of complete waves of output from the photocell, small movements in terms of the amplitude of the signal.

2.5 Accelerometer systems

For sensing the quantities of acceleration, velocity and distance travelled, systems based on accelerometers are used. The basis of all accelerometers is the action of acceleration on a mass to produce force, following the equation $F = Ma$ where F is force measured in newtons, M is mass in kilograms and a is acceleration in units of m/s^2. The use of a mass, often termed an *inertial mass*, in this way is complicated by the effect of the acceleration of gravity which causes any mass to exert a force (its *weight*) that is directed towards the centre of the Earth. The mass that is used as part of an acceleration sensor must therefore be supported in the vertical plane, and the type of support that is used will depend on whether accelerations in this plane are to be measured.

- Accelerometers are widely used in vibration and shock measurements, particularly in studies of the effectiveness of packaging. This applies also

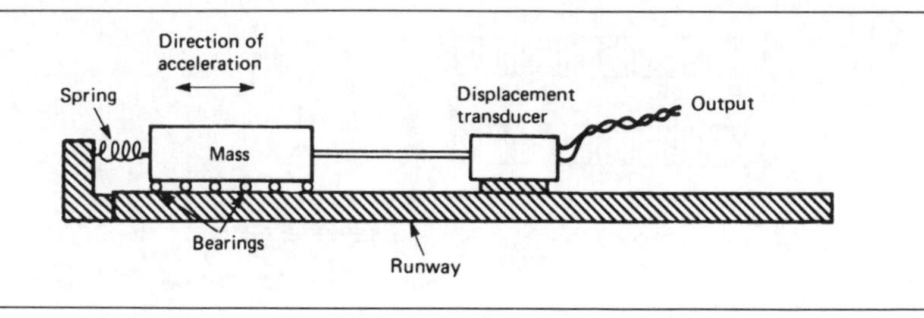

Figure 2.15 Measuring acceleration in one direction. The acceleration of the mass causes a force equal to mass × acceleration. This in turn will stretch or compress the spring, and the amount of this displacement can be measured by any of the usual methods, usually by an LVDT.

to car crash testing and to metering systems used to assess the performance of racing cars.

If the acceleration to be measured is always in a horizontal direction, then the mass can be supported on wheels, ball bearings or air-jets, depending on the sensitivity that is required. Since a force on the mass is to be sensed, the mass will also have to be coupled to a sensor. The method of sensing force is to measure the displacement of the mass against the restoring force of a spring, so that the outline system for a horizontal accelerometer is as sketched in Figure 2.15. The weight of the mass is supported on ball bearings, and the mass is held in the horizontal plane by springs. In Figure 2.15, acceleration in only one direction is to be measured, so that only one spring is illustrated, and it is assumed that sideways movements of the mass are restrained in some other way, by another set of springs or by guide rails.

When an acceleration in the chosen direction affects the mass, it will be displaced against the springs, extending one spring and compressing the other. The amount of linear movement will be proportional to the force, so that any sensor for linear displacement can be used to give an output that is proportional to acceleration. Suitable transducers include potentiometers, capacitive distance gauges, inductive gauges and LVDTs for the larger ranges of movement that are possible. Sensors for acceleration that use this spring and displacement principle are generally intended for the measurement of very small accelerations, usually in one plane.

If only a single dimensional acceleration can be measured, the acceleration in any other direction will produce a false reading, equal to the component of acceleration in that direction. Figure 2.16 shows how this component of acceleration is related to the true value and to the angle between the true acceleration and the measured acceleration directions.

- If the mass can be supported in a cradle of springs and three displacement sensors connected, one for each axis of motion (two horizontal and one vertical), then the outputs can be used to compute the magnitude and direction of an acceleration that can be in any direction.

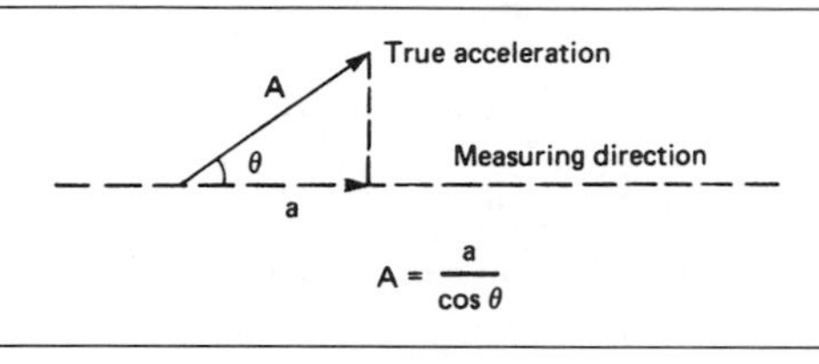

Figure 2.16 The relationship between acceleration measured out of line and true acceleration, if the correct direction is known.

A strain-gauge sensor can be connected to an inertial mass, with or without spring suspension in order to measure acceleration, but the most common type of accelerometer uses an inertial mass coupled to a piezoelectric crystal. The piezoelectric effect has been known since the end of the 19th century, even though in 1960 one major University textbook of physics did not mention it. The principle is that some crystalline materials such as quartz (silicon oxide), Rochelle salt and barium titanate are composed of charged particles (ions) which do not move uniformly when the crystal is stressed. Because of this non-uniformity, movement of the ions produces a difference in charge between opposing faces of the crystal, and if these faces are metallized, a voltage can be measured. The voltage is proportional to the strain of the crystal and can be very large, of the order of kilovolts for some ceramic crystal types, if the strain of the crystal is large. This principle is used to provide a spark for cigarette lighters, gas fires, and gas or oil boiler ignition systems.

The use of an inertial mass bonded to a piezoelectric crystal therefore provides an accelerometer that requires no springs or special supports for the mass. It is even possible to obtain two-dimensional signals from one crystal, and the system will respond to a very wide range of accelerations. Although the unit of acceleration is m/s^2, acceleration is very often measured relative to the 'standard' value of the acceleration of gravity, which is $9.81\,m/s^2$. This leads to figures of acceleration in 'g' units, from which the value in scientific units can be calculated by multiplying by 9.81.

Piezoelectric transducers can cope with acceleration values from a very small fraction of 'g' to several thousand 'g', a huge range compared to those that can be obtained by using spring and displacement systems. The snag is that the piezoelectric crystal is, from the circuit point of view, a capacitor, and the signal is in the form of a charge. The connection of a resistance to the contacts on the faces of the crystal will therefore allow this capacitance to discharge with a time constant equal to CR seconds, where C is the capacitance between the crystal faces in μF and R is the resistance between the faces in MΩ. If, for example, the capacitance is 1000 pF (= 0.001 μF) and the resistance is 10000 MΩ (the input of a FET DC amplifier, perhaps), then the time constant is only 10 s. This makes the piezoelectric type of sensor more suitable for measuring changes of acceleration that occur over a short time, perhaps a fraction of a second, than for measuring fairly constant values of acceleration for a long period.

As it happens, most accelerations are not sustained. Newton's first law states in effect that the natural state of any object in the universe is uniform motion in a straight line with no form of acceleration, and acceleration comes about only because of force. The only steady and constant value of acceleration that we normally encounter is the acceleration of gravity, and most of our acceleration measurements are on accelerations that are caused by short-duration forces, such as those encountered when one object hits another. To put this into perspective, an acceleration of only 1 g for 10 s corresponds to falling a distance of about 500 m in a vacuum. In general, unless you are working with propulsion systems for outer space, the measurement of small accelerations that are applied for long periods will be of no practical interest.

When an accelerometer of any type produces an electrical output, this output can be used for computing other quantities. One of these quantities is speed or, more correctly, *velocity* because an acceleration exists when a change of direction or a change of speed, or both, take place. The relationship between speed or velocity value and acceleration is shown in Figure 2.17, so that if the starting speed of an object, its acceleration (assumed constant) and the time of acceleration are all known, then the final speed can be calculated.

The mathematical action needed to find change of speed from acceleration and time is called *integration*, and analogue computers can carry out this action on a voltage signal from an accelerometer. The initial speed can be set in the form of a voltage applied from a potentiometer, and the output of the analogue computer is proportional to final speed. A second integration of the voltage output (the speed output) will produce a signal proportional to distance, so that this quantity can also be found by using an analogue computing action on the output of the accelerometer. If the starting point of the motion is rest (no starting speed), then no constants need to be fed in. The analogue computer can consist of little more than a pair of operational amplifiers if a simple one-dimensional motion is being sensed. If quantities are expressed in digital terms, then a digital computer can be used to calculate the integral values.

2.6 Rotation

There are very few machines that do not include a rotating shaft at some place, and the sensing and measurement of rotational movement is therefore important. The quantities that are used to measure rotation correspond to the quantities that are used in the measurement of linear motion in such a way that the same types of equations can be used, substituting the rotational quantities for the linear ones. The quantity that corresponds

For uniform (steady) acceleration:

$$\text{acceleration} = \frac{\text{change of speed}}{\text{time taken}} \quad \text{(direction unchanged)}$$

For non-uniform acceleration:

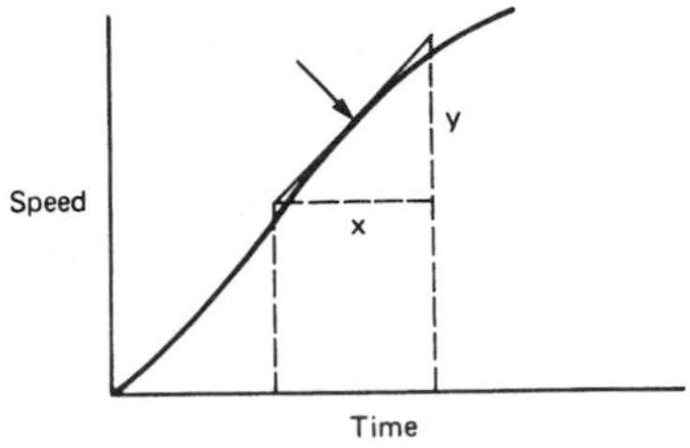

y
Speed
x
Time

Acceleration value at arrowed point $= \frac{y}{x}$

Note: change of direction also constitutes acceleration. For circular motion, the revolving object has acceleration equal to $\frac{v^2}{r}$, directed to the centre of rotation, where v = linear velocity at any instant, r = radius of circle.

Figure 2.17 The relationship between acceleration and velocity (speed) if the direction is constant.

to distance for a rotation is the angle rotated. For a complete rotation of a shaft, this angle is 360°, and so, in terms of degrees, the total angle turned by a shaft is 360 × the number of complete turns. The degree, however, is an artificial unit that is not used in calculations, and most textbooks show the relationships between rotational quantities in terms of *radians*. The definition of the radian is illustrated in Figure 2.18, and it leads to the angle for one complete rotation being 2π radians. To convert from degrees to radians, divide the angle in degrees by 57.3; to convert from radians to degrees, multiply by 57.3.

Because angle in rotational motion corresponds to distance in linear motion, rotational speed is defined as the angle through which a shaft turns per second, and corresponds to linear speed. If the rotating object is a wheel and is in contact with a surface, then Figure 2.19 shows how the linear distance and velocity are related to the angle turned and the angular velocity. The angular acceleration is defined as the rate of change of angular velocity, and the equations that relate these angular quantities, along with the linear counterparts, are illustrated in Figure 2.20. The sensors for angular motion are also very closely related to those that are used for linear motion.

Taking angular velocity first, the simplest form of sensor, which can also act as a transducer, is the AC or DC generator. For sensing and measurement purposes only a minimum of power must be used, so that a miniature AC generator called the *tacho-generator* is normally used. The construction

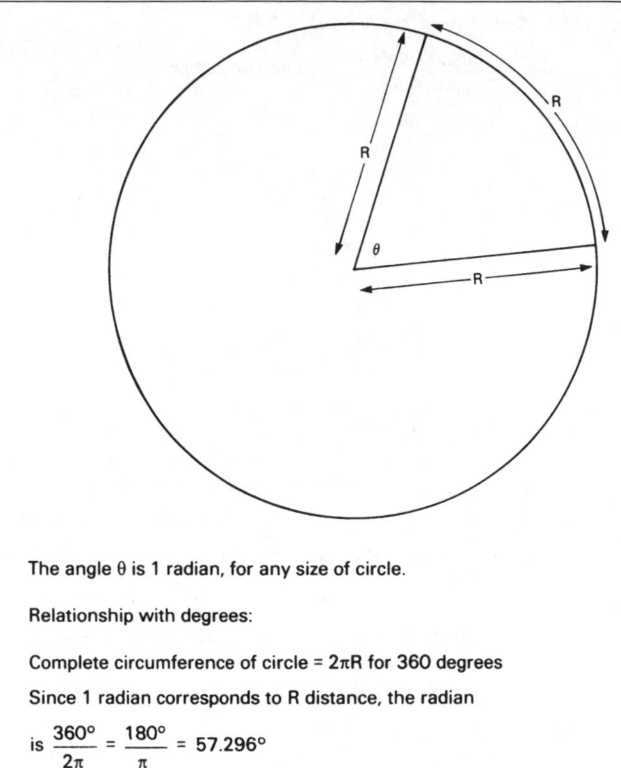

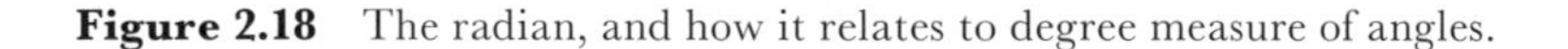

Figure 2.18 The radian, and how it relates to degree measure of angles.

of the tacho-generator, usually abbreviated to tacho, is a more precision-built version of an AC generator and usually has rotating magnets with output from stator coils so as to avoid the need for slip-rings. The frequency of the output signal is proportional to the revolutions per second of the shaft that is coupled to the tacho, so that a frequency-sensitive detector can be used to give a DC output proportional to angular frequency. The tacho can be used over a wide range of angular velocity values, and if the frequency detector is reasonably linear then the readings can be of high precision. The drawback for some applications is the need to make a mechanical coupling between the tacho and the revolving shaft.

Another form of measurement depends on 'drag systems'. A drag system involves a frictional coupling between the rotating shaft and some object that is restrained and whose displacement can be measured. One version of this is the drag cup, whose principle is illustrated in Figure 2.21. The end of the vertical rotating shaft dips into oil contained in a cup, and the cup is

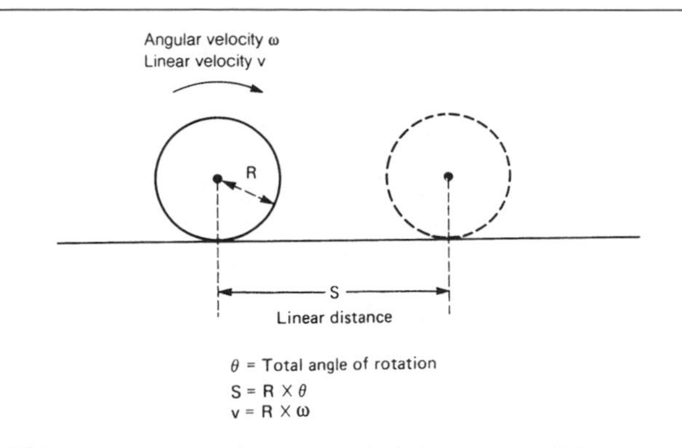

Figure 2.19 Linear and angular distance and speed.

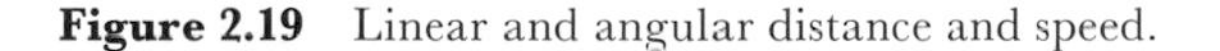

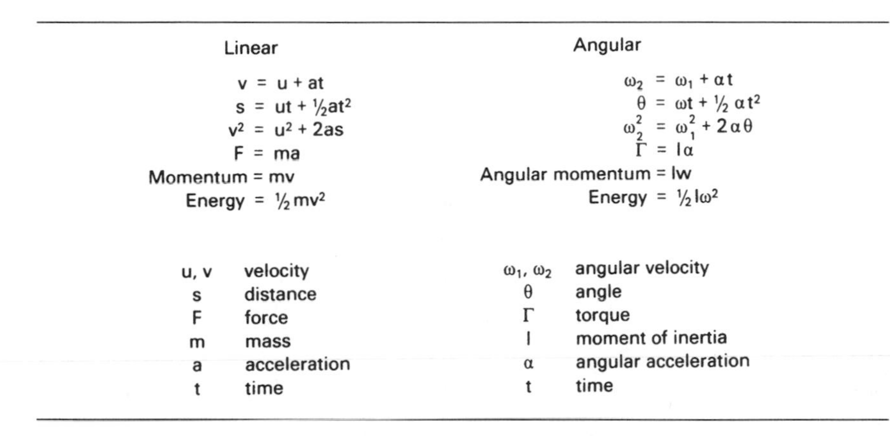

Linear	Angular
$v = u + at$	$\omega_2 = \omega_1 + \alpha t$
$s = ut + \frac{1}{2}at^2$	$\theta = \omega t + \frac{1}{2}\alpha t^2$
$v^2 = u^2 + 2as$	$\omega_2^2 = \omega_1^2 + 2\alpha\theta$
$F = ma$	$\Gamma = I\alpha$
Momentum = mv	Angular momentum = Iw
Energy = $\frac{1}{2}mv^2$	Energy = $\frac{1}{2}I\omega^2$

Symbol	Meaning	Symbol	Meaning
u, v	velocity	ω_1, ω_2	angular velocity
s	distance	θ	angle
F	force	Γ	torque
m	mass	I	moment of inertia
a	acceleration	α	angular acceleration
t	time	t	time

Figure 2.20 The relationships between linear and angular quantities illustrated by corresponding equations.

held in a spring mounting so that its rotation can be measured by any type of sensor, which can be a potentiometer, capacitor system, rotary LVDT or digital encoder (see later). The motion of the shaft is communicated by the viscosity of the oil to a turning force (torque) on the cup, and the displacement of the cup against the spring is measured by the sensor. Since displacement should be proportional to torque, which in turn should be proportional to rotational speed of the shaft, output from the sensor is proportional to rotational speed over a small range of speeds. The range is small because the assumption that torque is proportional to rotational speed holds good for only a small range of speeds, and the system is best suited for slow rotations.

Another version of this method that is much more versatile is the magnetic disc type. A magnet on the end of the shaft will cause a rotating

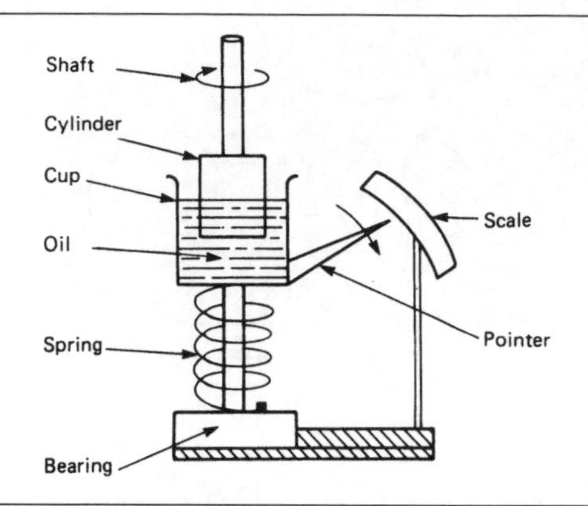

Figure 2.21 The principle of the drag cup method of measuring angular velocity. The method can be converted to electrical output by dispensing with the pointer and mounting the cup and its retaining spring on to the shaft of a potentiometer.

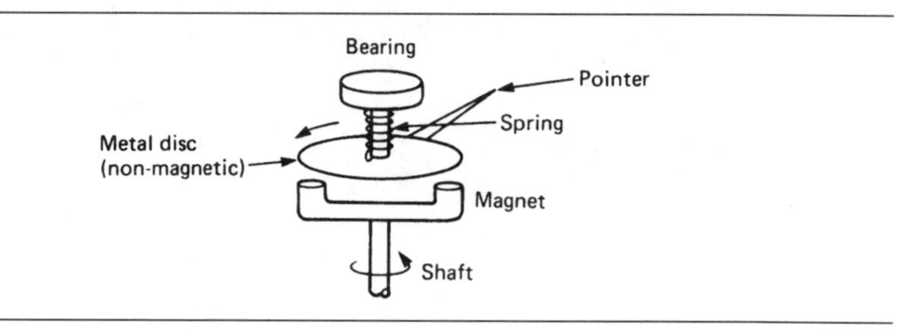

Figure 2.22 The principle of magnetic drag, using induced currents. This has been extensively used in car speedometers, and can be adapted to electronic measurement by replacing the disc and its mountings by a sensing coil.

field as the shaft turns, and if a metal disc (which need not be magnetic) is held close to this magnet (Figure 2.22) then the torque on the disc (caused by the interaction of the magnet and the eddy currents that are induced in the disc) will be proportional to the angular speed of the shaft. This is the scheme that has been used for many years for car speedometer heads, and it operates best at a medium range of angular speeds. One particular advantage is that no contact is needed, though there must not be any metal between the magnet and the disc.

For some purposes, a signal that is sent out for each revolution of a wheel or shaft is sufficient for angular velocity sensing, and this can be achieved by the use of piezoelectric or magnetic pulsing. A piezoelectric pulse can be operated, as indicated in Figure 2.23, by a cam on a shaft that will cause the piezoelectric crystal to be compressed on each rotation of the shaft. Since the signal from the piezoelectric crystal can be of several volts amplitude, this type of sensor often needs no amplification, but the output

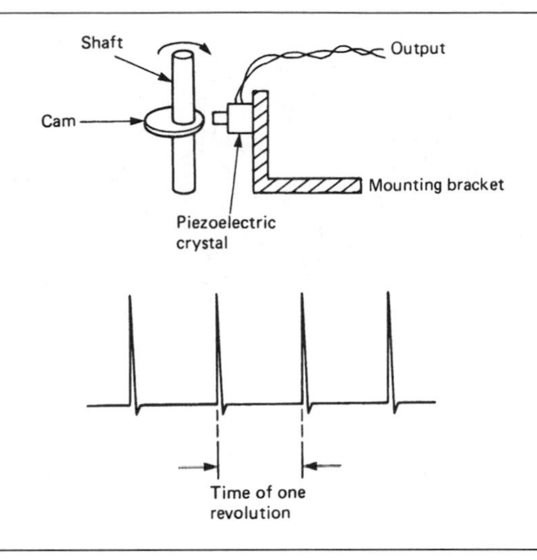

Figure 2.23 Rotational speed sensing using a cam and a piezoelectric crystal. The disadvantages are that there is only one pulse per revolution, and that the contact between cam and crystal causes a frictional drag on the rotation of the shaft.

is at a high impedance. In addition, the friction on the shaft is fairly large compared to the alternative system of magnetic pulsing. The magnetic pulse system uses a permanent magnet mounted on a wheel or shaft and passing over a coil at one part of the revolution. As the magnet moves, a voltage is induced in the coil, so providing a signal for each revolution. The magnetic pulse system is not ideal if the rotation is slow, because the amplitude of the pulse is affected by the rotational speed, and there must not be any magnetic metal between the magnet and the coil.

Angular displacement measurement methods fall into two groups: those for very small angular displacements that are temporary, and those for large displacements. Temporary small displacements are of a degree or less, and the shaft or other rotating object returns to its original position following the rotation. Such small displacements are sensed by variants of the methods used for small linear displacements, such as capacitive, strain gauge and piezo sensors.

For larger displacements, inductive (particularly LVDT) methods are useful, and Figure 2.24 shows a typical method. Potentiometric methods are also useful, but if the normal type of potentiometer construction is used, the angle of rotation is restricted to about 270°. The rotary digital encoder can be used to provide a direct digital output, using a wheel of the form shown in Figure 2.25 that has transparent and opaque sectors drawn in a pattern that provides as many channels of digital bits as will be needed for the required resolution.

The digital encoder can also be used to measure angular velocity and acceleration, and this makes it particularly useful for a very wide range of

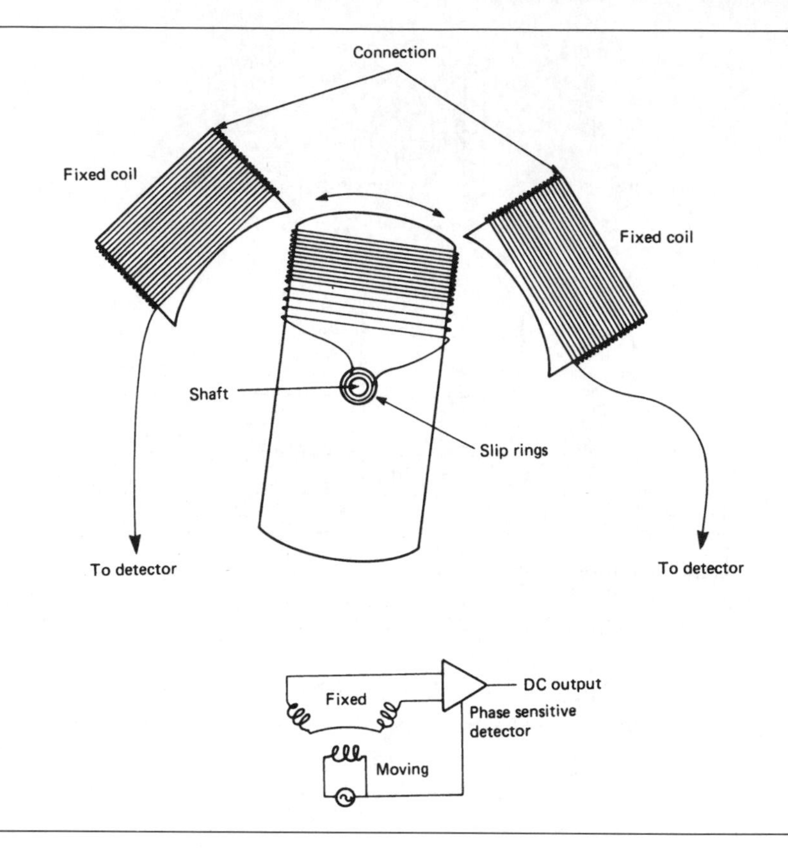

Figure 2.24 A rotary form of LVDT that can be used to sense larger displacements.

applications, assuming that suitable equipment is available to obtain the velocity and acceleration values from the angular position digital codes. This usually implies input to a computer that can run suitable software. If digital methods cannot be used, the changing output from another type of angular velocity measurement can be differentiated by means of an analogue computer stage using an operational amplifier.

In practice, optical encoders often use methods that are closer to analogue than to strictly binary digital methods, employing a disk on which the sectors are equally spaced (Figure 2.26). Such encoders can be obtained in the open form, allowing the disk to be attached to any rotating object, or as sealed units with a shaft that must be coupled to the rotating object. The number of output pulses per revolution is typically 1000–2500, depending on the type, and the signal levels are arranged to be suitable for digital TTL circuits, with a low level at a maximum voltage of 0.4 V and a high level with a minimum of 2.4 V. Maximum permissible disk speeds for the larger enclosed types range from 2400 rpm for the high-resolution disks

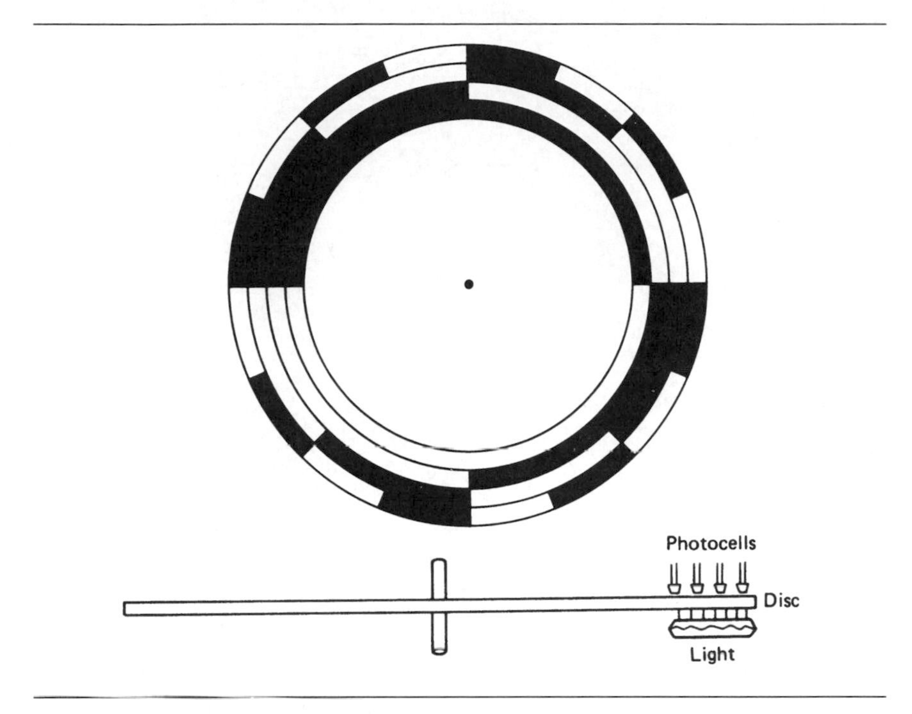

Figure 2.25 A digital rotary optical encoder disk. In this example, only a 16-position disk is shown, but practical examples would use eight tracks to obtain 256 positions.

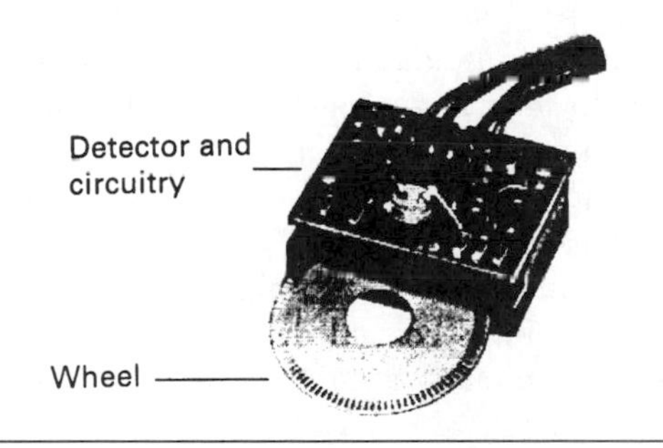

Figure 2.26 A rotary encoder of the type using a single channel, but with two-phase output.

to 6000 rpm for the lower resolution types, corresponding to a maximum operating frequency of 100 kHz. Miniature varieties can be used up to speeds of 30 000 rpm, and the unenclosed disks can typically be used up to 12 000 rpm.

The outputs of these optical encoders are not as a set of binary channels, but as two-phase signals, with the phase angle maintained at 90°.

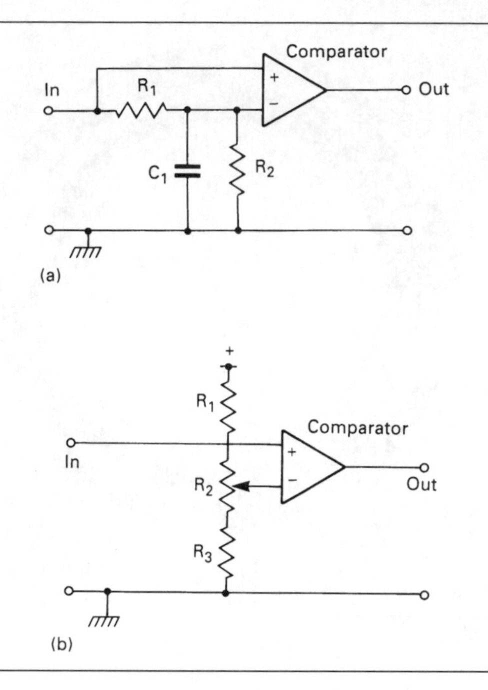

Figure 2.27 Circuits for the rotary encoder (a) using a low-pass filter, not effective at DC, (b) using a biased operational amplifier (op-amp) for a wider frequency range.

Although the signals are square-waves at the lower speeds of rotation, they become approximately sinusoidal at higher speeds. In addition to the phase signals, a synchronization signal is available which allows a phase-sensitive detector to find the direction of rotation.

For the simplest applications in sensing shaft speed, either of the phase outputs can be fed into a comparator, Figure 2.27(a), with the AC signal fed to the positive input and the DC component (because of the low-pass filter) to the negative input. This circuit is applicable only to rotational speeds for which the frequency is high enough to allow the filter to be effective, and for response to zero frequency the circuit of Figure 2.27(b) is better, using a potentiometer to set the input level at the negative input of the operational amplifier to a level midway between the signal peaks. A counter can be connected to the output of the comparator in either case.

Where the direction of rotation must be sensed, the phase differences can be used in a circuit such as that of Figure 2.28, in which the output from a J-K flip-flop is used to indicate direction (other circuits being used to indicate speed). In this circuit, the OR gate (which can be made using two NOR gates) forms a clock input to the J-K flip-flop, and the A and B inputs from the encoder are each taken by way of a comparator to square the signals to an acceptable degree.

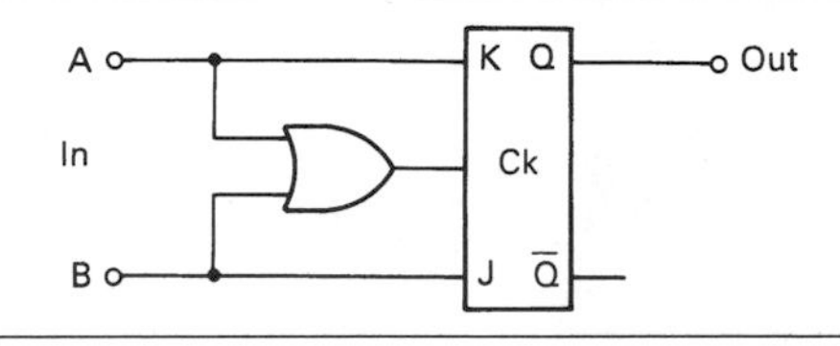

Figure 2.28 A circuit that uses the two-phase output of the rotary encoder to sense the direction of rotation.

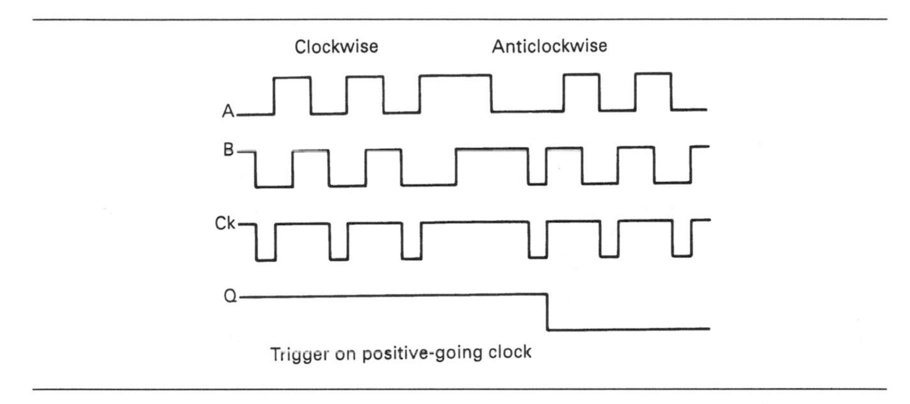

Figure 2.29 The waveforms for the circuit of Figure 2.29, showing how the change of phase is sensed.

The waveforms for this circuit (Figure 2.29) show the action. The J-K flip-flop in this case is one that triggers on the positive-going edge of the clock, making use of the voltage levels at the J and K inputs just prior to triggering. When the rotation is clockwise, the A signal is high and the B signal low just before the rising edge of the clock pulse (the delay in the OR gate will ensure that the clock is slightly delayed with respect to the signals at J or K), and this makes the Q output high. For anticlockwise rotation, the clock occurs when A is low and B is high, reversing the output of the flip-flop.

The resolution of an encoder which can produce as little as 100 pulses per second can be considerably improved by making use of the two-phase output, and a precision of 1° or better is possible by using a dual monostable in the circuit of Figure 2.30.

One particularly useful sensor for angular displacement is the *synchro*, also known as the *selsyn*. This makes use of inductive principles, and a simple synchro system is illustrated diagrammatically in Figure 2.31. The aim is to sense an angular displacement, convert it to phase changes in electrical signals, and reproduce the same angular displacement at a receiver. As the diagram shows, a rotor is fed with an AC signal, typically a 1 kHz sine wave. The rotor is encased by a three-phase stator, with the coils equally

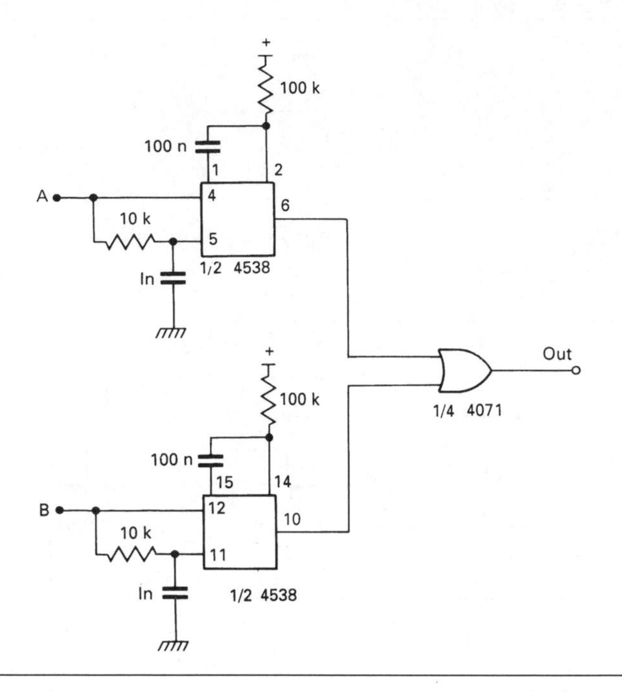

Figure 2.30 A circuit that makes use of both output phases to improve the resolution of the encoder.

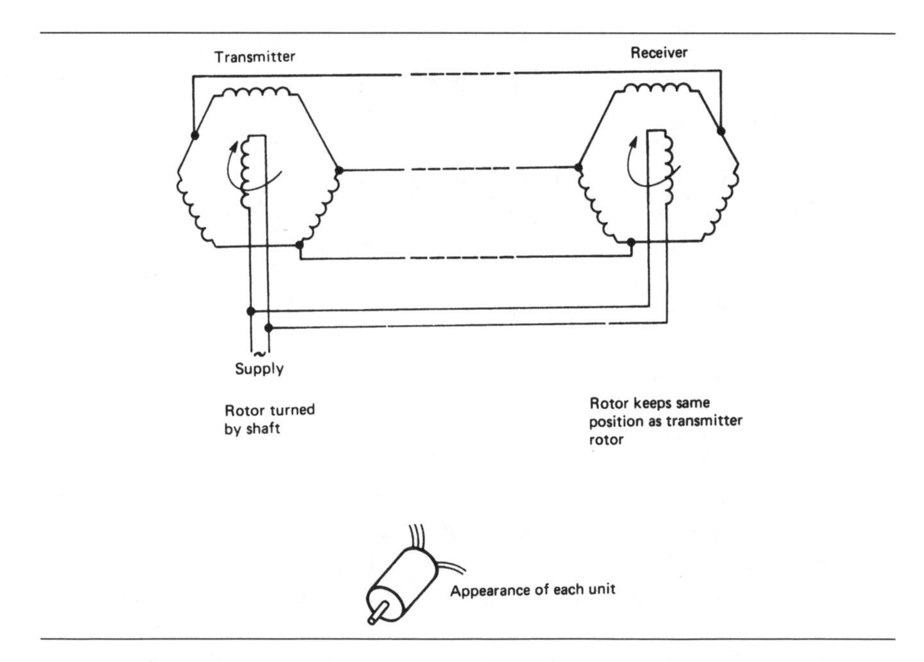

Figure 2.31 Principle of the synchro or selsyn. This is a form of rotary LVDT that uses three stator coils and one rotor. The rotor is fed with an AC supply (often at high frequency), and the connection of the stator coils ensures that when one rotor is moved, the other (slave) units will respond with an identical rotor movement.

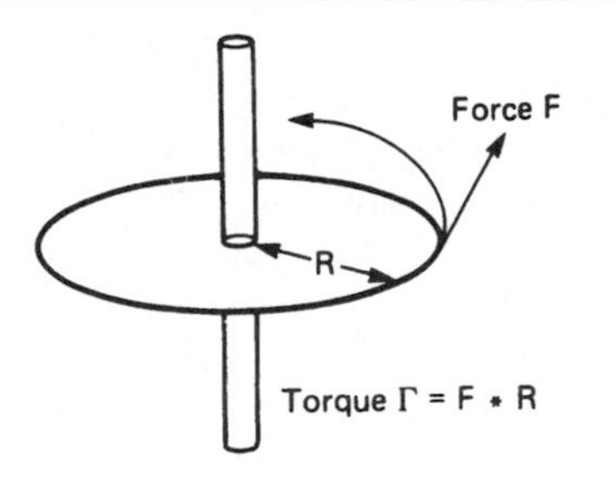

Figure 2.32 Relationship of force, torque and radius for a revolving object.

spaced so as to make the induced voltages at 120° to each other. The coils of this transmitter synchro are connected to the corresponding stator coils of a receiver synchro, whose rotor is fed with AC whose phase is locked to the phase of the AC used for the transmitter.

Any movement of the rotor of the transmitter synchro will cause a change in the amplitudes and phases of the voltages induced in the stator coils of the transmitter, and these same voltages and phases will exist across the receiver coils. The effect will be to cause a rotational force on the rotor of the receiver that will not reach zero until the rotor is at the same angular position relative to its stator coils as exists in the transmitter.

The device can have very low friction, and because the transmission is electrical, and because each unit is fed from a mains supply, the sensitivity can be high and the amount of torque fairly large. The device is used for diverse purposes, such as to transmit the angular position of a radar aerial, the direction of a wind vane, or the reading of a compass. The use of a synchro transmitter connected to a rotating radar aerial along with three-phase coils used for deflecting a cathode ray tube beam (with no rotor coil) was the main method of implementing the PPI (plan position indicator) type of radar for many years.

Finally, the most difficult of the rotational quantities to measure is torque, the rotational equivalent of force. As Figure 2.32 shows, torque can be obtained by using measurements of force or of angular acceleration. For static torque used, for example, to determine to what extent a nut or a bolt has been tightened, simple torque wrenches depend on a spring balance whose scale has been calibrated in terms of torque. Other static torque systems make use of a similar type of force measurement, using a strain gauge mounted on a shaft so as to measure strain in the direction of the circumference of the shaft. This is often the most practical arrangement for analogue measurements, since strain is proportional to stress caused by force. For a digital system, the use of an angular digital encoder and computer allows torque readings to be calculated.

Torque measurement on rotating shafts, as distinct from static measurement, is much more difficult. The conventional method of measuring torque on the shaft of a rotating motor uses a load with frictional coupling that can be adjusted so that the load remains still, but exerts torque on strain gauges. The measurement reliability can be improved by digitally processing the signals, but the method is not really useful for transient torque changes.

A more modern approach does not use mechanical coupling, and depends on the measurement of stator currents and voltages in an electric motor. The associated circuitry then computes the torque values, allowing for fast changes to be measured.

In addition to position, direction, distance and motion of an object, it may be necessary to sense the near presence of an object. This requires proximity detection, and the topic is discussed in Chapter 6.

Chapter 3

Light and associated radiation

3.1 Nature of light

Light is an electromagnetic radiation of the same kind as radio waves, but with a very much shorter wavelength and hence a much higher frequency. The relation between wavelength, frequency and the velocity of electromagnetic waves is given by:

$$\lambda\nu = c$$

where λ is wavelength in metres, ν is frequency in Hz, and c is the velocity in m/s. The velocity of light in free space is 3×10^8 m/s.

The place of light in the range of possible electromagnetic waves is shown in Figure 3.1, along with an expanded view of the portion of the range which we class as infrared, visible and ultraviolet. This range of wavelengths, i.e., the *spectrum* of electromagnetic waves, is the subject of this chapter, and the shorter wavelengths (higher frequencies) are considered separately from the longer wavelengths of electromagnetic (radio) waves for two important reasons. One is that this range of waves is not generated or detected by the conventional electronic methods that are used for waves in the millimetre to kilometre range.

The other point is that the very short wavelengths of this range cause effects that are not a problem when we work with radio waves. One such problem is *coherence*, mentioned earlier in conjunction with laser interferometers. Light from a small portion of a source, such as light passing through a pinhole, can be coherent over a short distance (a fraction of a millimetre), but only laser light is coherent over distances of a metre or more. Only coherent light exhibits the effects of interference which are so common with radio waves, though light is diffracted at edges and pinholes, causing spurious images.

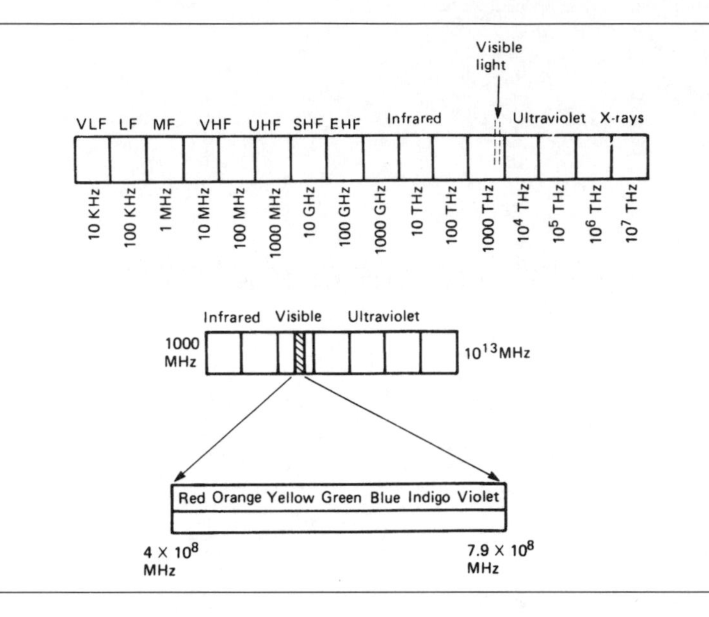

Figure 3.1 The spectrum of electromagnetic waves ranging from 10 kHz upwards. Light is a very small part of this complete spectrum, and the waves that can be used for radio communication are at the lower frequency end.

Like any other form of radiated wave, light can be *polarized*, and this is a topic that is of importance in some applications. Normal unpolarized light consists of waves whose direction of oscillation can be in any plane at right angles to the direction of motion (Figure 3.2). When such light is passed

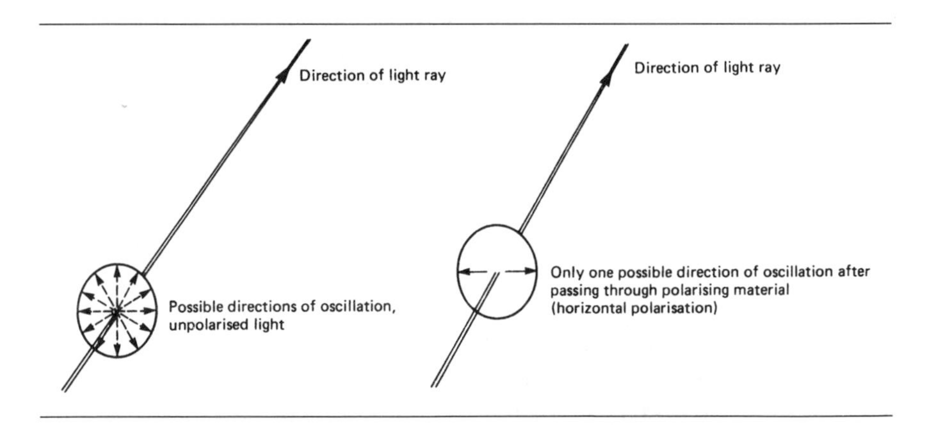

Figure 3.2 Light and polarization. Light is electromagnetic radiation that consists of an electric oscillation and a magnetic oscillation. The electric oscillation (a) can be directed anywhere at right angles to the motion of the light, but a polarized beam (b) has its direction of electric oscillation fixed. The magnetic oscillation is always at right angles to both the electric oscillation and the direction of the beam.

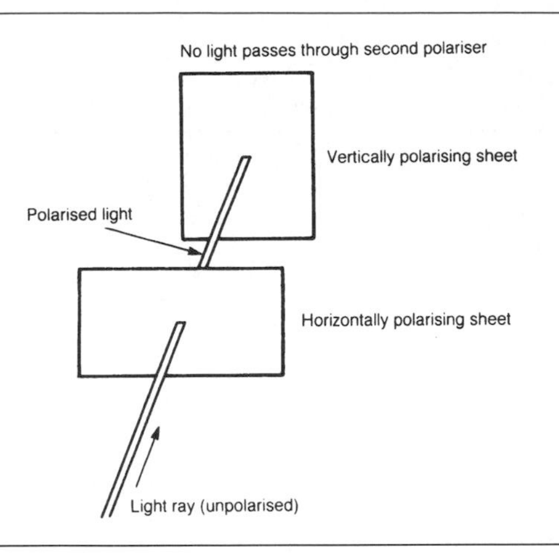

Figure 3.3 Polarizers and their effect. Several natural crystals, and some synthetic materials, can plane-polarize light, confining the direction of the electrical oscillation to one direction. When two sheets of such materials are placed in the path of a light beam and arranged so that their directions of polarization are at right angles, no light will pass through the second polarizer.

through a polarizing material, the light can become plane-polarized, meaning that the oscillation is in one plane of the material only.

Materials that polarize light in this way are generally crystals or other materials that contain lines of atoms at a critical spacing. If polarized light is beamed on another sheet of polarizing material, the amount of light that passes through the material depends on the angle of the sheet, because two sheets of polarizing material with their planes of polarization at right angles will not pass any light (Figure 3.3). Like radio waves, light can also be polarized in other ways (circular, elliptical polarization), but plane polarization of light is the most common.

Like the other waves in the electromagnetic range, light (considering the broad range of infrared to ultraviolet) has a velocity in space that is fixed at about 3×10^8 m/s, so that the usual relationship between wavelength and frequency (Table 3.1) holds good. When light travels through transparent materials, however, the velocity is reduced, and the factor by which the velocity is reduced is called the *refractive index*, a number that is always greater than unity. The velocity of light in any transparent material is therefore equal to the free-space velocity of 3×10^8 divided by the refractive index value for the material. For visible light, refractive index values for materials range from just above unity to about 2 – the highest values are for gemstones, notably diamond.

Refractive index values for infrared can be greater than this, and materials that are opaque to visible light may be quite transparent to infrared – an example is rock-salt. This illustrates how critical the effect of

Table 3.1 The relationship between wavelength, frequency and speed of waves. The figure of 3×10^8 m/s is for waves in free space (vacuum or air), and the speed can be significantly slower in other materials.

Speed of light $= f \times \lambda = 3 \times 10^8$ in free space

$$f = \frac{3 \times 10^8}{\lambda}$$

$$\lambda = \frac{3 \times 10^8}{f}$$

Note: The speed of light and other electromagnetic waves is lower in media such as glass

the wavelength of the radiation can be. In general, the interaction between an electromagnetic radiation and a material can be expected to be critical when the wavelength of the radiation is of a value similar to the distance between atomic particles in the material.

Light radiation carries energy, and the amount of energy carried depends on the square of the amplitude of the wave. In addition, the unit energy depends on the frequency of the wave. This concept of unit (quantum) energy is seldom considered when longer wavelengths are being used, but it determines, to a very considerable extent, what can be done using light waves and particularly the sensing and transducing actions. The quantum nature of light will be explained in more detail when we consider the action of the vacuum photocell, the effect of which was first discovered and explained at the turn of the century. The explanation of photoelectric emission, incidentally, was the achievement for which Albert Einstein won his first Nobel prize.

3.2 Colour temperature

The colour temperature (or equivalent temperature) of light and other radiation is often quoted, and for anyone unfamiliar with the principle it can be very misleading. Any hot object, meaning an object whose temperature is greater than that of absolute zero, will radiate energy, but the spectrum of that energy, meaning the relative percentage of the energy that is radiated at each detectable frequency, will depend on the temperature of the object. When an object and its surroundings are at the same temperature, each radiates to the other and the amount of energy leaving the object is balanced by the amount entering, keeping the temperature constant. The radiation is always over a wide band of frequencies, with a definite peak at one frequency. Figure 3.4 shows typical graphs of radiated power density (W/m^2) plotted against wavelength of light. The vertical

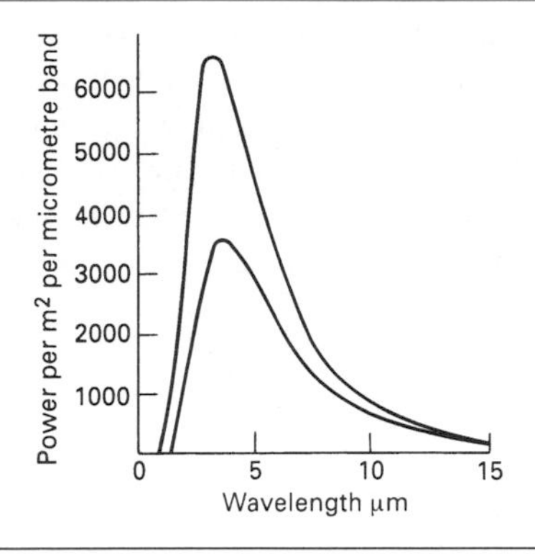

Figure 3.4 Radiated power and temperature. The upper curve is for an object at a higher temperature, radiating more power overall and with the peak at a shorter wavelength (higher frequency).

axis is scaled in terms of power radiated per m^2 of area per unit bandwidth; the horizontal axis is scaled in μm of wavelength.

Since any object hotter than absolute zero will radiate, the radiation temperature is measured in Kelvins (degrees absolute) which for all practical purposes are calculated by adding 273 to Celsius temperature. At low absolute temperatures, the radiation is predominantly in the far infrared, with no trace of near IR (which would be easier to detect) or visible light. As the temperature is raised, the spectrum shifts towards the higher frequencies and the absolute amount of energy at each frequency increases. In other words, as an object is heated, it radiates very much more energy and the energy is predominantly of a shorter wavelength. The amount of energy radiated is proportional to the fourth power of the absolute temperature, so that an object at 1000K is radiating 16 times as much as one at 500K.

At a temperature of around 700K, the spectrum of radiation has shifted sufficiently for some of the radiation to be in the visible region, so that we say that the object is *red-hot*. As the temperature increases, the spectrum continues to shift, with the simple relationship that the wavelength of the peak multiplied by the absolute temperature is a constant (double the absolute temperature and you halve the wavelength of the predominant radiation). In addition, the bandwidth of the spectrum increases. Temperatures of 3000K or so will provide light that is predominantly blue, approaching the spectrum of sunlight.

The spectrum of any radiation can therefore be described very compactly in terms of its *colour temperature*, meaning the temperature to which a perfectly radiating object (a *black body*) would need to be raised to radiate in the same way. To say that a light has a colour temperature of 3500K does not necessarily mean that it comes from an object at that temperature, only that it has the same composition in terms of predominant frequency and bandwidth as light from an object at that temperature. Fluorescent tubes, for example, can give light of a high colour temperature but are themselves cool to touch. In this example, the discrepancy is due to the small fraction of the contents of the tube that is actually radiating. There will usually be a discrepancy between actual temperature and colour temperature for any radiating material.

3.3 Light flux

The sensitivity of photocells can be quoted in either of two ways: as the output at a given illumination, using illumination figures in units of lux, often 50 lux and 1000 lux; or as a figure of power falling on the cell per cm^2 of sensitive area, a quantity known as *irradiance*. The lux figures for illumination are those obtained by using photometers, and a figure of 50 lux corresponds to a 'normal' domestic lighting level good enough for reading a newspaper. The level of illumination required for close inspection work and the reading of fine print is 1000 lux; on this scale, direct sunlight registers at about 100 000 lux. The use of mW/cm^2 looks more comprehensible to anyone brought up with electronics, but there is no simple direct conversion between power per cm^2 and lux unless other quantities such as spectral composition of light are maintained constant. For the range of wavelengths used in photocells, however, you will often see the approximate figure of $1\ mW/cm^2 \approx 200$ lux.

Another important point relating to the use of photocells is their wavelength at peak sensitivity. For many types of sensors, this may be biased to either the red or the violet end of the visible spectrum, and some sensors will have their peak response for invisible radiation either in the infrared or the ultraviolet. A few devices, notably some silicon photodiodes, have their peak sensitivity for the same colour as the peak sensitivity of the human eye. This makes the devices more suitable for use in automated processes that once involved visual inspection, but such replacements are not always successful. The reason is that although the peak sensitivity of the sensor may match that of the eye, the sensitivity at other colours may not follow the same pattern as that of the eye. In general, most sensors are more sensitive than the eye to colours at the extremes of the spectrum, and if a photosensor is to be used in applications such as colour matching, then filters will have to be placed in the light path to make the response curve over the whole spectrum match that of the eye more evenly. Figures 3.5 and 3.6 show some spectral response curves for selenium and for photoresistive cells, respectively.

3.4 Photosensors

Photosensors are classed by the physical quantity that is affected by the

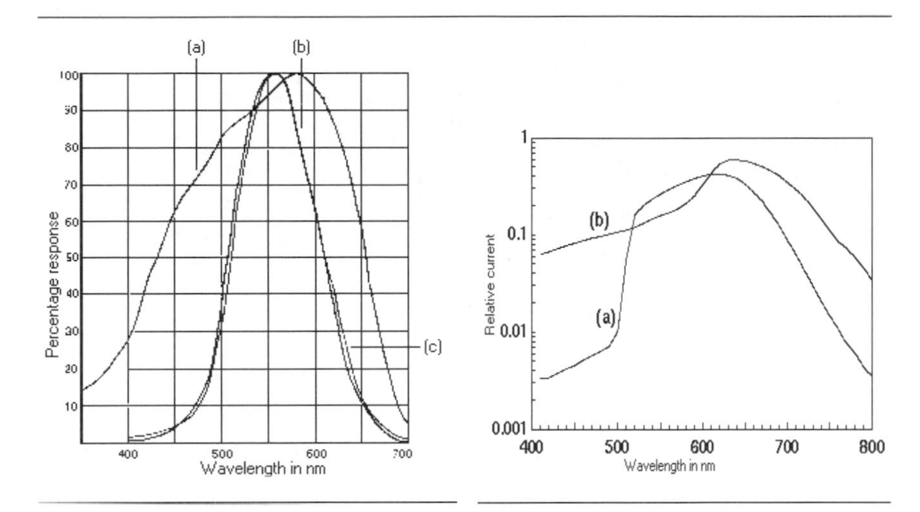

Figure 3.5 Response curves for selenium cells: (a) normal response, (b) modified response, (c) response of human eye.

Figure 3.6 Response curves for photoresistors: (a) cadmium sulphide cell, (b) cadmium sulphide and cadmium selenide mixture.

light, and the main classes are photoresistors, photovoltaic materials, and photoemitters. Historically, photoemitters have been more important in unravelling the theory of the effect of light on materials, but photovoltaic materials, notably selenium, were in use for some considerable time before the use of photoemission became practicable. Since photoemission allows us to combine the description of a usable device with the quantum effect, we will consider this type of sensor first, which can also be used to a limited extent as a transducer. The photoemissive cell, in fact, was the dominant type of photosensor for many years, and played a vital part in the development of cinema sound since it was the transducer first used for converting the film soundtrack into audio-electrical signals.

A simple type of photoemissive cell is illustrated in Figure 3.7. This is a vacuum device, because the photoemissive material is one that oxidizes instantly and violently in contact with air, and even in quite low pressures it will oxidize sufficiently to prevent any photoemission. The photoemissive action is the release of electrons into the surrounding vacuum when the material is struck by light. This release of electrons will take place whether the electrons have anywhere to go or not, but unless there is a path provided, all the electrons will lose energy and return to the emitting surface, often termed the photocathode. By enclosing another metal – a nickel wire, for example – which is at a voltage more positive than the photocathode, the cell can be used as part of a circuit in which current will flow when light strikes the photocathode. When light releases electrons, the current that flows is proportional to the amount of energy carried by the

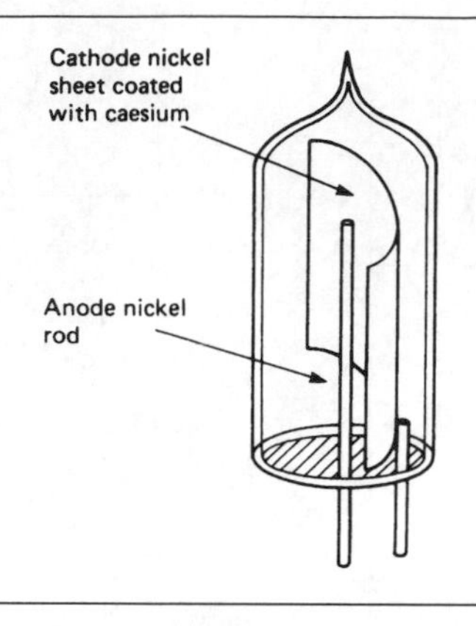

Figure 3.7 A typical photemissive cell. The cell must be contained in an evacuated enclosure, and consists of a nickel rod anode with a nickel sheet cathode coated with a photoemissive material such as caesium.

light beam. This, however, is true only if the light is of a frequency that will release electrons. For any photoemitting material there will be a *threshold frequency* of light. Below this threshold frequency, electrons will not be emitted no matter how intense the light happens to be, so that, in general, photoemitters do not respond to infrared, particularly the far infrared. The explanation of this threshold effect is due to Einstein, and makes use of an idea that was earlier put forward by Planck.

Planck's theory was that energy existed in units just as materials exist in atoms, and he named the unit the *quantum*. The quantum is a unit of *action*, a physical quantity that was not considered of practical interest prior to Planck's theory. The size of the quantum for a light beam is equal to the frequency of the light multiplied by a constant that we now call Planck's constant. We touched on this idea when considering laser interferometry, because the factor that prevents light from most sources from being coherent is that it is given out in these quantum-sized packets rather than as a truly continuous beam.

Einstein reasoned that the size of the quantum affected the separation of an electron from an atom in a photoemitter, and if the amount of energy carried in one quantum was less than the amount of energy needed to separate an electron, then no separation would take place. The total amount of energy carried by the beam was of no importance if the units of energy were insufficient to separate the electrons. The theory was confirmed by experiment, and this work also established Planck's quantum theory as one of the main supports of modern physics.

The practical effect as far as we are concerned is that photoemissive cells have a limited range of response, and whereas it is easy to make cells that sense ultraviolet light (high frequency), it is much more difficult to prepare materials that will sense infrared. Most photocells are noticeably much

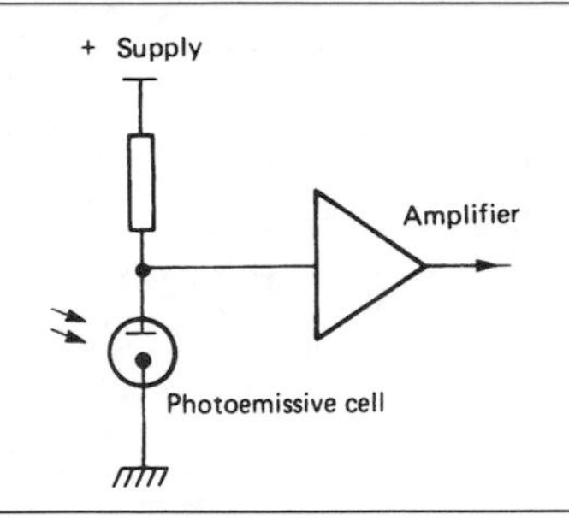

Figure 3.8 The use of a photoemissive cell in a circuit. The supply voltage is in the range +25 V to +300 V with a load resistor to provide a signal voltage to the amplifier.

more sensitive to light in the blue/violet end of the spectrum than in the orange/red end. Mixed photocathode materials containing antimony along with the alkali metals caesium, potassium and sodium, have been the most successful emitters in terms of providing an electrical output that is reasonably well maintained for the visible frequencies of light. This does not imply, however, that the output is by any means uniform.

The photoemissive cell is used in a circuit of the type shown in Figure 3.8, with a voltage supply that is often in the region of 25–100 V between anode and cathode. Some circuits make use of the current through the cell directly, but the most common circuit is as shown here, using a load resistor in series with the cell and amplifying the voltage signal. This is particularly useful when the incoming light is modulated in some way so that the electrical output will be an AC signal, such as that in use as a cinema soundtrack transducer.

Where a DC output is needed, the use of a photoemitter is less simple, and a current amplifier is more useful than a load resistor and voltage amplifier. The current through a typical cell is of the order of a microamp, so that fairly large load resistors and a considerable amount of amplification will be needed. This creates difficulties with both frequency response and noise level when the photocell is used to convert modulated light signals into AC signals.

For measuring purposes, any photoemitter detector will have to be calibrated for each light colour for which it will be used, and if a reasonably high precision is needed, this calibration will be a long and tedious matter. It is important to realize that 'white' light is a mixture of all the frequencies of the visible spectrum, and can also contain a proportion of invisible ultraviolet. This means that imperceptible changes in the composition of 'white' light which have no effect on its total energy will nevertheless have very large effects on the output from a photoemissive cell. This is, in fact, something that affects most photosensitive devices, as keen photographers will know.

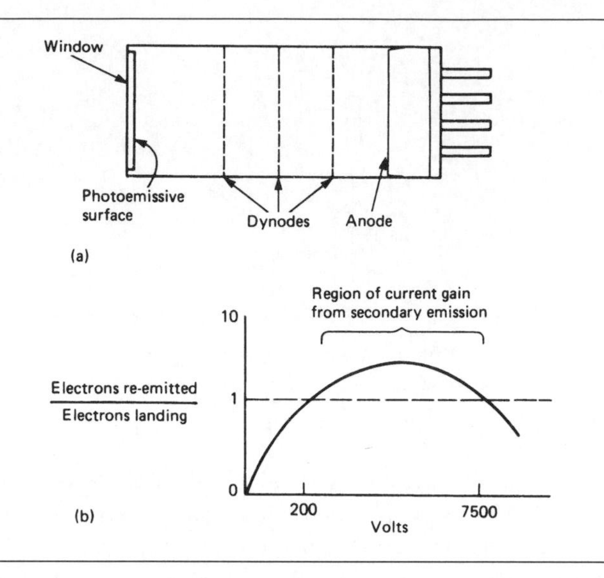

Figure 3.9 (a) A photomultiplier using three dynodes (or secondary multiplying electrodes). The electrons emitted from the transparent photocathode are multiplied in the dynode stages to provide a much greater output current than would be obtained from the original photocurrent. (b) The secondary emission characteristic for a typical dynode material. Materials that are good photoemitters are usually good secondary emitters as well.

Comparisons of light levels must therefore be made only when the composition of light is constant, and this is something that is very difficult to achieve, particularly for natural lighting. For artificial lighting using fluorescent tubes, constant light composition is much easier to achieve, but filament lamps give a light that contains a large fraction of red light and for which the light composition varies very sharply as the voltage is altered. A filament lamp which is run below its rated voltage gives light that is predominantly red; when run above its rated voltage it can give light that is biased to the blue end of the spectrum (with a greatly reduced life).

Photoemitters are still used where a fast response is needed, because many competing devices are solid-state rather than high-vacuum, and the speed of electrons in a solid is very much lower than the speeds that can be obtained in a vacuum. Even for cinema soundtracks, however, the vacuum photocell has now been replaced, and at the time of writing, vacuum devices are found only in old equipment and in instruments intended for specialized use. For these reasons, then, more detailed descriptions of vacuum photoemissive cells will not be given here. However, the photomultiplier obtains very much greater sensitivity from a photocell at very little cost in noise or time delay.

The principle, illustrated in Figure 3.9, makes use of *secondary emission*

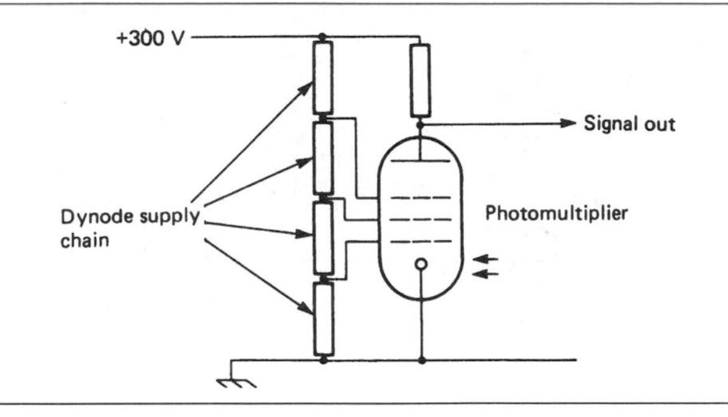

Figure 3.10 The electrical arrangement for the photomultiplier, using a chain of resistors (1–12 MΩ values) to supply the dynodes.

from the same type of materials as are used for photocathodes, notably caesium. Secondary emission occurs when a material is struck by electrons and releases more electrons than strike the surface initially. The effect is at its peak for electrons that have been accelerated by a voltage in the range of 30–200 V, and for surfaces of caesium the multiplication factor can be large, for example, 3–7. This means that an electron beam from a cathode can be directed to a secondary emitting surface and be re-emitted as a beam that contains a much larger number of electrons. In practical terms, this means a beam at a higher current.

By cascading stages, this multiplication effect can be very large, so that, for example, five stages that each give a multiplication of 5 will increase the current of a beam from a photoemitter by a factor of $5 \times 5 \times 5 \times 5 \times 5 = 3125$. This is unique among amplification methods in being virtually noiseless, and photomultipliers are used in detectors for very low light levels for specialized purposes. The secondary multiplying electrodes are known as dynodes, and each dynode must be run at a voltage level that is substantially greater than the one preceding it – Figure 3.10 shows a typical DC supply arrangement.

3.5 Photoresistors and photoconductors

Many materials have a resistance value that will change when light strikes the material. The theory behind this effect is that these materials are semiconductors that in their normal state have few free electrons or holes. The effect of light is to separate electrons from holes and so allow both types of particles to move through the material and carry current. Since a definite amount of energy is needed in order to separate an electron from a hole, the size of the light quantum is important, but it is not difficult to find materials for which the amount of energy is small, corresponding to a

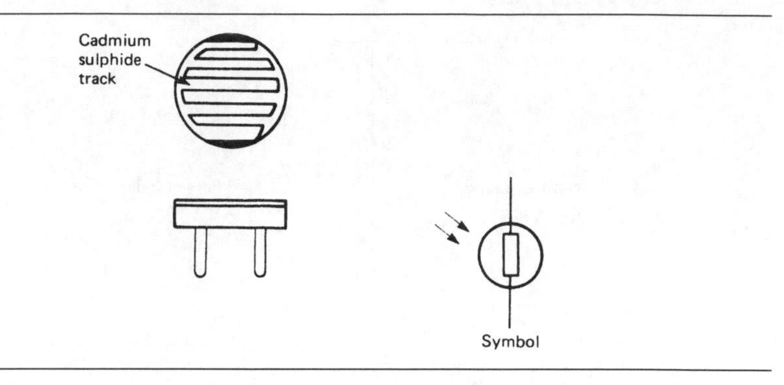

Figure 3.11 The photoconductive cell, using a cadmium sulphide track, is by far the most common type of photosensor used industrially. It is mechanically and electrically rugged, and has a good record of reliability. The alternative name is LDR (light-dependent resistor).

quantum of infrared light. Since, in addition, the materials are comparatively easy to prepare in a practical form, the use of photoresistive or photoconductive materials is very common. The two names should really be synonymous, but some lists show devices under one name or the other.

Because of the physical action, the effect is always that the material has high resistance in the absence of light, and the resistance drops when the material is illuminated. The effect of light on a photodiode is also one of reducing resistance, in this case the resistance of a reverse-biased diode, and the reason for the use of a separate name is that the manufacturing process for photodiodes is the same as is used for other semiconductor devices.

The most common form of photoconductive cell is the cadmium sulphide cell, named after the material used as a photoconductor. This is often referred to as an *LDR* (light-dependent resistor). The cadmium sulphide is deposited as a thread pattern on an insulator, and since the length of this pattern affects the sensitivity, the shape is usually a zigzag line (Figure 3.11). The cell is then encapsulated in a transparent resin or encased in glass to protect the cadmium sulphide from contamination by the atmosphere.

The cell is very rugged and can withstand a considerable range of temperatures, either in storage or during operation. The voltage range can also be considerable, particularly when a long track length of cadmium sulphide has been used, and this type of cell is one of the few devices that can be used with an AC supply. Table 3.2 shows typical specifications for a popular type of cell, the ORP12. The peak spectral response of 610 nm corresponds to a colour in the yellow–orange region, and the dark resistance of 10 MΩ will fall to a value in the ohms to kilohms region upon illumination. The sensitivity is not quoted as a single figure because the change of resistance plotted against illumination is not linear. The maximum voltage

Table 3.2 The characteristics of the ORP12 type of photoconductive cell (LDR, courtesy of RS Components Ltd).

Peak spectral response	610 nm
Cell resistance at 50 lux	2400 Ω
Cell resistance at 1000 lux	130 Ω
Dark resistance	10 MΩ
Max. voltage (DC or peak AC)	110 V
Max. dissipation at 25°C	200 mW
Typical resistance rise time	75 ms
Typical resistance fall time	350 ms

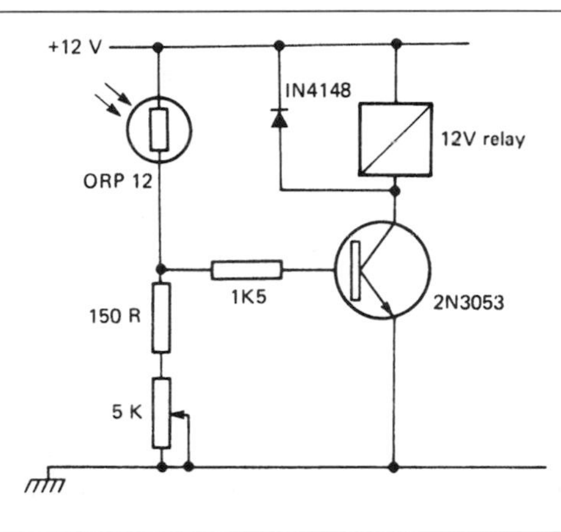

Figure 3.12 A typical ORP12 industrial application circuit (courtesy of RS Components Ltd).

rating of 100 V DC or AC peak allows a considerable latitude in power supplies, subject only to the 200 mW maximum dissipation. A typical spectral response graph was illustrated earlier in Figure 3.5.

The peak response at the yellow–orange region of the spectrum makes this LDR particularly useful for detecting the presence of a flame. The ORP12 finds its main application in control units, in particular for vaporizing oil-fired boilers (so that oil is cut off if the flame is extinguished) and in fire detecting equipment. In such applications, the main drawback of the LDR, which is a comparatively long response time, is not a drawback. The typical resistance fall time, meaning the time for the resistance to reach its final value when the cell is illuminated, is 350 ms; the resistance rise time when illumination is cut off is typically 75 ms. These times are reasonable when compared to the operating times of the relays that the ORP12 is so often used to operate, but they make any type of use with modulated light beams out of the question. A typical circuit using the ORP12 is illustrated in Figure 3.12.

Table 3.3 Characteristics of a typical general-purpose silicon photodiode (courtesy of RS Components Ltd).

Peak response wavelength	750 nm
Typical sensitivity	0.7 μA/cm^2
Dark current at −20 V	1.4 nA
Temperature coefficient of dark current	×2 per 10°C
Reverse breakdown voltage	−80 V at 10 μA
Temperature coefficient of change of signal current	0.35% per °C
Max. forward current	100 mA
Max. dissipation	200 mW at 25°C
Capacitance at −10 V bias	12 pF
Response time	250 ns

3.6 Photodiodes

A photodiode is a type of photosensor in which the incident light falls on a semiconductor junction, and the separation of electrons and holes caused by the action of light will allow the junction to conduct even when it is reverse biased. Photodiodes are constructed like any other diodes, using silicon, but without the opaque coating that is normally used on signal and rectifier diodes. In the absence of this opaque coating, the material is transparent enough to permit light to affect the junction conductivity and so alter the amount of reverse current that flows when the diode is reverse-biased.

Because this is a reverse current diode, its amplitude is not large, and the sensitivity of photodiodes is quoted in terms of μA of current per mW/cm^2 of incident power. For the normal range of illuminations, this corresponds to currents of 1 nA to 1 mA at a reverse bias of −20 V. The lower figure is the dark current, the amount of current that flows with no perceptible illumination. As with all semiconductor devices, the dark current of a photodiode will increase considerably as the temperature is increased, doubling for each 10°C rise in temperature.

Table 3.3 shows the characteristics of a typical silicon general-purpose photodiode from the RS Components catalogue. The peak spectral response is at 750 nm, which is in the near infrared, and the sensitivity is quoted as 0.7 μA/mW/cm^2. The typical dark current at a bias of −20 V is 1.4 nA, which means that the minimum detectable power input is of the order of 2 μW/cm^2.

The current plotted against the illumination gives a reasonably linear graph, and the response time is around 250 ns, making the device suitable for modulated light beams that carry modulation into the video signal region. Figure 3.13 shows the suggested circuit for using this type of photodiode along with an operational amplifier for a voltage output. The

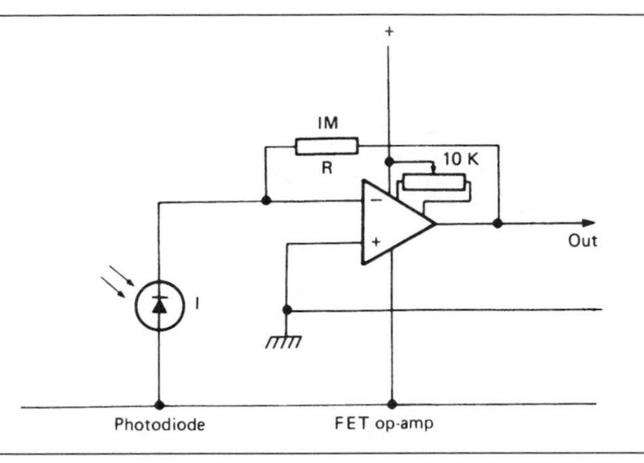

Figure 3.13 A circuit making use of the silicon photodiode (courtesy of RS Components Ltd).

feedback resistor R will determine the output voltage, which will be $R \times I$, where I is the diode current. Note that when an operational amplifier is used in this way, the frequency response of the system is determined more by the operational amplifier and the stray capacitance across the feedback resistor than by the photodiode.

3.7 Phototransistors

A phototransistor is a form of transistor in which the base–emitter junction is not covered and can be affected by incident light. The base–emitter junction acts as a photodiode, and the current in this junction is then amplified by the normal transistor action so as to provide a much larger collector current, typically 1000 times greater than the output current of a photodiode. The penalty for this greatly increased sensitivity is a longer response time, measured in μs rather than in ns, so that the device is not suitable for detecting light beams that have been modulated with high-frequency signals. Phototransistors are used to a smaller extent nowadays because it is just as simple to make a chip containing a photodiode integrated with an operational amplifier, and better response times can usually be obtained in this way. The phototransistor is still found in conjunction with an infrared emitter used for such purposes as punched tape readers, end of tape detectors, object counters and limit switches.

3.8 Photovoltaic devices

The first form of photovoltaic device was the selenium cell, as used in early types of photographic exposure meters. The principle is that the voltage across the cell is proportional to the illumination, and since for the selenium cell the voltage was of an appreciable size (of the order of 1 V in bright illumination), an exposure meter using this type of cell needed no amplification and could use a meter of reasonably rugged construction. The use of selenium is now only of historical interest – the properties of the metal were discovered by Wheatstone (of Wheatstone Bridge fame) when he was trying to construct high-value resistors to act as standards in a bridge circuit to measure the resistance of a trans-Atlantic telegraph cable.

Modern photovoltaic devices are constructed from silicon, and the construction method is as for a photodiode. A silicon photovoltaic device is a silicon photodiode with a large area junction and used without bias. It is connected into a large load resistance, and the typical voltage output is of the order of 0.25 V for bright artificial illumination of 1000 lux. The main application of such cells is in camera exposure controls, because with suitable filtering the peak response and the response curve can be made very close to that for the eye. For the same reasons, the device can be used for the monitoring and control of light levels in critical manufacturing processes, and in instruments used for checking light levels. The photovoltaic cell can also be used as a photodiode, and if amplification is to be used, this mode is generally preferred to its use as a photovoltaic device.

3.9 Fibre-optic applications

The increasing use of optical fibres along with photosensors and optical transducers has led to the development of a range of devices that are specially intended for use with fibres. The detecting devices are photodiodes, so that the principles are the same as have already been discussed, but the physical form of each device must be suited to coupling to the fibre-optic cable. Fibre-optic links have become a practical proposition only since suitable terminations and connectors were developed, and it is futile to expect to be able to couple a general-purpose type of photodiode or emitter to a fibre cable. Several manufacturers offer complete fibre systems, composed of emitters, detectors, couplers and cables for either experimental work, assessment, or production use.

3.10 Light transducers

The conversion of energy between electrical and light forms has been of considerable importance for more than a century. The conversion of electrical energy into light has been, and still is, performed mainly by using the electrical energy to heat a coil of wire to a high temperature so that it gives out light. The light spectrum that is obtained in this way depends on the temperature of the object, and practical limitations usually result in the light having a spectrum that is biased to the red, so that the light from sources of this type, i.e., incandescent sources, is unsuitable for colour matching. This type of transducer is, however, the simplest to manufacture and use. It also has the lowest efficiency of conversion.

Early designs of light bulb used carbon filaments, which were fragile (some, astonishingly, have survived and are still working), and the bulbs were evacuated to avoid oxidation of the carbon. The improvements made by Swan to the original Edison design included the use of the metal tungsten as a filament material, and the use of argon gas in the bulb. The inert nature of the gas avoided oxidation, and the pressure of the gas (although less than atmospheric pressure) avoided problems due to evaporation of the metal of the filament. Early tungsten-filament lamps failed in an hour or less because of the blackening of the glass caused by the tungsten vapour condensing. Although blackening is still the main cause of failure, the time has been greatly extended through the use of the inert gas in the bulb.

The development of gas-discharge lamps has resulted in a variety of light sources of a very different type, all very much more efficient than the incandescent lamp (usually at least five times the conversion percentage of the filament lamp). The original types of gas-discharge devices used gases such as neon at low pressure, so that high voltages in the range 1–15 kV were needed to operate the devices. These were, and are, the familiar neon signs, in which the gas is nowadays seldom neon but more usually a mixture of gases that have been chosen to achieve some particular colour. The colour of the light from discharge tubes is the colour of the predominant spectral lines of the gas or vapour that is used, and this leads to the light being in many cases almost monochromatic, or of one single colour. Low-pressure discharge lamps are, therefore, unsuitable for general application to illumination and they remain used predominantly for advertising signs.

Vapours of liquids and solids have also been used in discharge lamps, and the most familiar of these are the mercury and sodium lamps that are used for street illumination. Used at low pressures, these vapours give strongly coloured light in which colour discrimination is almost impossible, but when higher pressures are used, the light spectrum broadens to give results that are more acceptable. The sodium high-pressure lamps are considerably more visible in fog, due to the orange–red predominance in the light, and are used extensively for motorway and main road illumination.

All varieties of discharge lamps require control gear, in the sense that they cannot simply be connected to a voltage source. Figure 3.14 illustrates the normal current–voltage characteristic of a discharge lamp, and shows that no current passes until a critical voltage is reached across the terminals. From that point (the ignition voltage), the voltage across the gas drops as the current is increased, a negative-resistance characteristic. This would cause the lamp to burn out if there was no way of controlling the current.

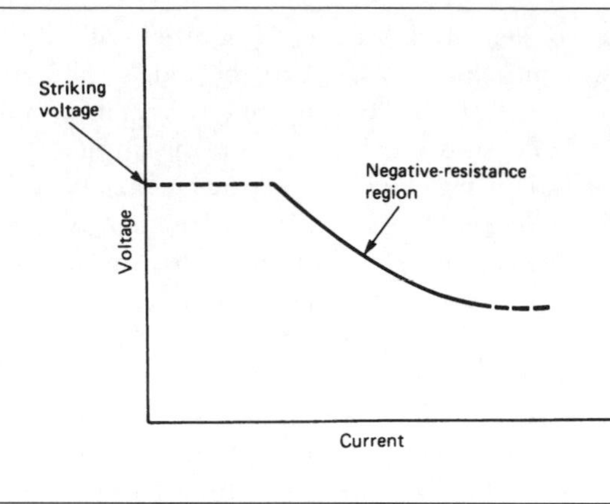

Figure 3.14 A typical gas-discharge characteristic, showing the striking voltage and the negative-resistance region.

These lamps are therefore normally operated with AC, using an inductor (choke) in series with the supply if the running voltage is less than the supply voltage, or a transformer with current-limiting inductance if the running voltage is higher than normal supply voltage. This additional equipment reduces the overall efficiency of the lamp, but is an essential part of the system.

For domestic use, the familiar fluorescent tube has been available for some considerable time now. This is not such a physically simple device as the older types of discharge tube, however, because it incorporates two energy conversions in series. The conversion from electrical to light energy is made by a discharge through mercury vapour at a pressure less than atmospheric, but this results in light that is mostly in the violet and ultraviolet range. The tube is coated on the inside with a phosphor material that acts as a frequency-changer. The violet/ultraviolet light from the mercury discharge causes energy changes in the atoms of the phosphor material, and as these changes are reversed, the energy is given out again as light in the normal range of visible light.

This system is very flexible, because the colour and composition of the light that is given out depends on the phosphor coating rather than on the mercury discharge. The phosphors are metal silicates with controlled amounts of impurity, and very small changes in composition can cause marked differences in both conversion efficiency and spectrum – the materials are essentially the same as those used for cathode ray tube coatings. Because of this, fluorescent tubes can be bought in almost any colour, and most importantly in a 'sunlight' version whose spectral make-up is that of mid-day sunlight, ideal for colour matching. For domestic purposes, tubes labelled as 'warm' give a light that is more biased to red than the daylight type so that the light output is rather closer to that of the

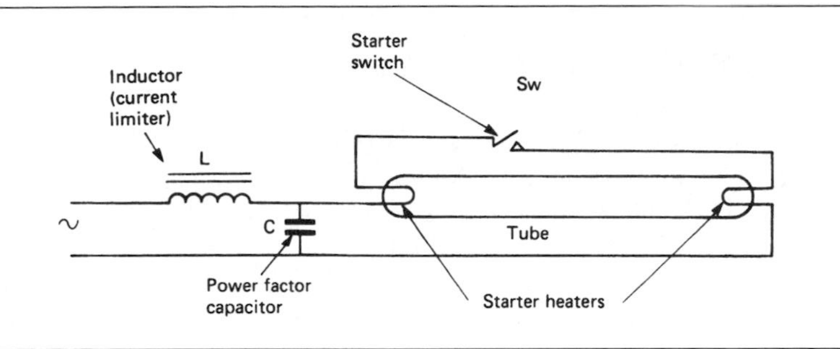

Figure 3.15 The conventional switch starting circuit for a fluorescent tube, along with the inductor and power factor capacitor.

incandescent lamp. Even with the double conversion, the fluorescent is almost five times as efficient as an incandescent lamp.

The control gear of the fluorescent will be either of the switch-start or the quick-start variety. The switch-start type uses the conventional inductor (choke) in series with the tube, and a capacitor for power-factor correction. The starting voltage of the tube is higher than the supply voltage, and the mercury must be vaporized in order to achieve sufficient vapour pressure for starting. This makes starting difficult in cold conditions, which is why a fluorescent in a garage or a loft often fails to light in the winter. The starting circuit (Figure 3.15) contains a thermal timeswitch that will pass current through the heating filaments when the tube is first turned on, and then break this circuit. At the instant when the switch breaks contacts, the back-EMF generated in the inductor causes the voltage across the tube momentarily to exceed the ignition voltage, so that the discharge starts and will then continue if the mercury vapour pressure is high enough. The current is then limited by the choke.

The alternative arrangement uses a transformer with separate windings for the heater filaments, and a high-impedance output greater than the ignition voltage for the main discharge. At switch-on, the filaments are heated and the ignition voltage is applied, so that the tube lights without the flickering or long delay that can be an annoying feature of switch starters. The high impedance of the secondary transformer then controls the tube current.

A more modern variation on this scheme uses an electronic converter to generate the heater and main voltages, so that the tube can be made with its control gear in one unit, and plugged directly into a domestic lamp socket. Tubes of this type are now readily available for domestic use, and combine long life with high efficiency. Other tubes with incorporated control gear still use the switch-start principle and so exhibit the disadvantages of slow flickering starts. All domestic fluorescent lamps of the compact type have a slow warm-up, so that the normal light output is not attained until the lamps have been in action for several minutes.

3.11 Solid-state transducers

Since the invention of the transistor, work on electronic devices that made use of current in solid semiconductors has resulted in the development of many useful devices, among which are the current-to-light transducers classed as LEDs (light emitting diodes). More recently, the place of the LED has been taken to some extent by a device that is, strictly speaking, neither a sensor nor a transducer but a light modulator, the LCD (liquid crystal display).

LEDs are semiconductor diodes in which the voltage drop for conduction is comparatively large. When an electron meets a hole in such a junction, the two combine and release energy that can be radiated if the junction is transparent. The conventional germanium and silicon diodes release only a small amount of energy that corresponds to infrared, and which is absorbed in the non-transparent material (and also by the paint or other coatings on the diode). Compound semiconductors, however, of which the first was gallium arsenide, can give radiation in the visible range and are themselves transparent and allow the light to escape.

The majority of LEDs in use are of gallium phosphide or gallium arsenide phosphide construction, and all are electrical diodes with a high forward voltage of about 2.0 V and a very small peak reverse voltage of about 3.0 V. In use, then, great care is needed to ensure correct polarity of connection so as to avoid burning out the LED. This has been made more difficult by the modern practice of identifying the cathode lead by making it shorter, because once the leads are cut to length for fitting into equipment, the identification is lost. The older form of identifying the cathode by a small flat area on the body of the diode was at least independent of the leads.

The intensity of illumination from the LED depends fairly linearly on the forward current, and this can range from 2 to 30 mA, depending on the physical size of the unit and the brightness required. All but the smallest types can sustain a power dissipation of the order of 100 mW, the main exceptions being the low-current LEDs that are used for battery-operated equipment. The current through the LED is governed by using a resistor in series, although when the LED is controlled by an IC, the resistor is usually incorporated into the IC and need not be wired separately.

The colour of the LED is determined by the material, and the two predominant colours are red and green. The use of red and green sources close to each other gives yellow light, in accordance with the rules on mixing of light colours, so that all three of these colours can be obtained from LED sources. The use of twin diodes in a package can allow switchable red, green and yellow lights to be obtained, and when LEDs are combined in a package with IC digital units, flashing LED action can be obtained.

Single LEDs can be obtained in a considerable variety of physical forms, of which the standard dot and bar types predominate. The bar type of

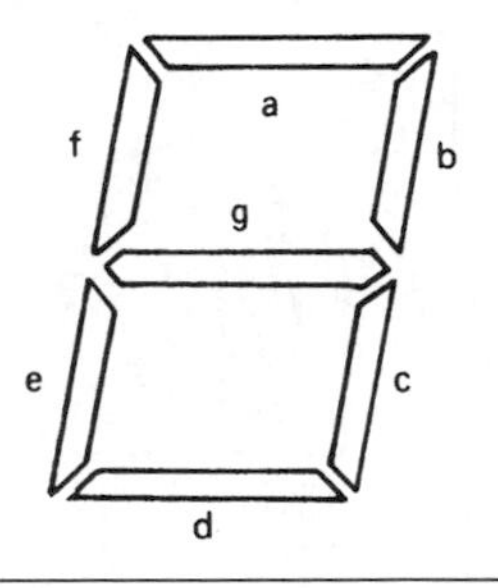

Figure 3.16 The shape and labelling of the sectors of a seven-segment display. Many displays add an eighth segment, a decimal point to the right or left of the main sections. This requires a decoder-driver IC to convert from binary representation into the correct pattern of bars for each number.

LED can usually be obtained in intensity matched form, in which the intensity at a specified current rating is matched closely enough to allow the units to be stacked to be used as a column display. In a display of this type, a diode whose intensity is higher or lower than the others is particularly obvious, hence the need for intensity matching.

The other shapes of LED are not supplied as single units but as preformed patterns. Bar-graph modules of five to 30 segments can be bought as single units, but the most common preformed types are the alphanumeric displays. Of these, the seven-segment type of display, as illustrated in Figure 3.16, is by far the most common. This uses seven bar-shaped segments to display numbers (and a few letters), and despite the name, most of the displays use eight segments because a decimal point is normally included and can be specified on the right or on the left of the digit. The use of seven-segment displays requires a decoder-driver IC to convert from binary representation into the correct pattern of bars for each number.

The alternative to seven-segment display is the dot matrix display. The dot matrix type of display can be used for numbers and letters, and if sufficient dots per unit are used, the display can be of any type of alphabet, not simply European alphabets. Dot patterns of 4 × 7 are used where only digits and capital (English) letters are needed, but 5 × 7 matrix patterns allow more versatility. Matrix displays are almost always sold as display units with controlling circuits built in, because there is seldom any requirement for a matrix display other than for the normal range of numerical and alphabetical characters (alphanumeric display).

Driving circuitry is also usually built in when multi-digit displays are used. Multi-digit displays are used on a multiplexed basis, meaning that display bars in identical positions on the different units are connected in parallel (Figure 3.17). This would mean that a binary-coded signal at the

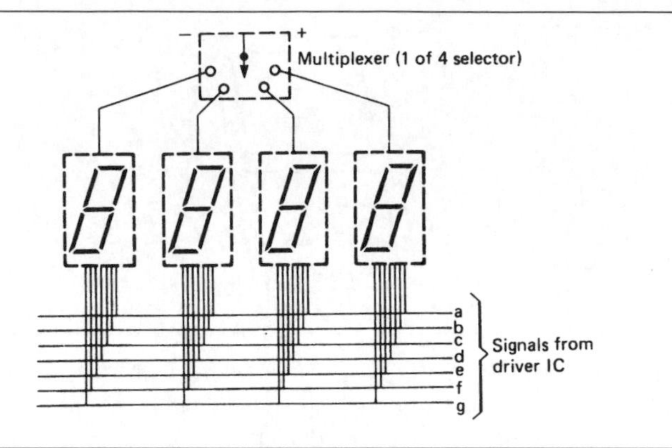

Figure 3.17 Multiplexed displays. The signals from the driver IC are applied to the bars of all the displays in a set, and each digit is switched on in turn by a multiplexer, a form of IC switch that selects any one out of the four as illustrated here.

inputs would activate the same number on all of the displays, but the multiplexing switches isolate the cathodes of all the diodes except on one unit. On a four-digit display, for example, as might be used for a digital voltmeter, the digits are displayed one at a time, and the binary number data has to be altered at the input for each unit. This means that the rate of changing data at the inputs must be synchronized to the rate of switching from one display to another. This usually calls for a display which is supplied without multiplexing circuitry, and an alternative is to use a display which incorporates a counter and multiplexer so that the pulses to be counted are input directly to the display, which also has a reset and a blanking input.

One very considerable disadvantage of LEDs, used in any quantity, is power consumption. Even a small seven-segment display can use 15–20 mA per segment, so that the worst-case consumption for a four-digit display showing the number 8888 would be at least 420 mA. This order of current is reasonable enough for mains-powered equipment, but is out of the question for battery-operated circuits, unless secondary cells are being used (as on car displays, although these are usually turned off when the ignition switch is off). The display type that is much more likely to be used for battery-operated equipment, ranging from pocket calculators to portable computers, is the LCD (liquid crystal display).

3.12 Liquid crystal displays (LCD)

Liquid crystals are not necessarily liquids, and definitely not crystals, but the name has stuck and is considerably easier to remember than any more precise title. The name of liquid crystal was given to certain types of organic materials (chemicals found in living organisms) in the early part of this century because of their remarkable chemical structure. A crystal is an arrangement of atoms in which the precise order governs the behaviour of the material; it is hard, has a high melting point, and often polarizes light. Each atom in a solid crystalline material affects each of its immediate neighbours very strongly, and the materials are of comparatively simple chemical structure, often composed of just two types of atoms.

There are materials, however, in which some sort of order exists, but which are not crystals in the classic sense. These materials consist of units that contain hundreds of thousands of atoms, but there is enough interaction to arrange the units into a structure, particularly if the material is a viscous (thick) liquid. The units in such materials, one of which is cholesterol (the fatty material in the bloodstream), are in the form of long chains, and when these chains are arranged in line, the material polarizes light very strongly. The types of liquid crystal material which are of interest in electronics are those whose long chains can be aligned in an electrostatic field obtained by placing a voltage between two conductors. The liquid crystal materials are non-conductive, so that making the chains align in a field like this causes no current to flow. A liquid crystal display therefore consists of sets of electrodes with the liquid crystal material covering them, and with one transparent enclosing wall and a reflective backplate (Figure 3.18).

The liquid crystal display cell is used with a sheet of polarizing material over the transparent wall. External light will pass through the polarizing sheet, and if the liquid crystal material is not aligned, will pass also through this and be reflected from the back-plate. If a voltage is applied between two electrodes, however, the material between these electrodes will become polarizing, and the already polarized light will not pass and cannot be reflected. This makes the affected area look dark, and the stronger the illumination of the cell the greater the contrast between this dark area and the brighter parts that have not been polarized. Although some displays are made with a built-in backlight for viewing in the dark, LCDs generally are intended for use in well illuminated areas. The use of DC on the electrodes causes irreversible changes to the liquid crystal material, so that the usual practice is to invert the DC supply to low-frequency (30–60 Hz) AC and use this as the supply to the electrodes. The use of electronic conversion can ensure that no trace of DC is present, and since the power consumption is so low, relatively simple inverter circuits can be used.

Because the current requirement of the LCD is so low, of the order of 8 μA for a four-digit display operating at 32 kHz, there is no problem about using battery operation, even when a complex driving IC is incorporated with the LCD. It is quite rare to find LCDs offered alone – practically every manufacturer supplies only the packaged units with inverters and driving logic built in. More recently, the ‘supertwist’ type of LCD has

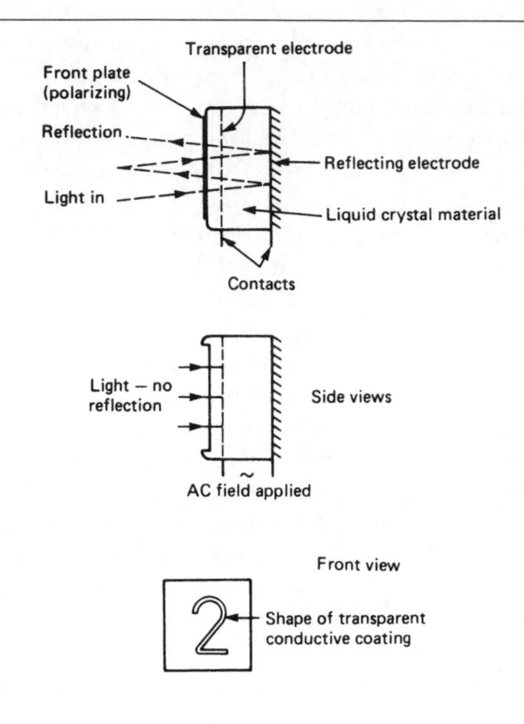

Figure 3.18 The LCD principle. The shape that is to be displayed is deposited in the form of a transparent conductive coating on the front face, and the reflective rear face is the other electrode. When power is applied, the material between these electrodes polarizes light, and the shape of the polarized zone accurately follows the shape of the electrode. Where the material is polarized, no light is reflected, causing a dark display.

become available, which offers much greater contrast along with faster operation, and this latter type is being used extensively for the screen displays of portable computers. The most recent development is the TFT screen, in which a set of thin-film transistors (TFT) controls each LCD cell of the screen. This allows fast-acting colour displays to be manufactured.

Other principles have been used, and are still used, for alphanumeric display. Filament displays, for example, can be obtained in seven-segment form, often with a better ratio of brightness to consumption than LEDs. Gas plasma displays, which were very common (in the form of the Ericsson Dekatron) prior to the emergence of LEDs, were used for some portable computer screens because they allowed viewing in poor illumination together with very low power consumption. The LED and LCD types, however, are by far the dominant types. Modern laptop computers use colour LCD displays in which each point on the screen is controlled by a thin-film transistor.

3.13 Light valves

A *light valve* is a device that is the electronic equivalent of a camera shutter. The LCD is one form of light valve, but the term is generally reserved for very high-speed devices that are capable of nanosecond timing. Light valves of this specialized type are used for ultra-high-speed photography, particularly for the photography of explosions and very fast chemical or nuclear reactions. The device is, like the LCD, a light modulator rather than a transducer, but some mention is due here.

The original type of light valve was a spin-off from TV camera techniques, and relied on the use of an electron beam. The principle is illustrated in Figure 3.19, with a photocathode at one end of an evacuated tube and a screen at the other. With a suitable accelerating voltage between the photocathode and the screen, and possibly a magnetic field applied to maintain focus, the image on the photocathode can also be obtained on the screen. The electrodes near the photocathode can affect the electron beam, usually by deflecting it so that it does not reach the screen. This is most easily done if the beam is allowed to cross over or follow a bowed path (Figure 3.20) on its way from the photocathode to the screen. High-speed electronic shutters of this type have been used to photograph events in nuclear explosions and are used extensively to analyse actions such as flame propagation in cylinder heads of car engines and in the ignition tubes of gas turbines.

The principle of photocathode and phosphor screen can also be used in *image intensifiers*. By using a large accelerating voltage between the photocathode and the screen, the light energy from the screen can greatly exceed the intensity reaching the photocathode. There can also be frequency conversion, because the photocathode can be an infrared-sensitive type and the screen a normal white one, allowing images to be seen in apparent darkness. This principle was used in World War II as a gunner's night

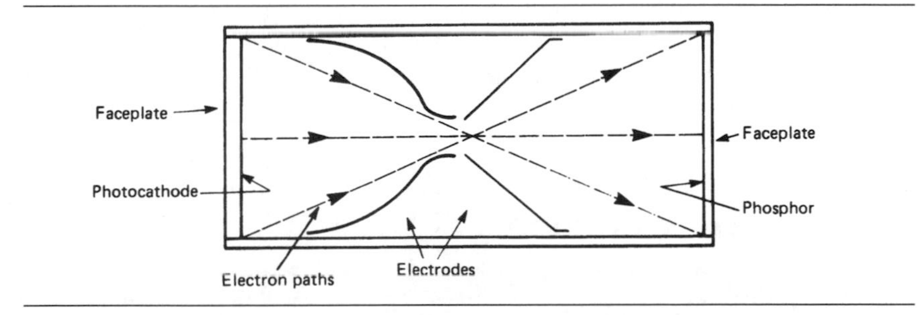

Figure 3.19 A simple light valve, which makes use of an electron image. The electrons from the photocathode are accelerated to the phosphor screen to form a bright image. This can be turned off by changing the voltage on one or both electrodes, and the rate of turning the image on and off can be very fast. The same arrangement can be used as a light amplifier if the accelerating voltage is high enough.

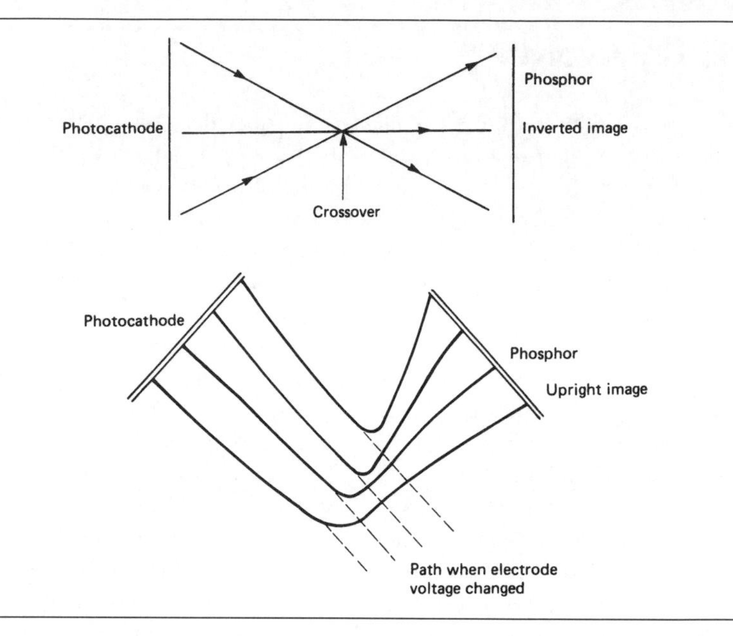

Figure 3.20 Electron paths in light valves. In the crossover type, the image is inverted, and the action of the electrode is to turn off the beam. In the bent-beam type, image distortion is greater, but much less power is needed to restore a straight beam shape which means that no electrons reach the phosphor.

sight, and has been considerably refined since then in terms of sensitivity, size and power consumption, and clarity of image.

Another well-established form of light switch is the *Kerr cell*. The Kerr cell uses an effect similar to the LCD – the effect of an electric field to rotate the plane of polarized light in a transparent material. Early types of Kerr cells used liquid suspensions, but later types have been able to use solid crystals. One application of modern Kerr cells has been to modulate laser light, allowing large-screen colour TV displays to be obtained from three laser beams. Another old-established effect, the Faraday effect, is used to deflect the light beams.

3.14 Image transducers

An image transducer is a device that will transform an optical image into electrical signals. The simplest image transducers are parallel types, in which a matrix of photodiodes supplies signals to a set of wires. Until recently, this type of transducer was used only for small-scale applications (such as optical character recognition) or coarse images, but modern developments in integration have led to renewed interest in their use. For the most part, however, image transducers are overwhelmingly of the serial output type in which the image is broken up into a set of repetitive signals.

The method of breaking up the image is *scanning*, a technique invented more than a century ago by Nipkov, and in its original mechanical form employed by Baird in his 'Televisor' of 1928. Nowadays, scanning means the deflection of an electron beam over a set of charges that are proportional to the illumination of parts of an image. However, a serial-register action that gives a similar signal output is used in modern solid-state TV cameras and is also termed scanning. The basis of TV scanning is the analysis of the picture into lines and fields. Samples of the picture brightness are taken at a large number of points along a line at the top of the picture. Each sample provides the amplitude of a brightness signal, which is an analogue signal for each line of the picture. As the line is scanned, vertical deflection of the scanning is also applied so that the next line to be scanned is a short distance below the first.

Although the picture could be completely analysed in one set of scan lines that traced parallel lines from the top to the bottom of the picture, TV scanning follows a more complex pattern. The principle of interlace is used, meaning that in a set of 600 lines (for example), the odd numbered lines 1, 3, 5, 7, etc. would be scanned, ending half way along line 299 and then returning to half way along the top of line 2 to scan the even numbered lines. This was historically used as a method of reducing the bandwidth of the electronic form of the TV picture while retaining a vertical repetition rate of 50 per second in order to reduce flickering.

In this book we are not concerned with TV methods except inasmuch as they affect the operation of the transducers, and this outline of scanning has been intended to explain the construction and use of the vidicon and the CCD pick-up devices. Computer monitors do not use interlace, and operate at higher horizontal and vertical scan rates.

The *vidicon* principle is illustrated in Figure 3.21. The sensitive material is a photoconductor, now traditionally lead oxide (antimony trisulphide was formerly used). The photoconductor is laid onto a transparent conductive coating which extends through the glass seal to metal contacts on the outside of the tube – it is from this connection that the video output signals are taken. The tube is evacuated, and the electron gun can produce a finely focused beam that can be deflected in the usual line and field pattern. Where the beam strikes the photoconductor, the connection will make the potential at the photoconductor equal to the potential at the cathode of the electron gun.

A light image is projected so as to focus on to the faceplate of the vidicon. This will have the effect of lowering the resistance of the photoconductor where the image is bright, allowing this portion to charge to the potential of the transparent conducting layer. The photoconductor behaves like a combination of capacitor and resistor (Figure 3.22), so that between the times when the beam scans a point, the effect of the light is to build up an

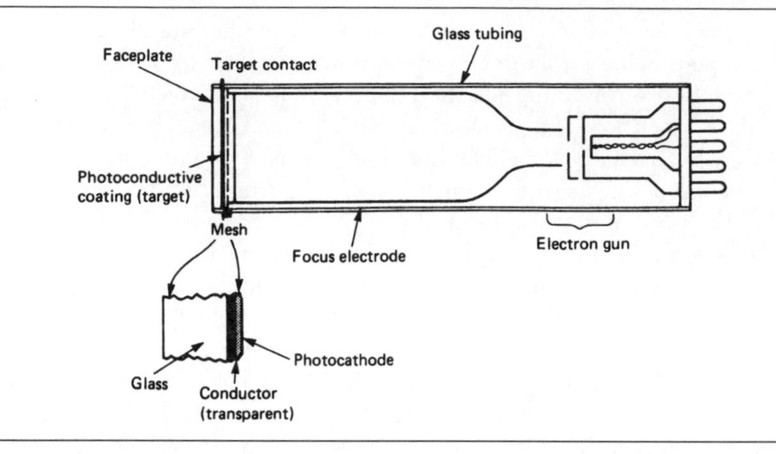

Figure 3.21 Cross-section of a vidicon. The glass faceplate is coated with a transparent conducting material (tin oxide) and this in turn is covered with a layer of lead oxide photoconductor. The scanning beam brings the beam side of the photoconductor to cathode voltage, and it rises to target voltage by leakage through the photoconductor in the time between scan intervals. The voltage that is attained is proportional to the brightness of the illumination, and the discharge current when the beam scans across constitutes the output signal at the target electrode.

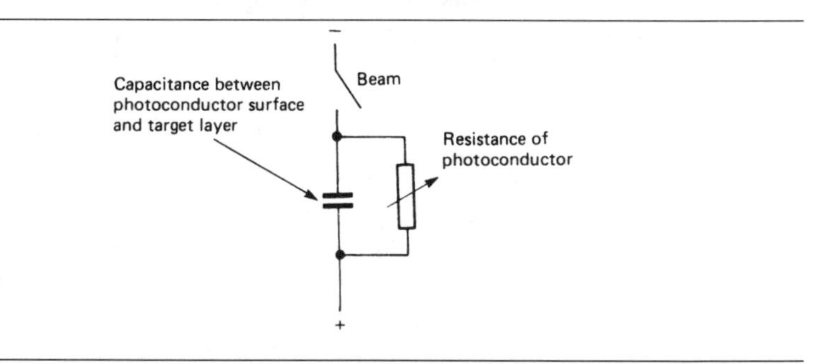

Figure 3.22 The equivalent circuit of the photoconductor and target electrode arrangement.

amount of charge that is proportional to the brightness of the light at that point. When the scanning beam strikes this spot, it will be discharged, and the capacitive discharge current will flow in the transparent layer, forming the video signal. This signal is at a comparatively high level, so that the vidicon is well suited to producing good signals in comparatively low illumination. Colour signals are obtained by using three vidicons, with filters used to make the images consist of the primary light colours of red, green and blue.

The solid-state image transducer is considerably more complex, and is made in IC form. The principle is to use the light intensity to charge a

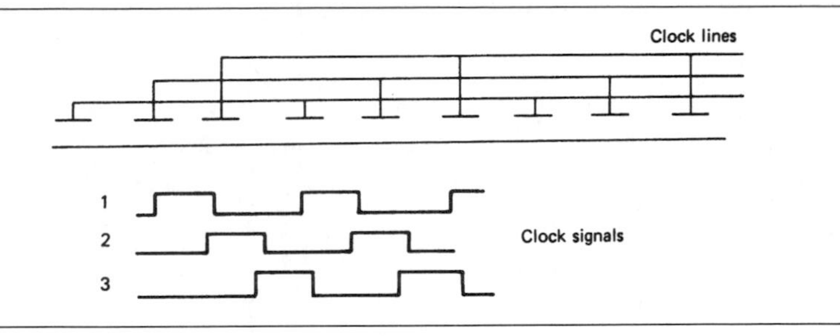

Figure 3.23 CCD principles. The small contacts are charged by conduction through the underlying photoconductor. At scan time, a set of pulses on the clock lines will transfer the charges from contact to contact in the same direction, so that a serial output signal is obtained from the last contact, forming a video waveform for this line.

capacitor and then to read the capacitor voltage by shifting it through a register. The digital type of shift register is unsuitable, and the device that is used is the CCD, charge-coupled device. This is a form of MOS device in which a clock pulse can pass a charge from one plate to another (Figure 3.23), like an analogue form of shift-register. The CCD type of TV camera pick-up device uses a large number of such devices that are arranged in lines. The lines are read out, giving a signal amplitude proportional to light brightness for each cell in the line, and a switching register ensures that the lines themselves are read in sequence. The picture quality does not approach that of the full-sized vidicon camera, but for very small cameras, or cameras that are used with low-resolution video-recorders (camcorders), the CCD type of device has the advantages of long life, high reliability, small size, and immunity to damage from bright lights.

3.15 Radio waves

The range of electromagnetic waves in the frequency range from around 100 kHz to several thousand GHz are radio waves in the sense that we use transducers that are recognizably of the type that we have always associated with radio and radar, such as long wires, dipoles, yagis and dishes. Aerials are unusual among transducers in being reversible, so that an aerial can be used either for transmission or for reception, although its characteristics for one task may not be so favourable for the other task.

The purpose of the receiving aerial is to intercept some of the energy of the modulated wave radiated by the transmitting aerial. Wave theory indicates that this can be done most efficiently by an aerial whose length is any exact multiple of half wavelengths of the signal itself. The smallest effective length of an aerial is thus one half wavelength of the desired signal. Since the wavelength, in metres, of the signal is equal to

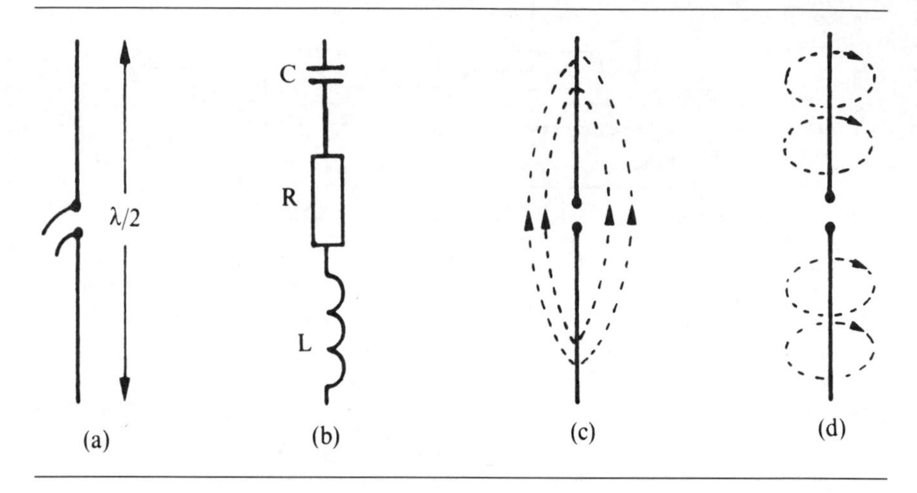

Figure 3.24 A half-wave dipole and its characteristics: (a) half-wave dipole aerial, (b) equivalent circuit, (c) the electric field, (d) the magnetic field.

$$\frac{3 \times 10^8}{\text{frequency in Hz}} \quad \text{or} \quad \frac{300}{\text{frequency in MHz}}$$

and the length in metres of an aerial is given approximately by the expression:

$$\frac{150}{\text{frequency in MHz}} \text{ (for a half wave aerial)}$$

Although approximate, this is close enough for many purposes. A small error occurs because 3×10^8 m/s is actually the velocity of electromagnetic waves in free space. The velocity in the metal conductors of the aerial is somewhat less. An aerial that uses a dipole of half-wavelength dimensions is a tuned aerial, and such tuning is essential if the aerial is to operate at high efficiency. We can often tolerate lesser efficiency figures, particularly for radio as distinct from TV applications, so that radio aerials in the past have used random lengths of wire (a long-wire aerial) or single (monopole) vertical rods as aerials. The older radio frequencies of the long-, medium- and short-wave bands would require long lengths of wire or rod for tuning, so that untuned aerials were a necessity, particularly for long wavelengths of several kilometres.

The most common tuned aerial element is the dipole shown in Figure 3.24(a), consisting of two equal-length rods connected to the transmitter or receiver via a feeder cable to the midpoint. The total length is approximately a half wavelength. Consider first a dipole of this type connected to a transmitter. If the transmitter frequency matches the frequency to which the dipole is tuned, then the dipole behaves as if it were a resonant tuned circuit as shown in Figure 3.24(b) with an impedance that is completely

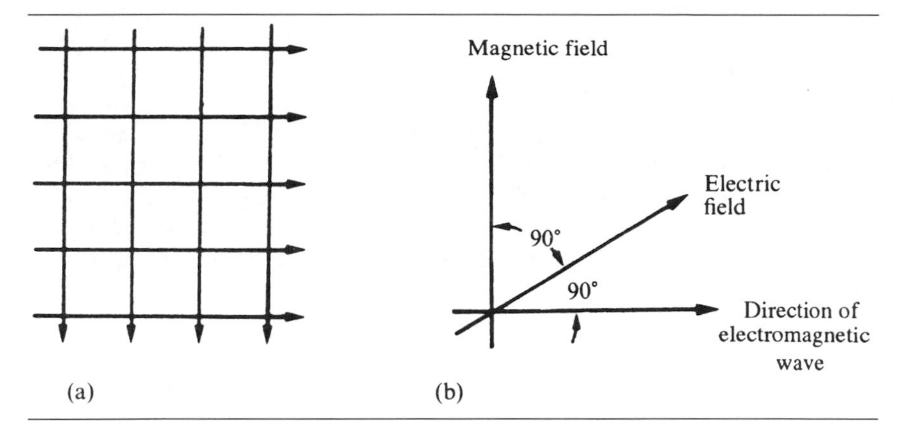

Figure 3.25 Electromagnetic waves in space: (a) wave front, (b) mutual relationship between components of propagation.

resistive. At frequencies on either side of this, the dipole acts either as an inductive or capacitive load to the transmitter, and a mismatch occurs, but there is a reasonable range of frequencies (bandwidth) for which the load is almost completely resistive. The resistance R of the equivalent circuit is usually called the *radiation resistance* for the aerial. R is not a physical resistance, only an equivalent one. If the aerial were replaced by a physical resistance of the same value, it would dissipate the same amount of heat energy as the aerial radiates electromagnetic energy into space.

The AC voltage fed to the aerial causes each rod to take up opposite alternating polarities which give rise to the lines of force of electric field, shown in Figure 3.24(c). In a similar way, the AC current also produces alternating magnetic fields around the rods that generate the lines of force of electromagnetic field, shown in Figure 3.24(d). These forces act at right angles to each other and together cause the energy to be radiated away from the dipole. Figure 3.25(a) shows an electromagnetic wavefront approaching an observer, while Figure 3.25(b) depicts the mutual relationship between the two components of the electromagnetic radiation and the direction of the propagation of electromagnetic waves.

If a complementary receiving dipole is to collect the maximum signal from the radiated energy, then its conducting rods must cut the lines of the magnetic field at right angles. That is, the receiving dipole must occupy the same plane as the electric field of the radiation. If the angle is less than 90°, the induced signal voltage will be reduced proportionally. For instance, if the relative angle is 45°, the signal is 0.707 of maximum (3 dB down). The direction of the electric field of the wavefront is referred to as the plane of polarization of the wave. As shown above, this occupies the same plane as the transmitting and receiving dipoles. At very high frequencies and ultra high frequencies, vertical, slant and horizontal polarizations are all in common use.

- The angle of polarization of the transmitted wave can be changed if the signal is reflected from a substantial object, such as a hill, a gas-holder, or a block of buildings; and in some difficult reception areas a small change in the polarization angle to which the receiving aerial is set can give a significantly better signal level. Similar polarization adjustments can also be used, in the right conditions, to provide an additional way of discriminating against unwanted signals.

The radiation or feed-point resistance of the simple dipole is typically 75 Ω. Up to a point, an increase in rod diameter will increase the bandwidth over which the aerial gives acceptable reception. The overall length of the dipole is normally cut for a frequency lying in the middle of the band of signals for which operation is wanted. Folding the dipole as shown in Figure 3.26(a) gives certain mechanical advantages but increases the radiation resistance to around 300 Ω. This figure can be reduced to some extent by increasing the cross-sectional area of the conductors.

A single dipole on its own seldom gives satisfactory signal pick-up unless the receiver is situated very close to the transmitter. By adding to the aerial a *reflector* placed behind the dipole and a number of *director* rods placed in front of it, relative to the direction of the incoming signal, a much larger signal amplitude can be gathered. The aerial then becomes more highly directional (it has higher *directivity*), accepting only signals arriving from a limited angle in front of the array, i.e., from the direction in which the director rods lie. The addition of these extra elements lowers the aerials radiation resistance to nominally 75 Ω.

An aerial consisting of a dipole, reflector and one or more directors (Figure 3.27) is termed a *yagi*, and is the usual form of TV aerial for a fixed site. The design of a yagi array is quite complex. Not only does it have to have high directivity, a definite value of feed-point resistance and a high gain relative to a dipole, but these parameters have to be maintained over a sufficiently wide bandwidth. Not unexpectedly, aerials that are inadequately designed or which are made using poor materials often give very poor results. The factors that make all the difference are often not apparent to the eye (will the insulating material, for example, still be an insulator at 800 MHz on a wet day?). Aerials should always be bought from a reputable manufacturer who can specify and measure the aerials' performance.

If transmissions are being beamed from a single fixed site, the yagi has to be properly mounted so that it is pointing directly at this transmitter. This can be done approximately using a compass bearing, or by using a field-strength meter connected to the aerial. A receiver itself cannot be used as a reliable guide to the received field strength unless its automatic gain

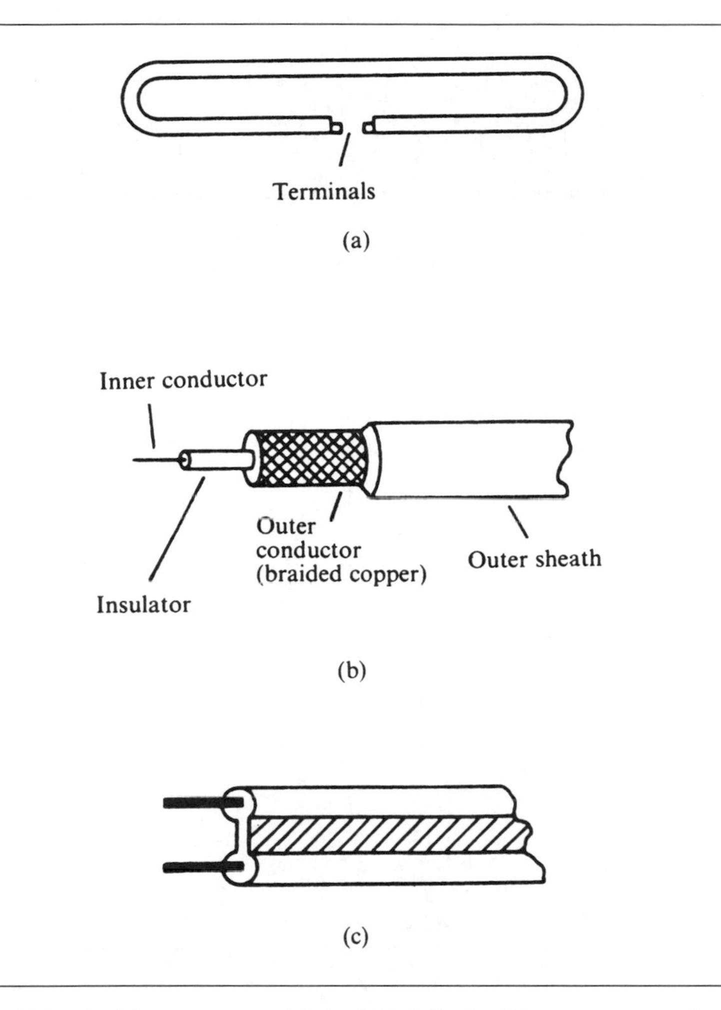

Figure 3.26 Aerial components: (a) the folded dipole, (b) a cut-away section of co-axial cable, (c) a section of twin-line cable.

control (AGC) is disabled, because AGC action will itself serve to mask the effects of signal variations.

Where two differently sited transmitters can be received by a single aerial array, two yagis can be connected to the receiver through a diplexer (a signal combiner) obtainable from the aerial manufacturer. Where several transmitters can be received from different directions (as happens in many parts of the USA), a rotating aerial assembly is often the only solution.

The feeder cables are designed to carry the signal with the least possible

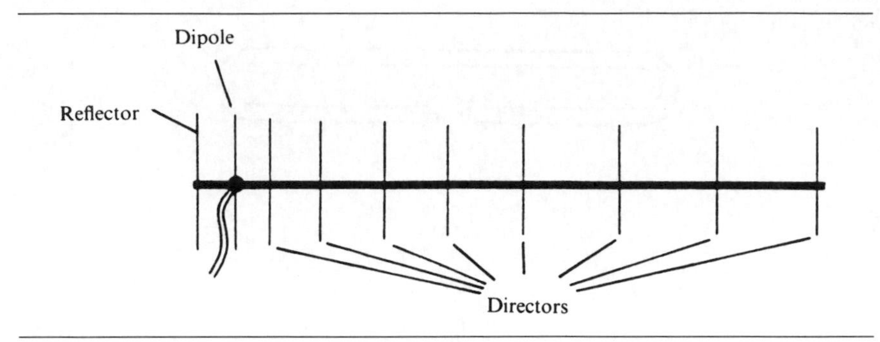

Figure 3.27 A yagi aerial.

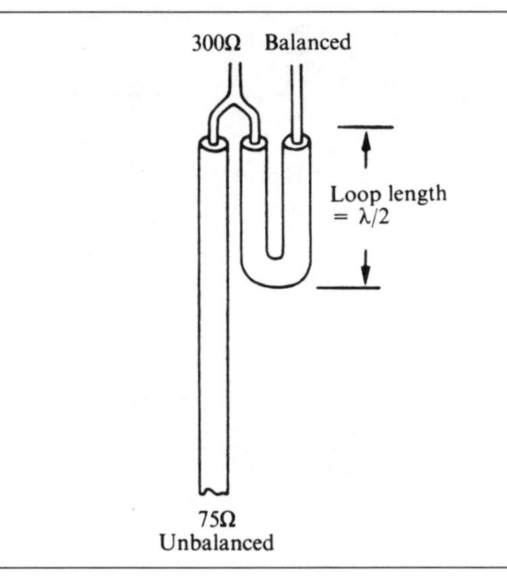

Figure 3.28 A typical balun capable of giving a 4 : 1 impedance conversion at its design frequency.

attenuation. Two common types of feeder are the twin-line of Figure 3.26(c), which is used extensively in the USA and has a 300 Ω impedance, and the coaxial cable of Figure 3.26(b) which is used almost exclusively in the United Kingdom and parts of Europe, and has a 75 Ω impedance. The impedances of aerial, feeder and receiver inputs must be correctly matched. An aerial system using 75 Ω co-axial cable, for example, must be used to connect a 75 Ω aerial and a 75 Ω receiver input.

Twin-line feeders are described as being *balanced*. With neither line earthed, the signals in the two lines are always anti-phase and of equal amplitude. On the other hand, co-axial cable is said to be *unbalanced*, the earthed outer conductor acting as a shield for the signal-carrying inner conductor. A form of transformer called a *balun* (balance to unbalance) is available (Figure 3.28) to match one cable system to the other.

Chapter 4

Temperature sensors and thermal transducers

4.1 Heat and temperature

The physical quantity that we call *heat* is one of the many forms of energy, and an amount of heat is measured in the usual energy units of joules. The quantity of heat contained in an object cannot be measured, but we can measure changes of heat content that take place when there is a change of *temperature* or a change of physical *state* (solid to liquid, liquid to gas, one crystalline form to another). In this sense, then, *temperature* is a measure of the level of heat for a material whose physical state has remained unchanged. The relationship between temperature and energy is very similar to that between voltage level and electrical energy.

The temperature sensors that we use all depend on changes that take place in materials as their temperatures change. Transducers for electrical to thermal energy make use of the heating effect of a current through a conductor, but transducers for thermal to electrical energy are not so direct, and in accordance with the laws of thermodynamics will require a temperature difference to operate, taking heat in at a higher temperature and discharging some heat at a lower temperature.

4.2 The bimetallic strip

Thermal sensing is important for the detection of effects as diverse as fire, overheating, or the failure of a freezer. The simplest type of thermal sensor is the *bimetallic* type, whose principle is illustrated in Figure 4.1. A compound strip is formed by riveting or welding two layers of metals, chosen so as to have very different values of *linear expansivity*. The linear expansivity (old name, expansion coefficient) is the fractional change of length per degree change of temperature and for all metals is positive,

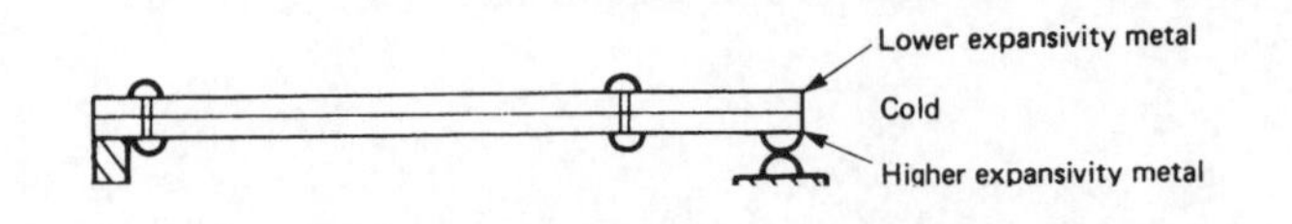

Figure 4.1 The bimetallic strip consists of two metal strips welded or riveted together. The strip can be extended into a spring shape for greater sensitivity, or can consist of two welded discs that will buckle when heated.

Table 4.1 Linear expansivity values for some metals – multiply figure shown by 10^{-5} for value. Because the metals do not expand by the same amount, however, the strip will bend as the temperature changes, as indicated in Figure 4.2.

Metal/Alloy	Expansivity	Metal/Alloy	Expansivity
Aluminium	2.4	Brass	2.1
Bronze	1.9	Chromium	0.85
Constantan	1.5	Copper	1.6
Invar	0.2	Iron	1.2
Magnesium	2.6	Manganin	1.6
Nickel	1.3	Platinum	0.90
Silver	1.9	Stainless steel	1.0
Tantalum	0.65	Tin	2.7
Tungsten	0.43	Zinc	2.6

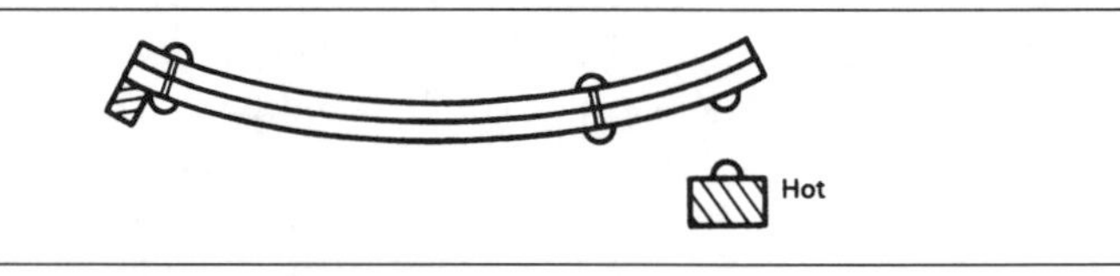

Figure 4.2 If one metal has a higher expansivity than the other, the strip will bend when heated, with the metal of higher expansivity on the outer side of the sector of the circle.

meaning that the strip expands as the temperature increases. Table 4.1 shows expansivity values for some metals in units of $K^{-1} \times 0^{-5}$.

This bending action can be sensed by a displacement transducer of any of the types discussed in Chapter 2, but is more often used to operate switch contacts, usually with the strip itself carrying one contact. The conventional type of bimetallic strip element is still to be found in some thermostats, although the strip is very often arranged into a spiral. This allows for much greater sensitivity, since the sensitivity depends on the length of the strip. The amount of deflection can be fairly precisely proportional to temperature change if the temperature range is small.

Thermostats of this type, however, have an undesirably large *hysteresis*, so that, for example, a thermostat set for a nominal 20°C might open at 22°C

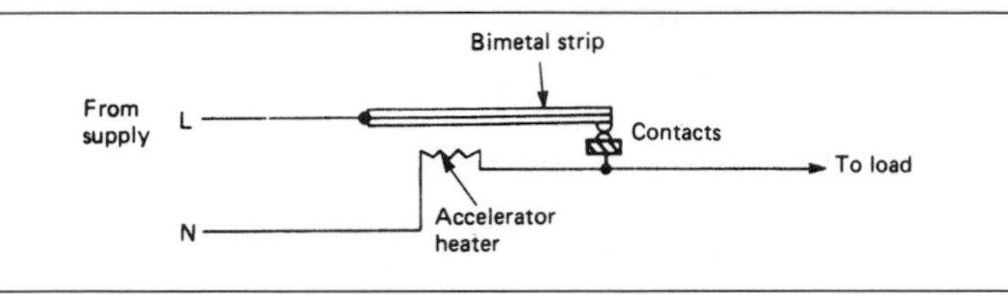

Figure 4.3 Using an accelerator with a bimetal thermostat. The accelerator ensures that the rate of rise of temperature at the thermostat is greater than that of the surroundings, so overcoming the hysteresis of the thermostat to some extent.

and close again at 18°C. This leads to undesirable temperature swings which can nullify the use of a thermostat. For example, with a bimetallic thermostat used to control room temperature the effect of the hysteresis is likely to be that the occupants of a room ignore the thermostat and turn the radiators directly on or off, or use the thermostat simply as an on/off switch. The hysteresis of the simple bimetal thermostat can be reduced by the use of an *accelerator*, consisting of a high-value resistor placed close to the element. The principle is that when the thermostat contacts close to switch on heating in a room, current is passed through the accelerator resistor (Figure 4.3) so that the rate of heating within the thermostat is faster than outside. See Chapter 10 for details of switch contacts.

This leads to the thermostat points opening before the same temperature is achieved in the room outside. The current through the accelerator resistor then switches off, and the thermostat will then cool more rapidly than the room so that the switch-on is more rapid than would otherwise be the case. The use of an accelerator, however, can lead to the desired working temperature being achieved very slowly or not at all in cold weather, and much too rapidly in hot weather. This has led to the use of more sensitive devices for thermostat use, based on thermistors (see later in this chapter).

The bimetallic strip exists in several physical forms, and one particularly useful form is the disc (Figure 4.4). For a change of temperature, a bimetallic disc will abruptly buckle giving a snap-over action that requires no form of assistance. This is the basis of the small thermal switches that are used for overheating protection in electronic equipment. These thermal switches can be bolted to heat sinks, small motors, transformers, jug-kettles, or other components that are likely to overheat and have a metallic surface.

Thermal switches can be bought as normally open or normally closed types, depending on whether they are to be used to detect rising or falling temperatures. The pre-set nominal temperatures have temperature hysteresis of the order of 3–5°C on each side of the set temperature since no accelerator is used. For more precise control, units that use long bimetal strips can be obtained with smaller hysteresis and variable setting temperature.

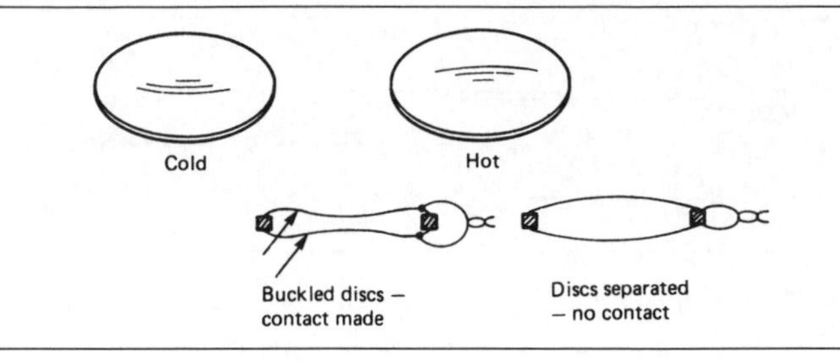

Figure 4.4 Bimetallic discs are used extensively as sensors for overheating components such as transformer windings and electric motors.

All types of long-element bimetal strip thermostats should be recalibrated at intervals, since the strip is subject to gradual changes (creep) that affect the thermostat setting.

4.3 Liquid and gas expansion

An older principle used in temperature sensing is liquid expansion in conjunction with a pressure switch, making use of the principles of the familiar mercury thermometer. The simplest sensor of this type is an adaptation of a mercury thermometer with two wire electrodes inset into the capillary (Figure 4.5). Since mercury is a conducting metal, a circuit will be made through the electrodes when the mercury level reaches the electrode whose

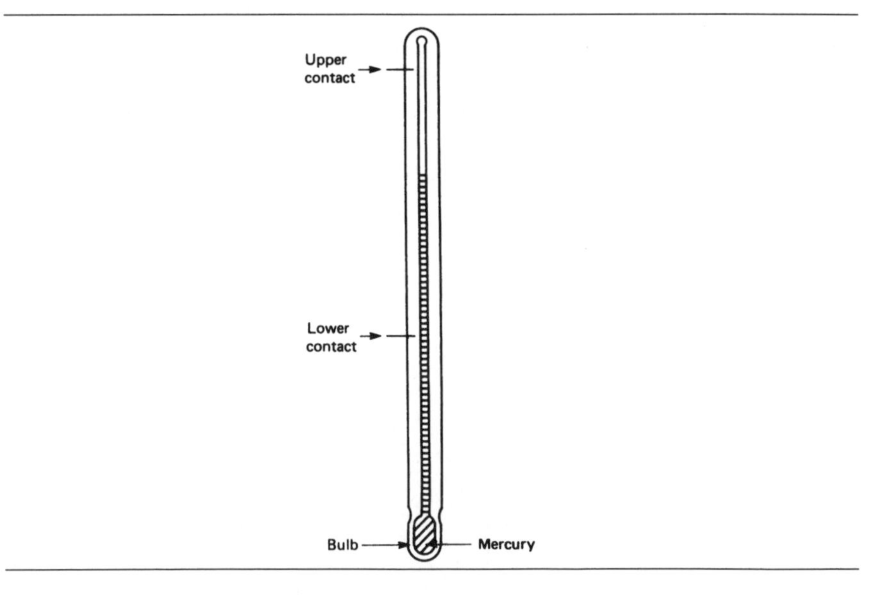

Figure 4.5 A temperature switch developed from an ordinary mercury thermometer using wire electrodes embedded in the glass tubing.

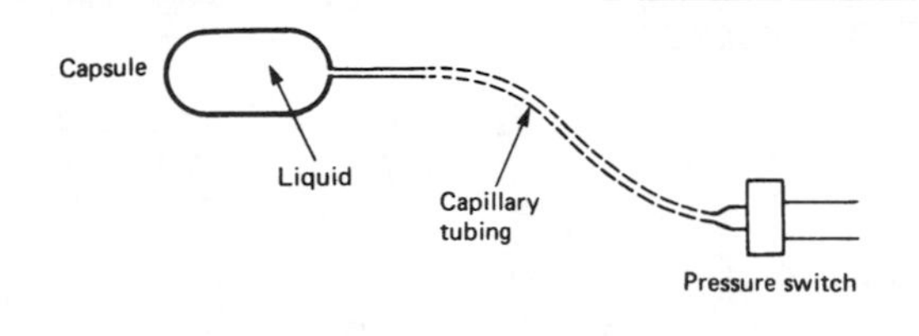

Figure 4.6 A liquid bulb and pressure sensor method for temperature sensing. The volume of liquid in the capillary tubing should be negligible compared to the volume of liquid in the bulb.

position corresponds to the higher temperature. This allows a predetermined temperature to be sensed, but for a switching action only and with no way of altering the temperature at which switching takes place other than by replacing the sensor with another one.

Although the mercury level can be used to change the frequency of an oscillating circuit and thus to provide a proportional sensing of temperature, this type of action is seldom used. The sensors that are used for temperature measurement as distinct from switching are mainly of the electronic type, including thermocouples and thermistors, and devices that make use of mechanical expansion are more likely to be used in switching circuits. The most common type is a development of the conventional bulb thermometer and has a sensing element (Figure 4.6) consisting of a capsule filled with liquid that is connected by a narrow-bore tube to the pressure switch. The liquid need not be mercury, and is nowadays more likely to be a form of synthetic oil.

Since the capsule can be remote, and involves no electrical connections, this is often a very useful method to use for hazardous environments, and the liquid can be chosen accordingly. The length of connecting tubing must be such that the volume of liquid contained in the tubing is only a small fraction of the total volume, since the temperature of the liquid in the tubing will also affect the pressure. The use of air or an inert gas in place of a liquid makes the device very much more sensitive, but the pressure switch needs to be able to respond to much lower pressures than are exerted by an expanding liquid.

One disadvantage of the system in general is that the sensing capsule needs to contain a reasonable volume of liquid and so cannot be small. In addition, since this volume of material has to be heated and cooled in order to follow temperature fluctuations, time is needed for the change, so that capsules cannot follow rapidly changing temperatures. The pressure sensor need not be a switching device, and the use of a diaphragm coupled to a potentiometer, LVDT or piezoelectric transducer can make the liquid/bulb type of temperature sensor into a fairly precise instrument, although this combination has few applications.

4.4 Thermocouples

THEORY

The thermocouple is frequently used as the sensing element in a thermal sensor or switch. The principle is that two dissimilar metals always have a (small) contact potential between them, and this contact potential changes as the temperature changes. The contact potential cannot be measured for a single connection (or junction), but when two junctions are in a circuit with the junctions at different temperatures, then a voltage of a few millivolts can be detected (Figure 4.7). This voltage will be zero if the junctions are at the same temperature, and will increase as the temperature of one junction relative to the other is changed until a peak is reached. The shape of the typical characteristic is shown in Figure 4.8, from which you can see

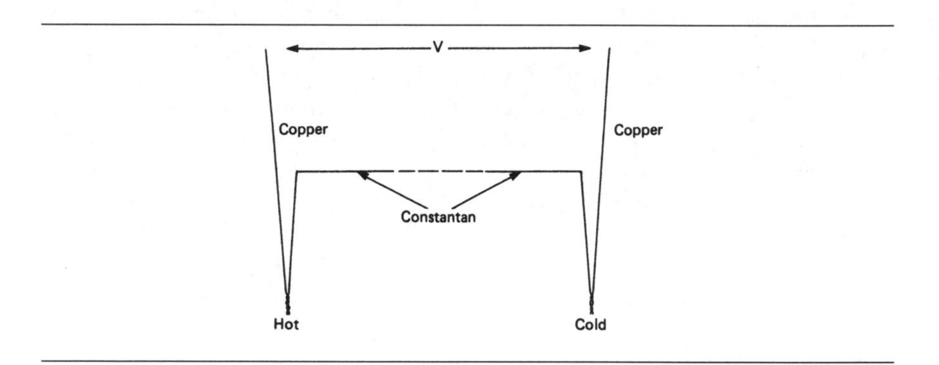

Figure 4.7 The construction of a thermocouple, in this example using copper and constantan alloy.

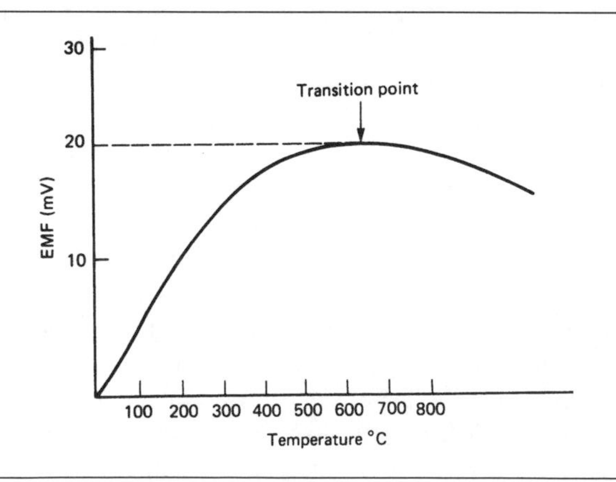

Figure 4.8 A thermocouple characteristic, showing the typical curvature and the transition point at which the characteristic reverses. A few combinations of metals (like copper/silver) have no transition, but have a very low output.

Table 4.2 The EMF (in mV) at 100°C difference for platinum/metal thermocouples.

Metal/Alloy	EMF in mV	Metal/Alloy	EMF in mV
Aluminium	0.4	Constantan	−3.3
Copper	0.75	Iron	1.88
Manganin	0.65	Molybdenum	1.2
Nickel	−1.5	Silicon	45.0
Silver	0.7	Tungsten	0.8

that the thermocouple is useful only over a limited range of temperature due to the non-linear shape of the characteristic and the reversal that takes place at temperatures higher than the turn-over point.

The thermocouple makes use of the Seebeck effect, and the theory of this leads to an equation for EMF:

$$E = a + b\theta + c\theta^2$$

where a, b and c are constants for the types of metals used in the thermocouple and θ is the temperature difference between them. If the cold junction is held at 0°C, then the EMF equation becomes:

$$E = \alpha T^2 + \beta T$$

where α and β are measured constants for the pair of metals, and T is the temperature difference. For temperatures below the transition point α is usually small, so that the EMF is almost directly proportional to temperature difference.

- The Peltier effect, see later, is the reverse of the Seebeck effect.
- The Kelvin effect is much less well-known, and it concerns EMF generated in a conductor with no junctions. In such a conductor, a temperature difference between two different parts of the conductor will cause an EMF to be developed.

When an electric current flows through a conductor whose ends are maintained at different temperatures, heat is released at a rate approximately proportional to the product of the current and the temperature gradient.

- Any practicable circuit including a thermocouple will contain more than two junctions of differing metals, and the circuits have to be designed so that only the intended junctions are at different temperatures.

The output from a thermocouple is small, of the order of millivolts for a 10°C temperature difference, and Table 4.2 shows typical EMF values for metals and alloys, using platinum as the other metal of the couple. Table 4.3 shows values of EMF and temperature difference for three commonly

Table 4.3 More detailed characteristics for three important thermocouple types, showing the useful range of temperature differences and EMF values in mV when the cold end of the thermocouple is at 0°C.

Temperature °C	Copper/ Constantan	Iron/ Constantan	Platinum/ Plat.Rhodium
−20	−0.75	−1.03	
−10	−0.38	−0.52	
0	0.00	0.00	0.00
10	0.39	0.52	0.05
20	0.79	1.05	0.11
30	1.19	1.58	0.17
40	1.61	2.12	0.23
50	2.04	2.66	0.30
60	2.47	3.20	0.36
70	2.91	3.75	0.43
80	3.36	4.30	0.50
90	3.81	4.85	0.57
100	4.28	5.40	0.64
200	9.28	10.99	1.46
300	14.86	16.57	2.39
400	20.87	22.08	3.40
500		27.59	4.46
600		33.28	5.57
700		39.30	6.74
800		45.71	7.95
900		52.28	9.21
1000		58.23	10.51
1200			13.22
1500			17.46

used thermocouple materials. Of these, the copper/constantan type is used mainly for the lower range of temperatures and the platinum/rhodium type for the higher temperatures. Because of the small voltage output, amplification is usually needed, unless the thermocouple is used for temperature measurement, together with a sensitive millivoltmeter. If the output of the thermocouple is required to drive anything more than a meter movement, then DC amplification will be needed, using an operational amplifier or chopper amplifier (see Chapter 12). The type of amplifier that is used needs to be carefully selected, because good drift stability is necessary unless the device is recalibrated at frequent intervals. This makes the chopper type of amplifier preferable for most applications.

If an on/off switching action is required, the thermocouple must be used along with a controller that uses a Schmitt trigger type of circuit which also permits adjustment of bias so that the switching temperature can be pre-set. The usual circuitry includes amplification, because the lower ranges of thermocouple outputs are comparable with the contact potentials (the same type of effect) in amplifier circuits, and attempting to use very small inputs for switching invariably leads to problems of low hysteresis and excessive sensitivity.

One particular advantage of thermocouples is that the sensing elements themselves are very small, allowing thermocouples to be inserted into very small spaces and to respond to rapidly changing temperatures. The electrical nature of the process means that the circuitry for reading the thermocouple output can be remote from the sensor itself. Note that thermocouple effects will be encountered wherever one metallic conductor meets another, so that temperature differences along circuit boards can also give rise to voltages that are comparable with the output from thermocouples. The form of construction of amplifiers for thermocouples is therefore important, and some form of zero setting is needed.

PRACTICAL USE

Thermocouples are very widely used industrially, making the thermocouple one of the most important of temperature sensors. Of the many possible combinations of metals that could be used for thermocouples, only a few are practical from the consideration of a reasonably linear scale and good resistance to high temperatures. Table 4.4 shows the most commonly used

Table 4.4 The most common thermocouple types, with their code letters.

Code	Metals	Range	mV @ 100°C	Notes
S	PtRh/Pt	0–1400°C	0.645	Needs ceramic sheath
R	PtRh/Pt	0–1400°C	0.647	Needs ceramic sheath
J	Fe/CuNi	0–800°C	5.268	Attacked by oxygen or acids
K	NiCr/NiAl	0–1100°C	4.095	Avoid reducing agents
T	Cu/CuNi	−200°C to +400°C	4.277	Low temperature use
E	NiCr/CuNi	0–800°C	6.137	High output

Notes: The S-type uses 90% platinum, 10% rhodium alloy with pure platinum as the other metal.
The R-type uses 87% platinum, 13% rhodium alloy with pure platinum as the other metal.
The J-type (or iron–constantan couple) uses copper–nickel alloy and iron.
The K-type (or chromel–alumel couple) uses nickel–chromium and nickel–aluminium alloys.
The T-type (or copper–constantan couple) uses copper and copper–nickel alloy.
The E-type (or chromel–constantan couple) uses nickel–chromium and copper–nickel alloys.

types. These fall into two groups, the *base metal* types such as iron–constantan and the *noble metal* types such as platinum–rhodium–platinum. The noble metal (originally named because of their resistance to all known acids) thermocouples are required for the higher temperatures, but have lower output levels and require ceramic sheathing to avoid oxidation damage. Thermocouples that use iron as one wire require protection against rusting and against oxidizing atmospheres generally.

The differences between thermocouple measurements and those made by other means are not always well appreciated, however. A thermocouple measurement is always a differential measurement, measuring the temperature difference between the cold or *reference* junction and the hot or *measuring* junction. When neither of the metals used in the thermocouple is the same as that used for the connecting cable, there will be two sets of junctions. Tables for thermocouple use are always made up in the assumption that the reference junction will be at 0°C. In industrial use this is very seldom true, and some form of compensation has to be used so that the output readings can be adjusted for the true temperature of the reference junction(s).

The usual method is *cold-junction compensation* incorporated into the amplifier/output portion of the instrument. A metal coil or a thermistor is used to sense the temperature at the reference junction or junctions, and the output from this sensor is used in an adding stage within the instrument to correct for the effect. This is most easily done in microprocessor-controlled equipment by using a table of correction values held in a ROM, but older analogue methods using an adding stage have been satisfactory in the past.

- Note that these methods apply a correction based on the cable that is supplied with the instrument. Changing to a different cable material (for example, extending the thermocouple connecting cable with copper cable) can make the built-in correction factor false, since two new junctions have now been added.

For precise measurements, although thermocouples are not ideal for such applications, it is more usual to have the reference junction(s) held in temperature reference units. The ice-point form of the reference unit is held at 0°C using Peltier cooling junctions (the reverse of the thermocouple effect) and precise sensors for the reference temperature, such as the bellows type which makes use of the expansion which occurs as water changes to ice. The traditional way of establishing the zero reference point was to use a vacuum flask filled with a mixture of ice and water, but this can cause large discrepancies unless used carefully.

The main objection is that ice which has been taken from a freezer will often be at 15°C or lower, and the water around it at around +5°C, so that the reference junction will quite certainly be at the wrong temperature and also at a temperature which will change considerably. The water–ice mixture is suitable if the water is demineralized (and the ice made from similar water), the ice is crushed, not in lumps, the ice has been in contact with the water for a considerable time and is constantly stirred, and the reference junction is not in contact with the ice.

The *hotbox* reference system uses a solid aluminium block with a drilled cavity into which the reference junction is placed. The block temperature is maintained constant, usually at a temperature well above ambient in the region of 55–65°C. The temperature of the block is raised quickly to its steady level by a heater which switches off when the temperature approaches the controlled level. From then on, the temperature is controlled by a thermistor and a small heating element operating in a loop with an amplifier. The instrumentation for the thermocouple must include circuitry that will correct the readings for the raised temperature of the reference junction by adding a small voltage to the output of the thermocouple.

- Another method, which is a passive system, consists of embedding the reference junction in a metal block that is well insulated, so that its temperature changes only very slowly. Another sensor in the block is connected to the instrumentation and generates the correction signal for the reference junction temperature.

The connections between the thermocouple and the reading system are important, as has been mentioned. When the distance between the thermocouple and the measuring instrument is considerable, extension or compensating cables should be used to connect these two. The difference between these two is that extension leads use the same materials as the thermocouples, and can be used at the same temperatures. Compensating cables use low-cost metals and can be used in ambient temperatures up to about 80°C only. The compensating cables must be matched to the type of thermocouple being used, and both extension and compensating cables must be connected in the correct polarity.

Cables made to the British standard (BS 1843:1952) all use codings in which the negative lead is blue, but the US ANSI cables use red for negative and the German DIN specifications use red for positive. In each case, the other colour is used for the opposite polarity (Table 4.5). Because the colours are not internationally standardized, it is important to know the country of origin of thermocouple extension or compensating cables and the market for which they were intended (cables of German origin might have been manufactured for sale in the USA and carry the ANSI colour coding).

Whatever type of thermocouple is used, many applications will require the measuring junction to be sheathed to prevent contact with such materials as molten metals, hot or corrosive gases and corrosive liquids. For some applications, the junction can be allowed to protrude outside the sheathing if a fast response is needed, particularly for gas temperature

Table 4.5 Colour codes for extension cables and compensating cables in the UK, USA and Germany.

Code		UK	USA	Germany
(a) *Extension cables*				
	Outer	Brown	Purple	—
E	Positive	Brown	Purple	—
	Negative	Blue	Red	—
	Outer	Black	Black	Blue
J	Positive	Yellow	White	Red
	Negative	Blue	Red	Blue
	Outer	Red	Yellow	Green
K	Positive	Brown	Yellow	Red
	Negative	Blue	Red	Green
	Outer	Blue	Blue	Brown
T	Positive	White	Blue	Red
	Negative	Blue	Red	Brown
(b) *Compensating cables – type U for noble metals, type VX for base metals*				
	Outer	Green	Green	White
U	Positive	White	Black	Red
	Negative	Blue	Red	White
	Outer	Red	Red	Green
VX	Positive	White	Brown	Red
	Negative	Blue	Red	Green

measurements, although this is not permissible if the gas is corrosive. The alternatives are to use the isolated type of sheathings, in which the junction is totally insulated electrically, or the grounded type, in which the junction makes contact with the sheathing. The latter type also gives good protection against corrosive materials, but with a considerably faster response.

Both the fully sheathed types must be used in high-pressure environments. Table 4.6 lists materials commonly used for sheathing thermocouples for industrial uses. The 27% chrome alloy is the most widely used type for low-temperature molten metal bath measurements, particularly for lead and zinc alloys. Stainless steel is better than nickel alloys in atmospheres that contain oxides of sulphur (exhaust gases from burning coal or oil, for example), and the ceramic materials must be used for sheathing noble metal thermocouples.

4.5 Metal-resistance sensors

All metallic conductors exhibit a change of resistivity when their tempera-

Table 4.6 Sheathing materials used for thermocouples.

Material	Max. °C	Comments
Mild steel	500–800	Depends whether cold drawn or solid drawn. Liable to oxidation
27% chrome–iron	1000	Used in molten tin or lead. Liable to oxidation
18/8 stainless steel	800	Good resistance to oxidation and corrosion
Inconel (nickel alloy)	1100	Must not be used in sulphur oxide atmosphere
Silicon carbide	1500	Outer sheath use, resists thermal shock. Can be oxidized
Alumina ceramic	1600–1900	Use for noble metals. Highly resistant to chemicals

Table 4.7 Resistance, resistivity and the change of resistance with temperature.

A wire with uniform cross-sectional area A, length s and resistivity ρ will have resistance R, given by:

$$R = \frac{\rho s}{A}$$

For a temperature rise of θ°C, the following changes occur:

length increases by $s\alpha\theta$, where α is the linear expansivity;
area increases by $2A\alpha\theta$, where A is area at 0°C;
resistivity increases by $\rho\alpha\theta$, where ρ is resistivity and α is temperature coefficient of resistivity.

For most metals, value of expansivity is of the order of $2 \times 10^{-5}\,K^{-1}$ and the temperature coefficient of resistivity is of the order of $4 \times 10^{-3}\,K^{-1}$, about 200 times larger, so that the changes in dimensions affect resistance only to a negligible extent. This allows us to use the temperature coefficient of resistivity as if it were the temperature coefficient of resistance. The formula for resistance change is therefore:

$$R_\theta = R_0(1 + \alpha\theta)$$

where R_θ = resistance at temperature θ, R_0 = resistance at 0°C, α = temperature coefficient of resistivity, θ = temperature difference.

ture changes and this change in the resistivity causes a change of resistance – Table 4.7 summarizes the relationships. The change of resistance is a more linear quantity over a large range of temperature than the output from a thermocouple. Although the characteristic shows deviation from a straight

Table 4.8 Temperature coefficients of resistance for some metals.

Metal	Coefficient ($\times 10^{-3}$)	Metal	Coefficient ($\times 10^{-3}$)
Aluminium	4.2	Copper	4.3
Iron	6.5	Nickel	6.5
Platinum	3.9	Silver	3.9

line at high temperatures, there is at least no reversal as is found in the thermocouple characteristic. The deviation is caused by the effect of the square and cube law components of the equation, and these effects are important only at high temperatures. For most metals, the first coefficient of resistance change (alpha) is close in value to the 1/273 (or 0.00366) figure for the expansivity of gases (see Table 4.8). A few metal alloys have a very low value of temperature coefficient, notably constantan with a value that is only about 10% of the average for pure metals, and manganin with an even lower value. Both of these materials are alloys of copper, nickel and manganese.

For comparatively small temperature ranges, up to 400°C or so, the resistance change of nickel or of nickel alloys can be used. For higher temperature ranges, platinum and its alloys are more suitable because of their much greater resistance to oxidation. For measurement purposes, the resistance sensor can be connected to a measuring bridge, along with a set of dummy leads whose temperature is also changed (Figure 4.9). A platinum resistance in this form can be used as a standard of temperature measurement. The National Physical Laboratory standard thermometer is a gas-expansion type, but this requires long and elaborate setting up, so

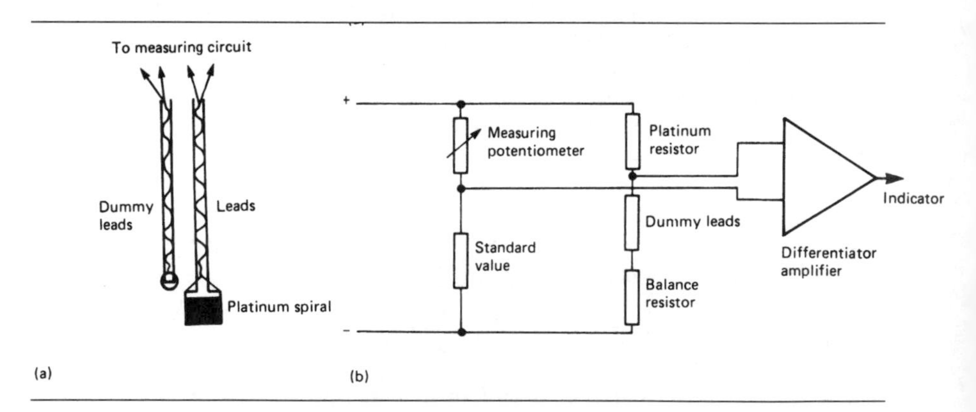

Figure 4.9 The arrangement of a platinum resistance thermometer using dummy leads to compensate for the change of resistance of the leads to the measuring element: (a) physical arrangement, (b) electrical circuit.

that platinum-resistance thermometers calibrated from the gas-thermometer standard are used extensively as secondary standards (often called, confusingly, sub-standards). The size of the sensing element and its heat capacity make the response slow compared to some of the purely electronic devices, such as the thermocouple.

When a switching action is required, the output from a bridge circuit connected to the platinum resistance thermometer can be used to operate a trigger type of circuit. This is seldom done, because the advantage of the resistance type of thermometer is its comparatively linear response, and switching can be carried out by a variety of less expensive devices.

RESISTANCE THERMOMETERS

The platinum-resistance thermometer was formerly used only as a laboratory standard, but advances in the construction of these and resistance thermometers generally have led to resistance thermometers being used in many applications for which only thermocouples were once considered. In particular, many industrial processes that at one time would have been considered as being tightly controlled if the temperature variations were held to 10°C, are now required to be held to much closer limits. The days of 'spit on it and see if it sizzles' are quite definitely over, and the days of 'near enough' are almost gone as well. Emphasis on quality control and uniformity of product now require temperatures in manufacturing processes to be maintained to much closer limits than was even thought possible at one time.

Although there are several materials that can be used in resistance thermometers, platinum has the considerable advantage of being the reference material for international standards, used in the range −270°C to +660°C. The laboratory form of platinum-resistance thermometer is used for calibrating other thermometers, but is a fairly bulky piece of equipment. Miniature versions are now available which combine the accuracy of the platinum-resistance principle with the ability of platinum to withstand corrosive atmospheres. Although nickel and copper can be and are used for some purposes in the lower ranges of temperature, platinum has the advantage that it can be prepared in a very pure state, is highly resistant to corrosion (like all noble metals) and has a resistance/temperature relationship which is almost perfectly linear over a wide range of temperatures. It is also a very stable material, both electrically and mechanically, so that drift of resistance value as the material ages, and with use, is negligible.

The other factors that have led to an increase in the use of platinum-resistance thermometers are the development of instrumentation that can match the high standards of the platinum-resistance system and the comparative ease of use. Connecting cables to a platinum-resistance thermometer can be ordinary copper cable, with no need for special extension

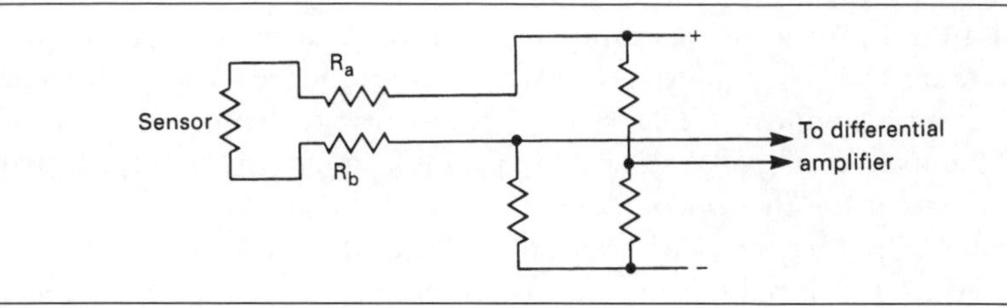

Figure 4.10 The simple Wheatstone bridge circuit used for a resistance thermometer. The cable resistance values R_a and R_b are measured in addition to the resistance of the sensor.

compensating cables. The calibration of a platinum-resistance thermometer is a once-only affair, with no need for cold-junction compensation or the use of thermostatic enclosures for a reference junction. The instrumentation portion of a thermometer can also be simpler, and adding stages to allow for compensation is unnecessary.

The construction for an industrial platinum-resistance thermometer is generally more rugged than the open coil construction used for a laboratory standard. The wire is wound in a spiral or small radius and the whole spiral sealed into a hollow alumina rod, using ceramic sealant where the wires enter the alumina. The sizes of such assemblies can be as small as 0.8 mm in diameter and 5 mm in length, and with careful monitoring of the winding resistance accuracy as close as 0.01% can be obtained. Another quite different form of construction is the use of thin films of platinum on ceramic substrates, which unlike the wound form can be mass-produced, at much lower prices. These are for many industrial purposes more suitable, although some care has to be taken with any platinum-film device to ensure that the device has no chemical catalytic action on any gases or liquids with which it is in contact.

The electrical circuit used for the platinum-resistance thermometer is a bridge circuit, and for many applications, a simple Wheatstone bridge circuit (Figure 4.10) is sufficient. In this type of circuit, the resistance $R_a + R_b$ of the two connecting leads is also measured, but if this resistance is negligible compared with that of the platinum coil (1% or less) then the error is also negligible. In some cases, however, the leads must be of a substantial length and cannot have negligible resistance, so that other circuits must be used. Although the full-compensation method of Figure 4.9 is used in laboratory equipment, a simpler system is illustrated in Figure 4.11, in which three leads are used, two carrying current and one acting as voltage sensor only. In this circuit, the added resistance is equivalent to the difference between the main connecting cables, and this quantity is, for practically all industrial measurements, negligible. The principle can be extended to a four-wire system in which two cables carry the bridge current and two others act as voltage connectors.

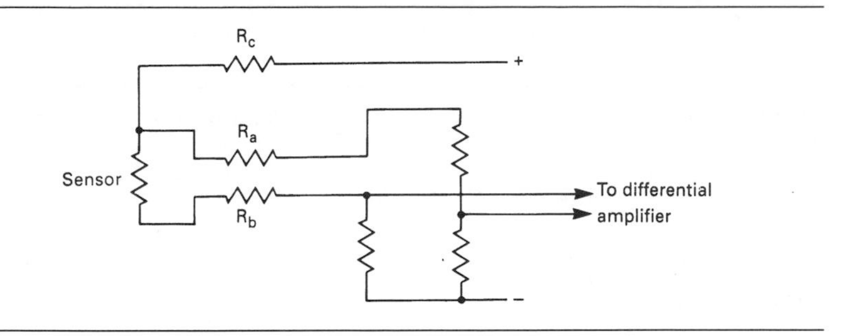

Figure 4.11 A three-wire compensating circuit in which the cable resistance is almost balanced out: only the difference in cable resistance is added to the sensor resistance.

For all resistance thermometer applications, the amount of current used in the bridge circuit must be so low that the self-heating of the platinum sensor is negligible. This creates a conflict between sensitivity and accuracy because the higher the current through the low resistance of the platinum, the greater the voltage to be measured and hence the ease of use of the bridge. Modern high-impedance electronic amplifiers make it possible to operate measuring bridges with very low currents without sacrificing sensitivity, so that this problem is much less significant than was formerly the case. Some care is needed, however, if a new platinum-resistance thermometer head is connected to an old measuring bridge, to ensure that the bridge current is low enough to avoid self-heating. The resistance of film-type sensors is higher than that of the older wire type, and lower currents will be needed for the films.

Table 4.9 compares and contrasts the merits of thermocouples and resistance thermometers as an approximate guide to specifying a temperature measuring system for industrial use.

4.6 Thermistors

Thermistors are a form of temperature-sensitive resistors formed using mixtures of oxides of exotic metals. The constructional methods are similar to those used for carbon composition resistors. Some of these mixtures have positive temperature coefficients, and in most cases it would be meaningless to quote a value for temperature coefficient, positive or negative, because the value is not a constant. The thermistors with a positive temperature coefficient are very non-linear, but the more common negative temperature coefficient types follow a roughly logarithmic law with no violent changes in resistance.

Given that the resistance of a thermistor is known at one temperature θ_2, it can be calculated for another temperature θ_1 by using the formula illu-

Table 4.9 The relative merits of thermocouples and platinum resistance thermometry compared.

Thermocouple	Platinum Resistance
0.5–5°C precision	0.1–1.0°C precision
−200°C to +1750°C range	−200°C to +650°C range
Price factor 1	Price factor 2.5
Sensitive at tip	Sensitive throughout stem
50 ms–5 s response	1–50 s response
Can be very small	Larger size
Reference zero needed	—
Can be used for surface temperature	—
Vibration resistant	Affected by vibration
No power supply needed	Needs power supply
No self-heating effect	Current must be limited
Long-term drift	Excellent stability
Very robust	Can be fragile
Special leads required	Uses copper cables
Output 10–40 μV/°C	Output 0.4Ω change/°C
Screening needed	Can be unscreened

Table 4.10 Thermistor formulae that relate resistance to temperature.

The temperature coefficient of a thermistor, is not a constant, but itself varies as temperature changes. A more useful quantity is the thermistor constant B which can be used to find the resistance at any temperature in the working range provided another pair of resistance and temperature values is known.

The equation is

$$R_2 = R_1 \cdot e\left(\frac{B}{\theta_1} - \frac{B}{\theta_2}\right)$$

θ and B values are in Kelvin (K) units of temperature.

For example, if the thermistor constant B is known to be 3200K, and the resistance at 30°C is 2 kΩ, then the resistance at 45°C can be calculated as follows.

The temperatures are 293K and 318K, so that the quantity in brackets is 0.8586. Using the exp function of a calculator, $R_2 = 2 \times 2.359 = 4.719$, about 4.7.

strated in Table 4.10. The use of θ rather than T for temperature in this formula is a reminder that the temperatures must be in units of Kelvins (absolute temperatures). The Kelvin or absolute temperature is obtained by adding 273 to the Celsius temperature. If you need to work to two places of decimals of temperature, use the figure 273.16.

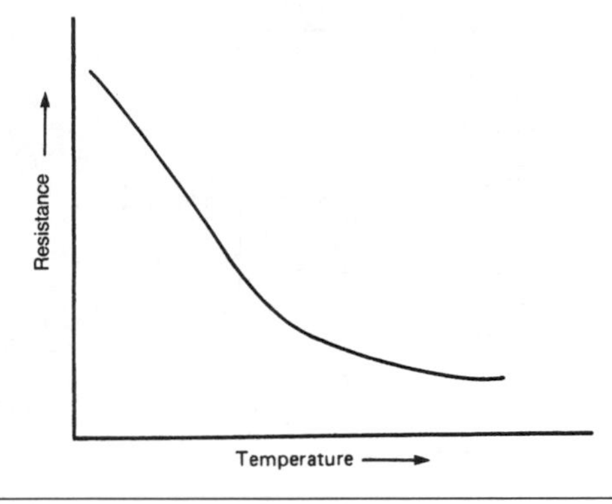

Figure 4.12 A typical resistance/temperature characteristic for an NTC thermistor. The resistance decreases as the temperature rises, and the shape of the characteristic is never linear.

Thermistors can be obtained in a variety of physical forms, as beads, miniature beads, plates, rods, and also encapsulated in metal containers. Negative temperature coefficient (NTC) thermistors are used for temperature control applications such as low-temperature oven controllers, deep-freezer thermostats, room temperature sensors and process controllers. Temperature limits range from 150°C to 200°C, with a few types able to withstand 600°C. The range of temperature that a thermistor can handle depends on the associated circuit, because the range of resistance will be very large compared to the range of temperature.

The typical NTC thermistor characteristic is shown in Figure 4.12, and it displays a negative change of resistance with increasing temperature. The shape of the characteristic is exponential rather than linear, and the useful temperature range is comparatively small.

In any of these applications, NTC thermistors have considerable advantages as compared to the old bimetal thermostat, notably the absence of any hysteresis effects (switching on at a different temperature than that for switching off). NTC thermistors can also be obtained in evacuated envelopes for use in such purposes as oscillator limiters and controllers for voltage-controlled amplifiers.

Thermistor circuits in general require the use of pre-set potentiometers in order to make working adjustments, but circuit costs can be reduced by employing curve matched thermistors whose resistance values are guaranteed to within close limits at each of a large range of temperatures. Thermistors of all types will also have quoted values of dissipation constant and time constant. The dissipation constant is the amount of power (in milliwatts) that is required to raise the temperature of the thermistor by 1°C above the ambient temperature. For the evacuated bulb type, the dissipation constant is very small, of the order of 12 μW/°C, so that the resistance

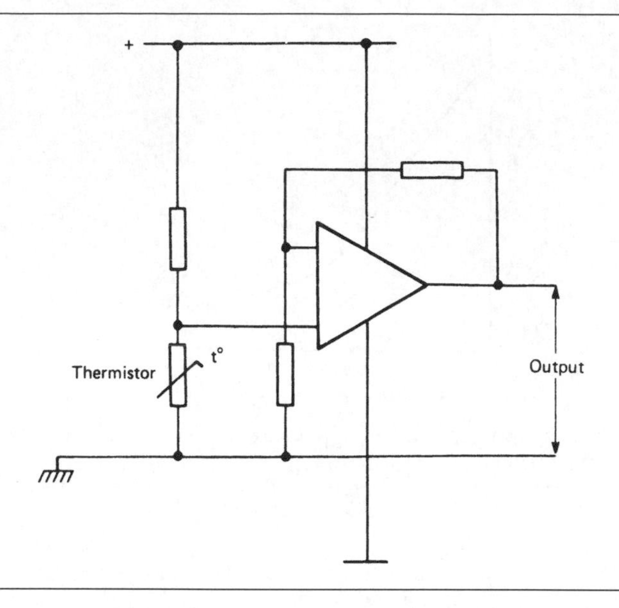

Figure 4.13 An NTC thermistor temperature sensing circuit that makes use of an operational amplifier. The sensitivity can be adjusted by altering the feedback ratio.

of this type of thermistor is substantially altered by only very small amounts of signal current. For temperature-sensing thermistors, values of dissipation constant in the range 70 = 500 μW/°C are typical.

The time constant for a thermistor is defined as the time needed for the resistance to alter by 63% of the difference between an initial value and a final value caused by a change of temperature. Time constant is measured with negligible current flowing, because otherwise the figure would be altered since part of the heating is internal rather than external. The figure of 63% may seem odd, but it corresponds to the definition of time constant for other networks, such as a capacitor and a resistor. By making the definition in this way, the value of time constant is genuinely constant over a large range of temperature changes. Time constants of 5–11 s are typical of the physically small thermistors (miniature beads and the evacuated bulb types), with much larger values for others, 18–25 s for the larger beads, and as high as 180 s for thermistors that have been assembled into temperature sensing probes.

NTC thermistors can be constructed from semiconducting materials with values of temperature coefficients that are usually much larger than the (positive) temperature coefficients of resistors. The phrase NTC resistor is used for devices with fairly small negative values of temperature coefficient, and the term thermistor is reserved for the types that have large negative values of temperature coefficient. Most of the thermistors that are incorporated into temperature-sensing circuits are of the NTC type.

PTC THERMISTORS

Positive temperature coefficient (PTC) thermistors are a more recent development, used mainly for protection circuits for sensing temperature or current. Unlike the NTC types, these PTC thermistors have a current–voltage characteristic that exhibits a change in direction, and two basic types are used, both depending on compounds of barium, lead and strontium titanates (ceramic materials). The over-temperature protection type of PTC device has a switchover point at a reference temperature (or trip temperature, T_r). At temperatures lower than the trip temperature the resistance of the PTC device is fairly constant, but around the trip temperature the PTC characteristic takes over and their resistance rises very sharply as the temperature rises. A typical graph of resistance plotted against temperature, along with a trip-sensing circuit, is shown in Figure 4.14. The sudden change in resistance can be used to operate an indicator or to switch other circuits for purposes such as motor protection or for preventing overheating of transformers.

The other type of PTC thermistor device is used for over-current protection circuits, and its resistance/temperature graph is illustrated in Figure 4.15. This characteristic follows an S-shaped curve which has two turnover points, one at a point of minimum resistance R_{min} and the other at the maximum resistance point R_{max}. Between 0°C and R_{min}, the temperature coefficient is negative, and the coefficient is also negative at the temperatures in the region higher than R_{max}. Between R_{min} and R_{max} the temperature coefficient is large and positive. In this PTC region, the change of resistance can be as large as 100% for each °C rise in temperature.

With one of these devices wired in series with a load, the load is protected against excessive current. At the working current, the PTC device is in its

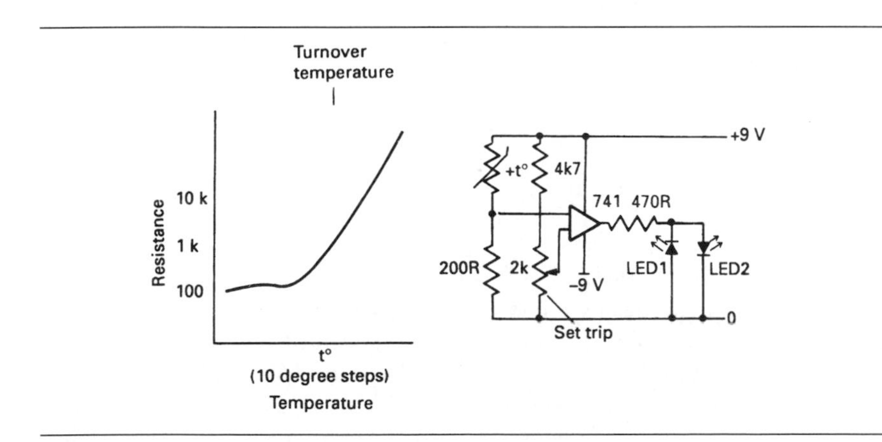

Figure 4.14 Characteristic for a PTC thermistor used for sensing excess temperature.

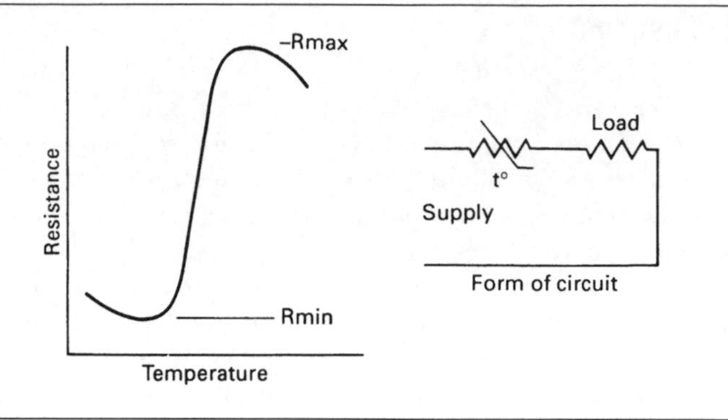

Figure 4.15 The type of characteristic used for current protection devices operated by a PTC thermistor.

low-resistance state, allowing most of the applied voltage to be across the load. When the current is increased, the thermistor will suddenly switch to its PTC mode, assisted by self-heating as more of the applied voltage will now be across the thermistor, until the current flowing in the whole circuit becomes very small. The circuit can be arranged so that it is self-resetting, returning to normal when the thermistor cools, or requires the circuit to be reset by switching off the current and allowing the thermistor to cool.

The change of resistance per unit change of temperature for a PTC thermistor can be so abrupt that circuit devices such as bridges and Schmitt triggers are sometimes not necessary. In a very few applications, the PTC thermistor can be used directly, but it is usually undesirable to have the controlled current passing through the thermistor, and more usually the thermistor is part of a transistor switch or an operational amplifier circuit. The output of such a circuit is not particularly linear, but the sensitivity can be high and the response can be rapid. One particular advantage is that the sensing element can be very small.

- Note that if an op amp with sufficient gain is used to amplify the voltage across a thermistor, any type of thermistor can be used to sense excessive temperature. The advantage of using a PTC thermistor is that many types of sensing applications can use circuits that require no op amp.

The connection of a thermistor to a switching circuit has the advantage, as compared to bimetallic strip devices, that it can be arranged so as to have zero hysteresis if this is a useful feature. For most switching purposes, however, some hysteresis is desirable in order to prevent rapid switching on and off as air currents strike the detector. The more elaborate temperature sensing systems that make use of thermistors are microprocessor controlled, and a form of time hysteresis is used. The temperature sensing output from the thermistor is monitored at short intervals, and a change registered only if the direction of temperature change is consistent. This allows much more rapid response to a temperature change than the conventional form of hysteresis, although the sensing intervals have to be adjusted to suit the type of use.

Transistors themselves can be used as sensing elements, because many transistor parameters follow an exponential negative temperature characteristic. The inherent amplification of a transistor makes the use of the temperature sensitivity of base–emitter voltage attractive as a sensing system, because the output can be taken in amplified form from the collector. The use of transistors as temperature sensors, however, is confined mainly to temperature compensation of transistor and IC circuits.

4.7 Radiant heat energy sensing

Radiant energy, which can be light, heat, radio waves or, in some instances, ionizing radiation, may need to be sensed, and a very large range of the electromagnetic spectrum is sensed by its temperature effect. The sensors for light have mainly been dealt with in the previous chapter, but there is one device, the *bolometer*, which has been held over to this chapter because its action is essentially thermal.

The bolometer principle is illustrated in Figure 4.16. A blackened material will absorb radiation well and so its temperature will rise when radiated energy falls on it. The change of temperature is then detected in the ways that have been described earlier in this chapter. The classic types of bolometers used in the nineteenth century were metal, and the effect of the temperature rise due to radiation was detected in sensitive measuring bridge circuits. Because the change of temperature due to radiated energy can be small, and the change of resistance correspondingly smaller, bolometers are usually connected into bridge circuits so that the output from an unradiated bolometer can be compared with the output from the bolometer which is deliberately exposed to radiation.

Modern bolometers can make use of semiconductor sensors, blackened for maximum absorption of radiation. The much greater change of resistance

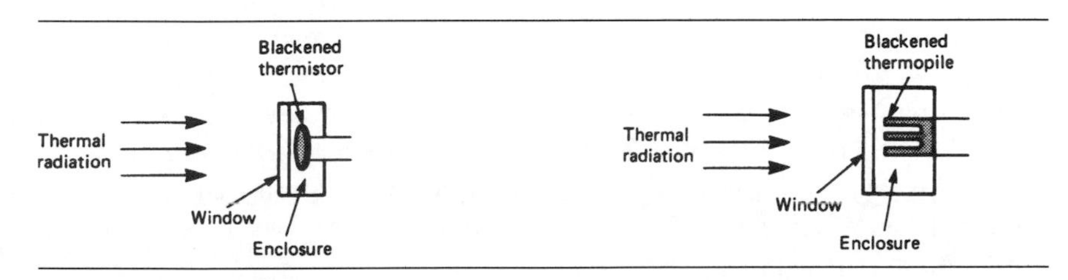

Figure 4.16 The bolometer consists of a temperature sensor whose surface is blackened and enclosed in a container (preferably evacuated). Thermal radiation passes through the window, heating the sensor. The sensor can be a thermistor or a thermopile.

of a thermistor for a small change of temperature makes this type of material ideal for bolometer use and permits much more sensitive detection than was possible with the older types. The non-linear nature of the thermistor is less important in this type of application, because the changes of temperature are usually small.

4.8 Pyroelectric detectors

Pyroelectric films are dielectric materials whose surfaces charge when radiated by infrared (IR). Plastics films have been used for this purpose, but the material that is favoured for modern passive infrared (PIR) detection systems is lithium tantalate. The construction of a detector is as a capacitor with one metal plate and the other a pyroelectric material with a conducting surface. Because of the effect of infrared in separating charges on the pyroelectric material (confusingly, but correctly, also called *polarization*) the charge and hence the voltage across the plates of a pyroelectric capacitor will alter as the amount of incident infrared radiation is altered.

The time constant is large, so that the response rate to alterations in IR is in the range 0.2–1 Hz. Because the detector is a capacitor, however, there is no DC response, so that a non-moving infrared emitter will not be sensed. In addition, the capacitor has a very high impedance, so that a practical pyroelectric detector consists of the capacitor and a MOSFET constructed as one unit with external connections to the source and drain of the MOSFET. The main applications for pyroelectric detectors are burglar alarms, automatic light-switching and door-opening equipment, and positioning systems.

The main parameters of a pyroelectric detector are noise-equivalent power (*NEP*), *responsivity* and frequency response. The NEP for a given source energy, signal rate of change and bandwidth express the lower limit for which the detector is useful, since signals below this limit will be below the noise level. For a typical detector for a source at a colour temperature of 500K, at a frequency of 10 Hz and 1 Hz bandwidth, the NEP figure of 10^{-9} W is quoted for a lithium tantalate pyroelectric material.

Responsivity can be quoted in terms of either voltage or current output as volts per unit radiant energy or current per unit radiant energy, at a given predominant wavelength or colour temperature of source. A typical figure of voltage responsivity is 3200 V/W. The frequency response of responsivity means the change of responsivity for different modulation frequencies (not radiated frequency), and this, as noted above, corresponds to a low-pass filter action with a peak at less than 1 Hz.

Figure 4.17 illustrates a typical PIR unit, using a DIL IC layout with four pins in a casing approximately 8 × 7.5 mm. The equivalent circuit is of two pyroelectric capacitors connected so that their voltages will add, and with the summed voltages applied to the gate of a MOSFET whose source and

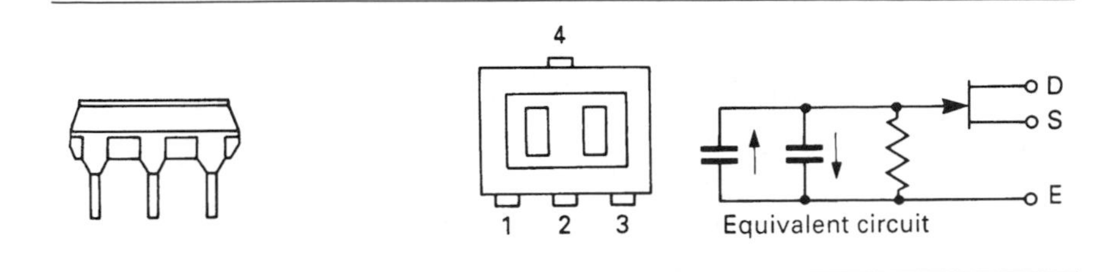

Figure 4.17 A typical pyroelectric passive infrared (PIR) unit and its internal circuit.

drain terminals are two of the pins of the package. A supply voltage in the range 3–15 V can be used, and the MOSFET is usually wired into a source–follower circuit with a resistor of 100–200 kΩ. The current consumption is 200 μA maximum. The relative response is a maximum for a frequency of 0.2 Hz, falling to 50% at 0.055 Hz and 0.7 Hz, making the unit well suited to respond to movement of humans and animals.

A unit of this type is used in conjunction with a faceted (Fresnel) lens which will focus infrared from an object on to the pyroelectric surfaces and also cause any movement of the object to sweep infrared beams across the detector. The addition of this lens is an essential part of the action, and its construction will also determine the angle over which detection of a moving warm object is effective.

Figure 4.18 shows a typical circuit, using a source–follower connection to the internal MOSFET with a 200 kΩ load. The output from the detector is amplified in two stages of a quad operational amplifier, with a voltage gain, overall, of around 3200 – the gain is controllable by the variable resistor in the feedback circuit of IC2. The output from IC2 is passed to a threshold circuit, using a 220 kΩ variable resistor to set the threshold voltage at which signals will be passed to the output transistor Tr1, which will amplify and shape the output from the detector. The output is shown operating an LED display with an output to an alarm circuit. In use, the alarm will be turned off when not needed, but the LED is usually kept in operation to indicate that the unit is working correctly.

4.9 Thermal transducers

The familiar transducer for electrical to thermal conversion is the heating element, and for most purposes this is composed of a nickel alloy wire such as *Nichrome*. This alloy of nickel, chromium and iron has good resistance to oxidation even when red hot, and its resistivity is high, so allowing a high resistance to be achieved without the need for very long lengths of narrow-gauge wire. The amount of energy transformed is given by Joule's equation (Figure 4.19), but the level of temperature that will be caused by

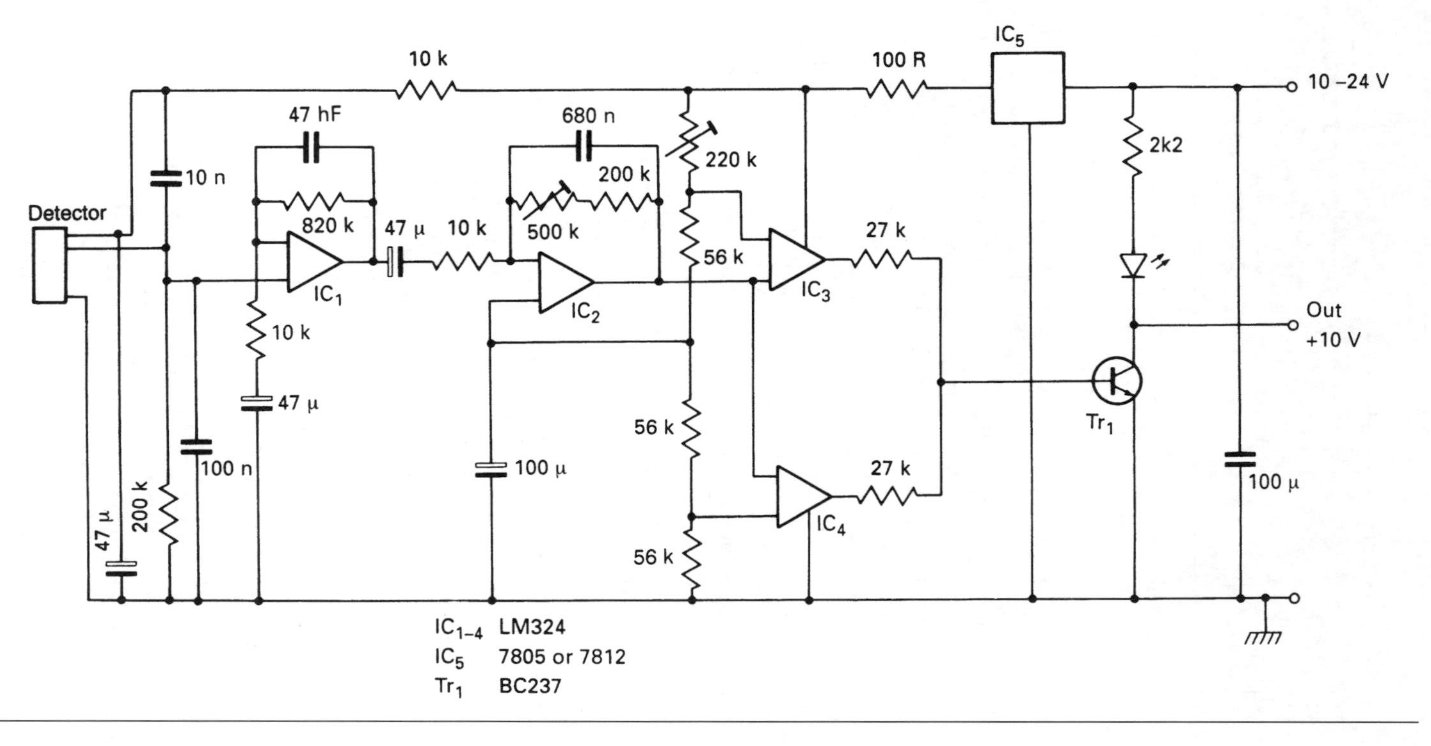

Figure 4.18 A suggested circuit for a pyroelectric detector unit.

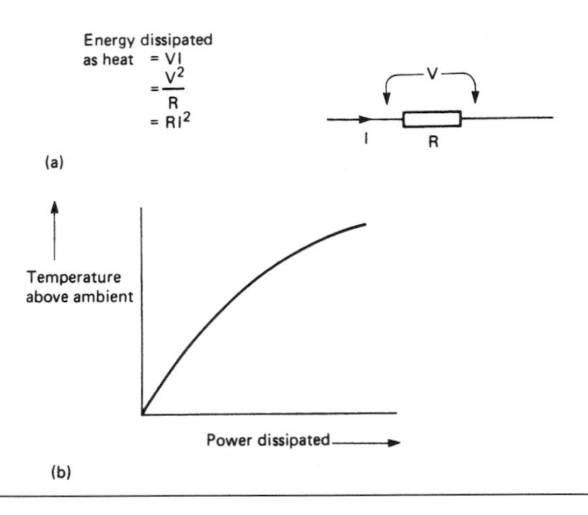

Figure 4.19 Heat dissipation. The Joule equations (a) show the conversion of electrical energy into heat energy, but there is no straightforward way of finding what the final temperature of the hot object will be. A typical temperature/power curve for a resistor is shown in (b).

a given current is less predictable. A material reaches a steady temperature when the rate of loss of heat energy is equal to the rate at which energy is input. The rate of loss depends on many factors other than the difference between the temperature of the material and the ambient temperature.

One point that is sometimes forgotten in making calculations on thermal conversion is that the resistance of a heating element is not constant. The current in a heating element is usually quoted as the operating current, the value of current for the hot element. Suppose, for example, that a heater has to operate at 240 V, 2 A. This implies that the resistance, when hot, will be 120 Ω. The resistance when cold will be less than this, and the difference between these values depends on the temperature difference and the material. If the heater of our example uses nichrome and runs at 400°C then the resistance at room temperature (taken as 25°C) is 119 Ω, because the temperature coefficient of nichrome is small. If the heating element had been constructed of pure nickel then the resistance at room temperature would be about 47 Ω, a very considerable change. Table 4.11 shows the basis of these calculations. The changes in resistance are particularly marked for incandescent lamps, because of the large difference between cold temperature and working temperature. Some care should always be taken if you need to work with the resistance values of such lamps, since the cold value will be only a small fraction of the hot value.

Any conducting material will act as a transducer of electrical energy to thermal energy, so that the use of materials that have a negative value of temperature coefficient needs special care. Passing current through such a material from a fixed voltage supply can cause heating that in turn lowers

Table 4.11 How resistance changes caused by temperature variations are calculated.

Resistance/temperature formula is:

$$R_\theta = R_0(1 + \alpha\theta)$$

where R_θ = resistance at temperature θ, R_0 = resistance at 0°C, α = temperature coefficient of resistance (taken as 7.7×10^{-4} in this example)

Example:

$$R_{400} R_0(1 + 7.7 \times 10^{-4} \times 400)$$

$$R_{25} = R_0(1 + 7.7 \times 10^{-4} \times 25)$$

so that

$$R_{400} = R_0(1.068) \quad \text{and} \quad R_{25} = R_0(1.00425)$$

and

$$\frac{R_{400}}{R_{25}} = \frac{1.068}{1.00425} = 1.06$$

If $R_{400} = 120\,\Omega$ then $R = \frac{120}{1.06} = 113.2\,\Omega$

the resistance and increases the current. This 'positive feedback' can cause fusing or other breakdown unless some part of the circuit limits the amount of current that can flow. The most notable examples of this effect are semiconductors and that curious mixture, glass.

The Peltier effect, discovered in 1822, is another form of electrical to heat transducer action, and is the reverse of the thermocouple (Seebeck) action. Once again, two junctions of dissimilar metals are used, and in this case, a current is passed around the circuit, resulting in one junction cooling and the other heating. The temperature changes for a given current are small for metal-to-metal junctions, but can be substantial for metal-to-semiconductor junctions, making this a useful way of temperature control for small spaces.

4.10 Thermal to electrical transducers

The conversion of electrical energy to thermal energy proceeds with almost 100% efficiency, but no conversions from thermal energy to any other form ever approach much more than 50% efficiency. The reasons for this are summed up in the laws of thermodynamics, and are founded on the principle that we do not know how much heat an object contains, and we cannot remove all of it. Any change from thermal energy to another form must involve heat taken in by a converter at a high temperature and a lesser amount of heat given out at a lower temperature. The efficiency cannot then be greater than the fraction given by:

$$\frac{\text{temperature difference}}{\text{high temperature}}$$

with the temperatures measured in the Kelvin scale whose zero point is equivalent to −273.16°C. This equation implies that 100% conversion is possible only if the converter exhausts its heat at 0K, which is not a practical proposition. In addition, this equation assumes that every other part of the conversion process operates at 100% efficiency.

The conversion from thermal to electrical energy is, on the large scale, carried out by way of steam generation, with the steam operating turbines that are coupled to alternators. The source of thermal energy can be nuclear. The use of the more direct gas turbine and alternator method is expensive and is used only for topping up the supply from conventional coal and nuclear power stations. One of the benefits of the inefficiency of the whole process can be the availability of large quantities of water at a useful domestic temperature of 40–60°C, and in some countries electricity generating stations also sell their waste heat in a type of scheme called CHP (combined heat and power). In the UK, the emphasis has always been on large-scale generation and grid distribution, and CHP has never been a practical proposition in such circumstances. The pioneer dream of a small nuclear station supplying electricity and heating for a self-contained community has never been realized, but it will have to become an option as the effects of burning half of the contents of the planet become more noticeable.

By comparison with the 40% efficiency that can be obtained from a well-designed and very large coal-fired power station, the efficiency of the generation of electricity from any other thermal transducers is extremely low. The most practical system in the past has used thermocouples stacked into large blocks that are heated at one end and cooled at the other, with the individual thermocouples connected in series so as to obtain a useful voltage. Several commercial units have been available in the past, and at one time the gas-fired radio receiver was, if not common, more than a museum piece. The Milnes converter was one of the most successful of these early units, and was manufactured in a factory in Tayport, Fife in the late forties, along with loudspeakers to the Milnes design. The efficiency of these thermocouple units, however, rarely exceeded 5%.

Other forms of thermal to electrical converters have achieved even lower efficiencies, and even with intensive development work, the efficiency figure of 10% is more of a dream than a reality. One method that was pursued for conversion of solar energy used the principle of thermal emission from a cathode heated by focused sunlight, with a large number of units connected in series in order to attain a reasonable voltage. This, however, proved to be as unreliable as any other solar source, and the ultimate comment on solar heating is provided by the advertisements in Australian papers offering to replace solar heating units by oil-fired boilers. If solar heating is uneconomic in Australia, its chances in the rest of the world are rather poor.

Chapter 5

Sound, infrasound and ultrasound

5.1 Principles

Sound and vibration are connected in the sense that any sound is associated with a mechanical vibration at some stage. Many sounds are caused by the vibration of solids or gases, and the effect of a sound on the hearer is to vibrate the eardrum. The sound wave is the waveform caused by a vibration and which in turn causes an identical vibration to be set up in any material affected by the sound wave. The mechanical vibration need not necessarily cause any sound wave, because a sound wave needs a medium that can be vibrated, so that there is no transmission of sound through a vacuum. Perhaps some day this will penetrate the minds of the makers of 'space' films, who always consider sound effects so important.

When sound is transmitted, the wave parameters are *velocity* (speed), *wavelength* and *frequency*. The frequency and the waveshape are determined by the frequency and waveshape of the vibration that causes the sound wave, but velocity and wavelength are dependent on the medium that carries the sound wave. The relationship between velocity, wavelength and frequency is illustrated in Figure 5.1. The velocity of sound in a given material depends on the density and the elastic constants of the material, and the velocity is highest in dense solids, lowest in gases at high pressure. Table 5.1 shows some examples of values of the speed of sound in common materials.

The perception of sound by the ear is a much more complicated business. Objective measurements of sound waves can make use of the intensity, measured as the number of watts of sound energy per square metre of receiving surface, or of the wave quantities of pressure amplitude or displacement amplitude. The ear has a non-linear response, and a sensitivity that varies very markedly with the frequency of the sound. The response of

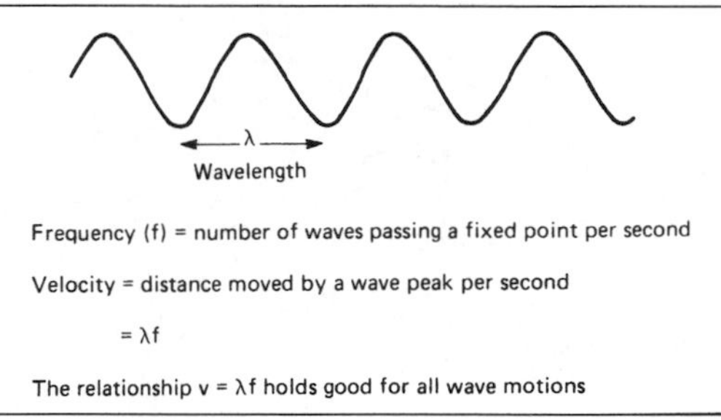

Figure 5.1 The relationship between wavelength, frequency and velocity for a wave.

Table 5.1 The velocity of sound in different materials.

Material	Speed of sound (m/s)	Material	Speed of sound
Aluminium	5100	Copper	3700
Glass	4500 (average)	Sea water	1510
Air	343	Helium	965
Hydrogen	1284	Nitrogen	334

Pressure amplitude (Pa)	Intensity in W/m²	Amplitude of displacement (m)
2×10^{-1}	10^{-4}	10^{-7}
2×10^{-2}	10^{-6}	10^{-8}
2×10^{-3}	10^{-8}	10^{-9}
2×10^{-4}	10^{-10}	10^{-10}
2×10^{-5}	10^{-12}	10^{-11}
2×10^{-6}	10^{-14}	10^{-12}

Minimum audible intensity (threshold of hearing)

10 100 1000 10 000

Frequency (Hz)

Figure 5.2 The minimum detectable sound level for the average ear plotted for different frequencies.

the ear is at its maximum to sounds in the region of 2 kHz, as the threshold curve of Figure 5.2 illustrates. This curve shows the amount of sound, measured in the three main ways, which is just discernible by the average human ear. The difference in the threshold intensity from the peak sensitivity frequency of about 2 kHz to a bass frequency of 100 Hz is very large, a ratio of about 10^5 in terms of W/m^2.

There is less variation in the threshold of pain (Figure 5.3), but this curve

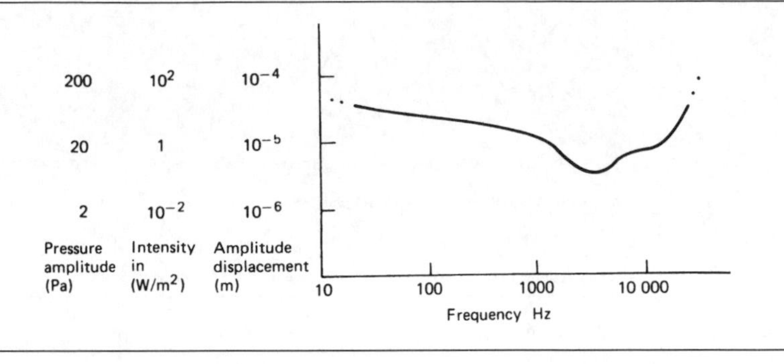

Figure 5.3 The pain level for the ear, plotted for the frequency range. This shows much less variation with frequency.

also shows the greater sensitivity to sound in the 2 kHz region. The pain threshold has been approached much more frequently in the latter part of the 20th century than in earlier times and frequent bombardment of the ears to energy levels near this threshold results inevitably in deafness. Legislation for health and safety at work is now having some impact on deafness caused by noisy machinery. Nothing seems to put an end to the excessive amplification used in the entertainment industry that is responsible for a much greater incidence of hearing problems. Unfortunately, this is a positive feedback problem, and the volume has to be progressively turned up as the audience becomes more deaf to it. If any factory imposed on its workers the sound levels that are now common in discos, the Health & Safety Executive would close it down.

The frequency range over which sound can be detected by the human ear is limited to the range of about 20 Hz to 20 kHz. In old age, the upper limit is reduced on average by about 1 Hz per day. The lower limit is determined by the sound-filtering effect of tissues in the ear, and avoids the unpleasant effect of the many low-frequency vibrations that exist around us. The transducers that are the subject of this chapter, however, are not necessarily constrained to these frequency limits, and some can be used with infrasound (very low frequencies) or with ultrasound (very high frequencies). Acoustic waves, in fact, can make use of frequencies in the MHz range, well above the upper limit of audible sound. In this region, waves become much more directional and much more subject to filtering effects that can be achieved by shaping the wave path. The topic of surface acoustic filters is based on the effects of surface shape on the transmission of acoustic waves (SAW) along a (solid) surface.

The effect of a sound wave on a material that it strikes is to vibrate the material, and in the course of this vibration every part of the material will be accelerated to and fro. The acceleration is in alternate directions, and there is no bodily displacement of the material, but an electrical output can be obtained from an accelerometer connected to the material. The sensors and transducers for sound to electricity are therefore of the same form as the transducers for acceleration and velocity, and the main differences are the ways in which these sensors and transducers are used. We make use of these devices mainly as transducers, since the aim of a microphone is to produce an electrical wave that is a faithful analogue of the sound wave that is striking the microphone. The power conversion, rather than detecting that a wave is present, is the important factor.

- Most microphones depend on a *diaphragm* as a transducer of sound wave amplitude to mechanical vibration, and it is the vibration of the diaphragm that is then used for sensing or transducing to an electrical form.

5.2 Audio to electrical sensors and transducers

The sound to electrical energy transducer is the *microphone*, and microphone types are classified by the type of electrical transducer they use. In addition to the transducer, however, a microphone will use acoustic filters, passages whose shape and dimensions modify the response of the overall system. These are needed because each transducer will have its own response that is determined by resonances in the materials as well as by the transducing principle itself. The correction to a more uniform response has to be made by means of the acoustic passages in the microphone housing. This type of compensation is preferable to the use of electrical methods, because acoustic filters can have much sharper effects with less impact on the rest of the frequency range.

- Microphones are used as sensors in sound-level meters. This calls for a microphone whose frequency response is carefully controlled, with electronic compensation for any substantial peaks or dips. The reading is usually in terms of decibels above the *threshold noise level*, i.e., the sound pressure level at which the ear can just detect that a sound is present.

The characteristics of a microphone are both acoustic and electrical. The overall sensitivity is expressed as mV or μV of electrical output per unit intensity of sound wave or in terms of the acceleration produced by the sound wave. In addition, though, the *impedance* of the microphone is of considerable importance. A microphone with high impedance usually has a fairly high electrical output, but the high impedance makes it very susceptible to hum pick-up, either magnetically or electrostatically coupled. A low impedance value is usually associated with very low output, but hum pick-up is almost negligible.

Another factor of importance is whether the microphone is directional or omni-directional. If the operating principle of the microphone is the sensing of the pressure of the sound wave, then the microphone will be

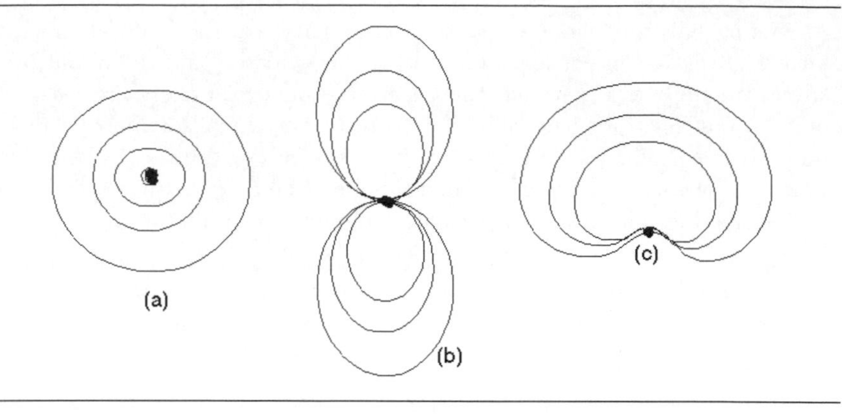

Figure 5.4 Simplified response curves for microphone types: (a) omnidirectional, (b) velocity-operated, (c) cardioid (heart-shaped).

omni-directional, picking up sound arriving from any direction. If the microphone responds to the velocity (speed and direction) of the sound wave, then it is a *directional* microphone, and the sensitivity has to be measured in terms of direction as well as amplitude of sound wave. The microphone types (see Figure 5.4) are known as pressure or velocity operated, omni-directional, or some form of directional response (such as *cardioid*).

The type of transducer does not necessarily determine the operating principle as velocity or pressure, because the acoustic construction of the microphone is usually a more important factor. If, for example, the microphone uses a sealed capsule construction, then the pressure of the sound wave will be the factor that determines the response. If the microphone uses a diaphragm or other moving element that is exposed to the sound wave on all sides, then the system will be velocity operated.

THE CARBON MICROPHONE

The carbon granule type of microphone was the first type to be developed for telephone use, and was still in use for telephones long after it had been abandoned for any other purposes. It has now been replaced by the electret capacitor type (see later) even for telephone use. The principle is illustrated in Figure 5.5, and uses loosely packed granules of carbon held between a diaphragm and a backplate. When the granules are compressed, the resistance between diaphragm and backplate drops considerably, and the vibration of the diaphragm can therefore be converted into variations of resistance of the granules. The microphone does not, therefore, generate a voltage and requires an external supply before it can be used.

The sole advantage of the carbon granule microphone is that it provides

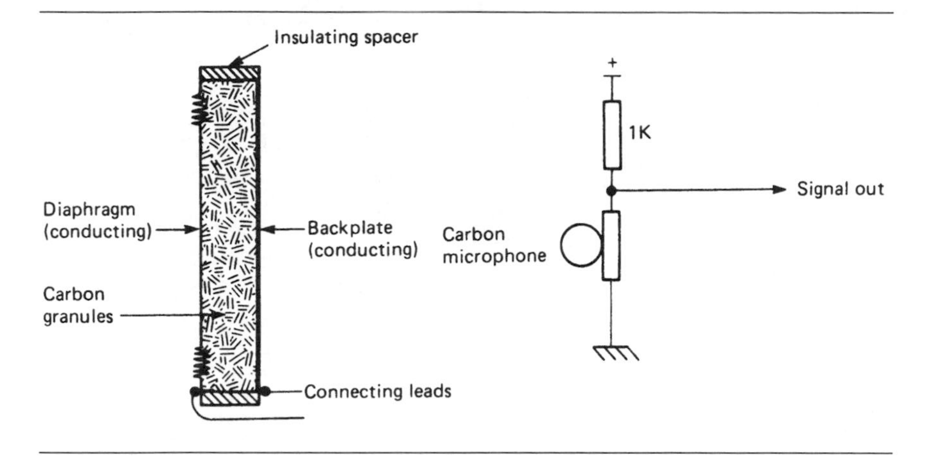

Figure 5.5 The carbon granule microphone principle and the type of circuit into which the microphone is connected.

an output that is colossal by microphone standards, with outputs of 1 V peak-to-peak possible. The linearity is very poor, the structure causes multiple resonances in the audio range, and the resistance of the granules alters in a random way even with no sound present, causing a high noise level. The predominance of the carbon microphone in the early days of telephony was due to its high output at a time when no amplification was possible, and the introduction of valve and, later, transistor amplifiers, caused the rapid demise of the carbon microphone for serious audio use.

THE MOVING IRON (VARIABLE RELUCTANCE) MICROPHONE

This type of microphone uses a powerful magnet that contains a soft-iron armature in its magnetic circuit, with the armature attached to a diaphragm. The principle is illustrated in Figure 5.6. The magnetic reluctance of the circuit alters as the armature moves, and this in turn alters the total magnetic flux in the magnetic circuit. A useful comparison is of a battery connected in series with a fixed and a variable resistor chain. A coil wound around the magnetic circuit at any point will give an EMF which is proportional to each change of magnetic flux. This makes the electrical wave from the microphone 90° out of phase with the sound wave amplitude, and proportional to the acceleration of the diaphragm. Most moving iron microphones are manufactured in the form of sealed capsules or with very limited access to one side of the diaphragm, so that the pressure of the sound wave is the predominant quantity that determines the action.

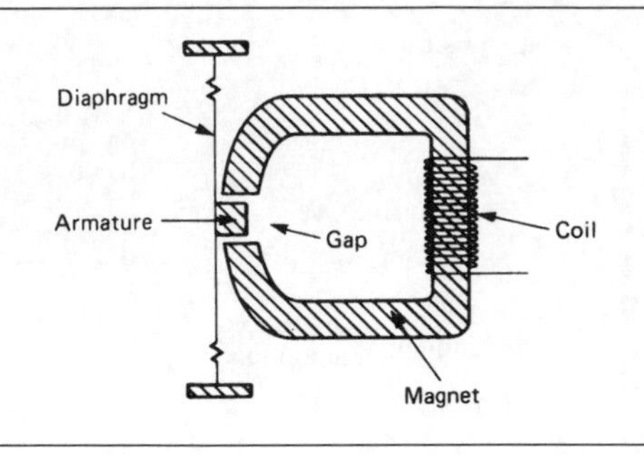

Figure 5.6 The principle of the moving iron (variable reluctance) microphone.

The linearity of the conversion can be reasonable for small amplitudes of movement of the armature, but is very poor for large amplitudes. The linearity can be considerably improved by appropriate shaping of the armature and careful attention to its path of vibration. These features depend on the maintenance of close tolerances in the course of manufacturing the microphones, so that there will inevitably be differences in linearity between samples of microphones of this type from the same production line.

The maximum useable output level from a moving iron microphone can be high, of the order of 50 mV, and the output impedance is also high, typically several hundred ohms. Because the flux path in the transducer is almost closed, external changes in the magnetic field will be very efficiently picked up, and the result is that the magnetic component of mains hum is superimposed on the output. This can be reduced by shielding the magnetic circuit, using mu-metal or similar alloys. The magnetic circuit that is the predominant feature of this type of microphone also makes the instrument heavier than some other types.

MOVING COIL MICROPHONE

The moving coil microphone uses a constant-flux magnetic circuit in which the electrical output is generated by moving a small coil of wire in the magnetic circuit (Figure 5.7). The coil is attached to a diaphragm, and the whole arrangement is usually in capsule form, making this pressure-operated rather than velocity-operated. As before, the maximum output occurs as the coil reaches maximum velocity between the peaks of the sound wave so that the electrical output is at 90° phase angle to the sound wave.

The coil is usually small, and its range of movement very small, so that linearity is excellent for this type of microphone. The coil has a low impedance, and the output is correspondingly low, but not so low that it has to compete with the noise level of an amplifier. The low inductance of

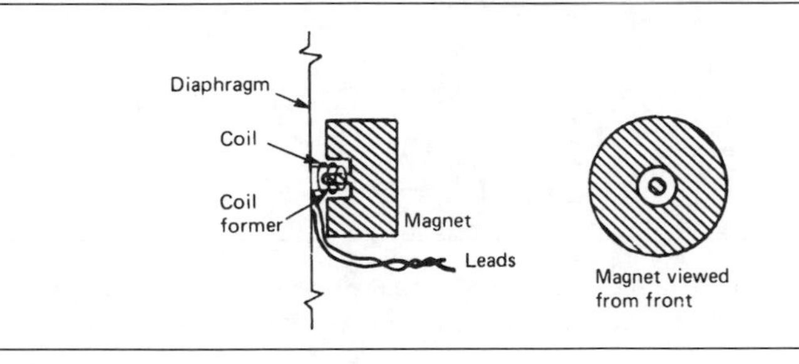

Figure 5.7 The moving coil type of microphone has a coil wound on a former attached to the diaphragm and moving in an annular gap of a magnet.

the coil makes it much less susceptible to hum pick-up from the magnetic field of the mains wiring, and it is possible to use hum-compensating (non-moving) coils, known as *humbuckers*, in the structure of the microphone to reduce hum further by adding an antiphase hum signal to the output of the main coil.

RIBBON MICROPHONE

The ribbon microphone is the logical conclusion of the moving coil principle, in which the coil has been reduced to a strip of conducting ribbon (Figure 5.8), with the signal being taken from the ends of the

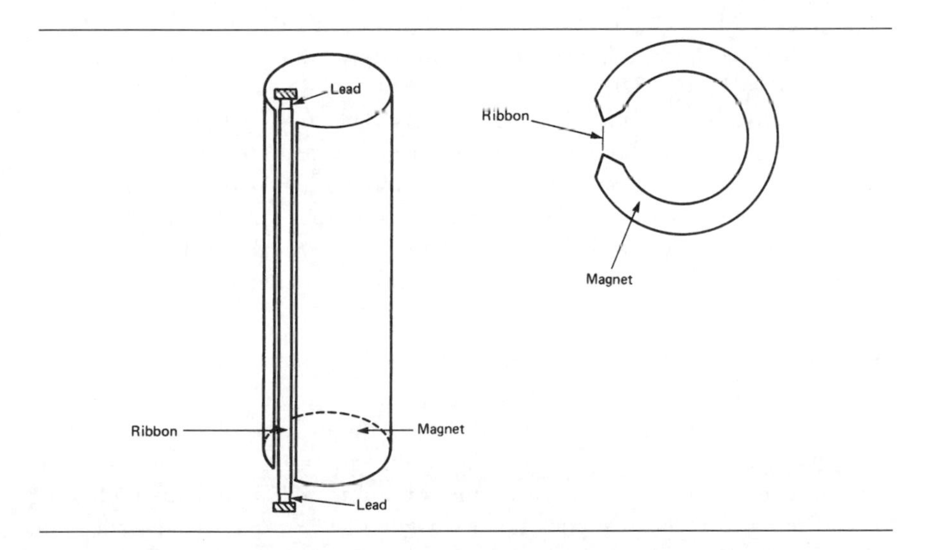

Figure 5.8 The ribbon microphone principle, which is ideal for velocity operation. The sensitivity is very low, and the output requires an amplifier built into the microphone housing, but the ribbon type of microphone provides the highest sound quality.

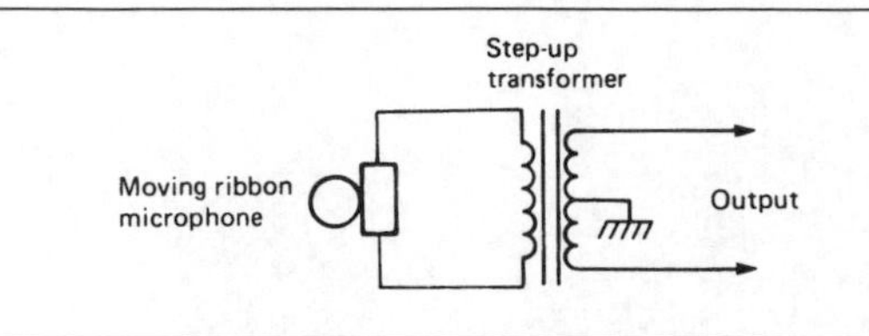

Figure 5.9 The use of a step-up transformer that also provides a balanced output as a match between a ribbon microphone and a line.

ribbon. An intense magnetic field is used, so that the movement of the ribbon cuts across the maximum possible magnetic flux to generate an electrical output whose peak value is as usual at 90° phase to the sound wave.

One of several features that make the ribbon microphone unique is the fact that it is a velocity-operated microphone, because the ribbon is affected by the velocity of the air in the sound wave rather than its pressure. This type of microphone is therefore used in situations where directional response is important, such as a voice commentary in noisy surroundings. The linearity is excellent, and the ribbon microphone is predominantly used where high quality reproduction is of paramount importance.

The construction of the ribbon microphone inevitably makes the output extremely low, and the microphones are usually equipped with a transformer to raise both the signal voltage level and the impedance level. The hum pick-up is also extremely low, and advantage can be taken of this to use a balanced output transformer to minimize hum pick-up in the microphone leads (Figure 5.9).

To be effective, a ribbon microphone has to be constructed to very precise limits, and good ribbon microphones are very expensive items, costing more than most domestic hi-fi users would contemplate spending on a complete sound system. The directional qualities are ideally suited to stereo broadcasting, although for some purposes the very directional response can be undesirable, and moving coil units have to be used.

PIEZOELECTRIC MICROPHONES

The piezoelectric transducer has the advantage over all the other types mentioned in this chapter of not being confined to use in air. A piezoelectric transducer can be bonded to a solid or immersed in a non-conducting liquid so as to pick up sound signals in any of these carriers. In addition, the piezoelectric transducer can be used easily at ultrasonic frequencies, with some types being capable of use in the high MHz region. All piezoelectric transducers require a crystalline material in which the ions of the crystal are displaced in an asymmetrical way when the crystal is strained.

The linearity can vary considerably with the type of material that is used, and from sample to sample.

The original types of crystal microphones used Rochelle salt crystal coupled to a diaphragm. This ensured very high output levels (of the order of 100 mV), with very high output impedance and very poor linearity. Rochelle salt was not used for long because of its unfortunate habit of changing to an inactive form when kept at a moderately high temperature and humidity. Many pioneer users of tape recorders were puzzled to find that their microphones (and gramophone pick-ups) refused to work after the machine was brought down from the loft or up from the cellar after a hot summer.

The types of piezoelectric transducers that are used nowadays are mostly synthetic rather than natural crystals. One such material is barium titanate, which is used in piezoelectric transducers for frequencies up to several hundred kHz. The original type of piezoelectric microphone which used a diaphragm coupled to the crystal is seldom seen nowadays. The sensitivity of modern piezoelectric materials to vibration is such that the impact of the sound wave on the crystal alone is enough to provide an adequate output. Most microphones of this type are made as pressure-operated types because one side of the crystal is normally used for securing the assembly to its casing.

The piezoelectric microphone has a very high impedance level and a much higher output than other types. The impedance level is of the order of several megohms, as distinct from a few ohms for a moving coil type. At this very high impedance level, electrostatic pick-up of hum is almost impossible to avoid, along with the problems of the loading and filtering effect of the microphone cable. For low-quality microphones, of the type that were once supplied with tape or cassette recorders, this is of little importance, but it rules out the use of a simple type of piezoelectric microphone for studio purposes. For such purposes, the crystal transducer can be coupled directly to a MOS preamplifier that can provide a low-impedance output at the same high voltage level as is provided by the piezoelectric transducer. The preamplifier operating voltage can be supplied from a built-in battery to avoid the problems of running supply cables along with signal cables.

CAPACITOR MICROPHONES

The capacitor microphone is a remarkable example of a principle that was comparatively neglected until another, equally old, idea was harnessed along with it. The outline of a capacitor microphone is illustrated in Figure 5.10. The amount of electric charge between two surfaces is fixed, and one of the surfaces is a diaphragm that can be vibrated by a sound wave. The vibration causes a variation of capacitance that, because of the fixed

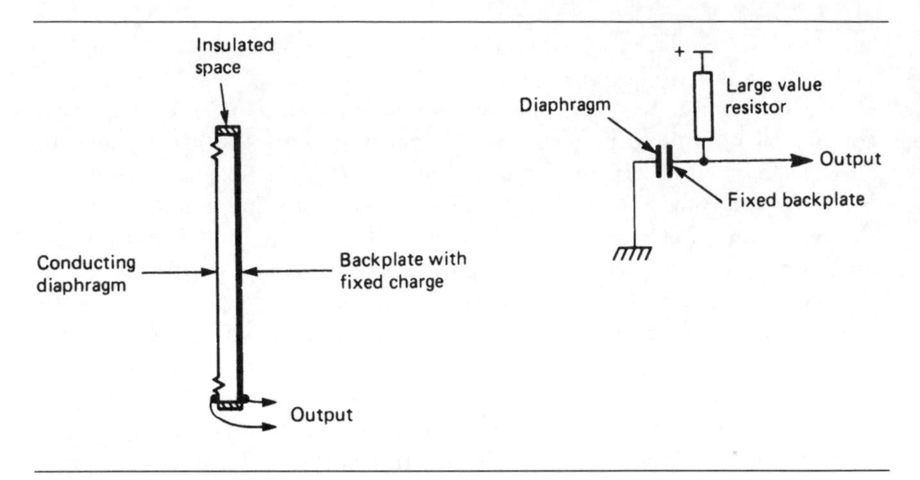

Figure 5.10 The capacitor microphone principle. The conducting diaphragm is grounded and the backplate is fed through a high-value resistor of several megohms so as to produce an approximation to constant charge conditions.

charge, causes a voltage wave. The output impedance is very high, and the amount of output depends on the normal spacing between the plates – the smaller this spacing, the greater the output for a given amplitude of sound wave. The construction of the microphone ensures that it is always pressure-operated.

The two main objections to the capacitor microphone in the past were the need for a high-voltage supply and the hum pick-up problems of the very high impedance. The high-voltage supply (called a polarizing voltage) was needed to provide the fixed charge; this was done by connecting the supply voltage to one plate through a very large value resistor. The high impedance made it difficult to use the microphone with more than a short length of cable (which added to the 'dead' capacitance), although it was suited to the valve input stages of the time.

The capacitor microphone can be very linear in operation and can provide very good quality audio signals without the need for elaborate constructional techniques. This was realized by a few users, notably Grundig, who always provided capacitor microphones with their tape recorders in the early days of domestic tape recording. There were also a few manufacturers who specialized in high-quality capacitor microphones for studio use, but capacitor microphones were always a rarity in comparison with moving iron and moving coil types.

The revival of the capacitor microphone came about as a result of a revival of interest in an old idea, the *electret*. An electret is the electrostatic equivalent of a magnet, a piece of insulating material that is permanently charged. The principle, which has been known for a century, states that if a hot plastic material (in the broadest sense of a material that can easily be

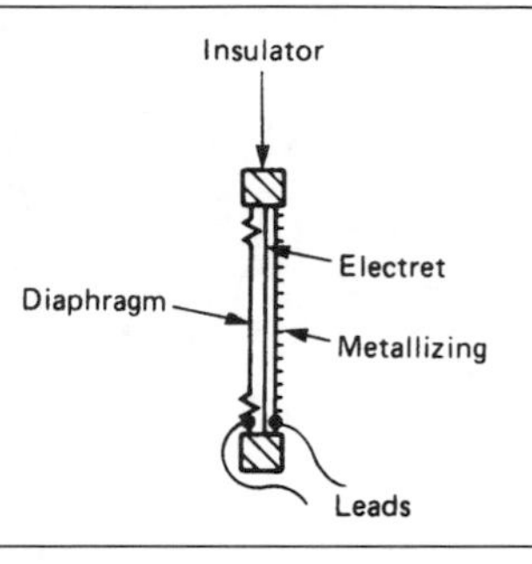

Figure 5 11 The electret capacitor microphone, which can be manufactured in very small sizes and which needs no polarizing supply.

softened by heating) is subject to a strong electric field as it hardens, it will retain a charge for as long as it remains solid. Materials such as acrylics (like Perspex) are electrets, and the idea was once considered for the manufacture of monoscope tubes, i.e., tubes that provide a TV test signal.

A slab of electret, however, is the perfect base for a capacitor microphone, providing the fixed charge that is required without the need for a polarizing voltage supply. This allows the very simple construction of a capacitor microphone, consisting only of a slab of electret metallized on the back, a metal (or metallized plastic) diaphragm, and a spacer ring (Figure 5.11), with the connections taken to the conducting surface of the diaphragm and of the electret. This is now the type of microphone that is built into cassette recorders, and even in its simplest and cheapest versions is of considerably better audio quality than the piezoelectric types that it displaced.

The high impedance of the capacitor electret microphone is no handicap now that MOS preamplifiers can be used, and for studio quality capacitor microphones the preamplifier can be battery operated so as to avoid the need for supply lines. The linearity of the microphone is independent of the electret, provided that the electret can supply a truly fixed charge. If the electret becomes leaky, as is possible through surface contamination, then the low-frequency response of the microphone will be degraded.

Another form of capacitor microphone uses a light film of pyroelectric material (see Chapter 4) whose polarization (separation of charge) is also changed by strain. Using such a film, metallized on one side, as one plate of a capacitor whose other plate is a perforated metal sheet creates a simple microphone whose output can be larger than that of most other capacitor types.

MICROPHONE PROBLEMS

Each specific type of microphone has its own problems and advantages, but

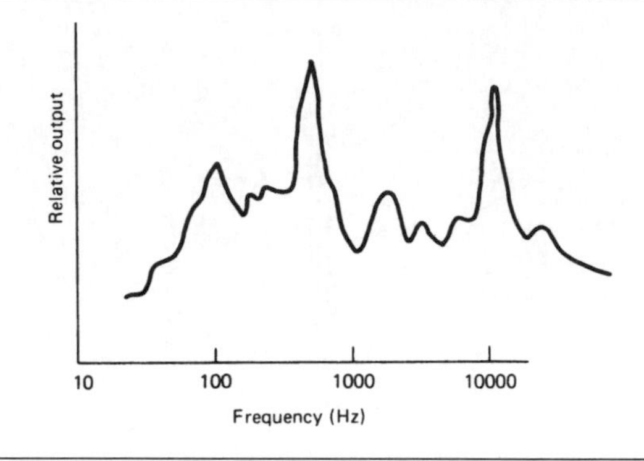

Figure 5.12 The type of response that can be obtained from an uncorrected microphone element. The peaks and troughs are due to resonances, mechanical and electrical.

there are problems that are common to all microphone types. The main problem of this kind is resonance, which will cause the output from the microphone to be distorted at some frequency, to form either a peak or a trough (Figure 5.12). These resonances can be electrical, but are much more likely to be mechanical and, as such, more difficult to deal with.

The two main techniques for dealing with mechanical resonances are shifting and damping. A resonance can be shifted by altering the mass of the resonating part, so that the resonance occurs outside the audio region. Reducing the vibrating mass will have the effect of shifting the resonance to a higher frequency, and when this technique is used, the aim is usually to shift the resonance to 30 kHz or higher. When the resonance is (unusually) at a low audio frequency, then adding mass can shift it to a lower, sub-audio frequency – this is more often a loudspeaker problem than a microphone problem.

Damping a resonance means that the energy of the resonating material must be dissipated, and this has to be done by using yielding materials rather then elastic materials like metals. Synthetic rubber materials can be made which have very high mechanical hysteresis (Figure 5.13). These make excellent damping materials to support diaphragms and other vibrating parts, and in general the use of damping calls for a considerable knowledge of materials of this type. Damping alone is seldom a cure for a bad resonance, and design effort has to be directed both to frequency-shifting the resonance and to damping it.

FREQUENCY AND WAVELENGTH

The use of microphones allows the frequency of sound waves to be measured

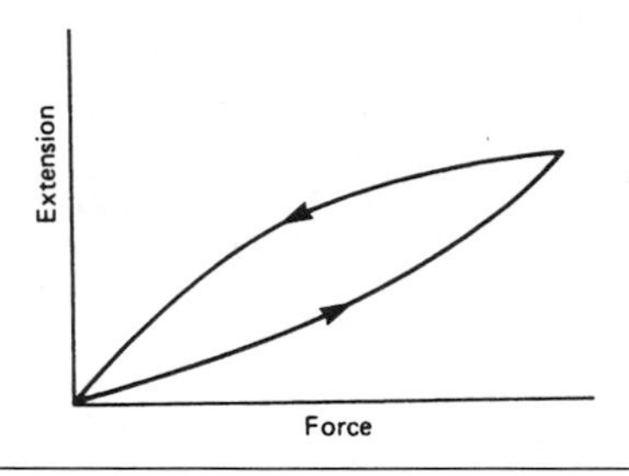

Figure 5.13 The extension/force characteristics for a typical high-hysteresis rubber material. The area enclosed by the loop is a measure of the amount of damping that can be obtained.

by connecting either an oscilloscope or a frequency meter to the microphone. It is more usual now to have a combined microphone and frequency meter. The wavelength of sound waves in air can be calculated from the frequency, using the relation:

$$\lambda = \frac{c}{f}$$

where λ is wavelength in metres, c is the speed of sound in air, and f is the measured frequency of the sound waves. Since the frequency of a sound wave can be measured very precisely, this is used in preference to direct measurements of wavelength. If the sound wave is being propagated in another medium, the value for speed of sound in that medium can be used in the equation.

Direct methods of measuring wavelength depend on setting up standing waves by reflecting a sound wave. When a wave is reflected, the forward wave and the reflected wave (Figure 5.14) can add to give wavepeaks (antinodes) and wave minima (nodes). If the reflecting surface is an integral number of half-wavelengths away, standing waves are set up with the nodes and antinodes in fixed positions (Figure 5.15).

The distance between two adjacent nodes is half of the wavelength, as is the distance between two adjacent antinodes. By measuring the distance between nodes (or between antinodes) we can find the wavelength directly, but such measurements can never be as precise as the calculation of wavelength from frequency.

- A microphone can distort the amplitude of the wave, in the sense of not providing a perfectly linear amplitude response, but it has no such effect on frequency.
- Hydrophones are a specialized form of microphone used under water, normally using piezoelectric transducers.
- Building acoustics are discussed in Chapter 7.

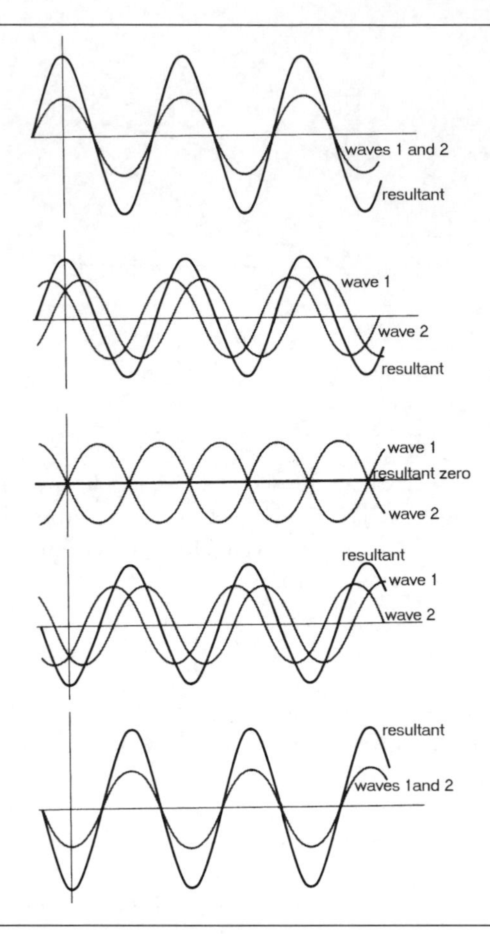

Figure 5.14 How waves interfere. Wave 1 is a wave travelling forward, and wave 2 is its reflection travelling backwards. The sum of wave amplitudes varies between zero and a maximum (sum of peak amplitudes) as the phase angle between the waves changes.

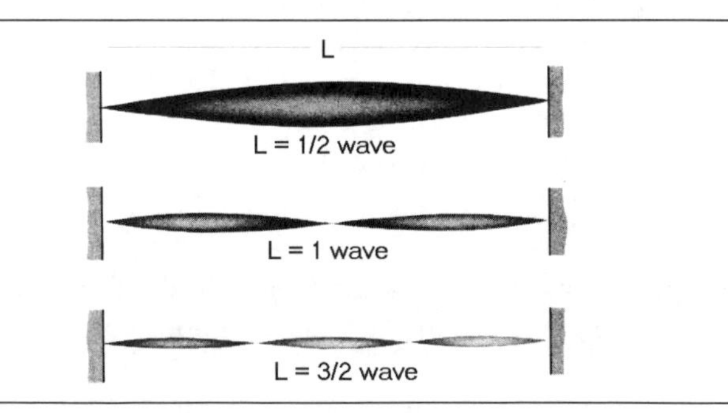

Figure 5.15 Illustrations of standing waves on a string, showing patterns for half-wave, full-wave and $1\frac{1}{2}$ wave.

5.3 Electrical to audio transducers

The microphone types that we have dealt with would be of little use unless we also had transducers for the opposite direction, and such transducers have been used for even longer than microphones. Earphones were used for electric telegraphs, in which the transmitter consisted of a Morse key, so that the earphone predated the microphone by a considerable number of years. The use of both earphone and microphone in a system is attributed to Bell, in trying to develop a hearing aid for the deaf (a passionate interest that he shared with his father, and a cause that has benefited greatly from the Bell fortune) created the first working telephone system. This also laid the foundation for a telephone system in the USA that makes that of any other country seem pitifully inadequate (and very expensive).

Until the use of thermionic valves became common in radio receivers, loudspeakers were a comparatively rare sight, although the basic principles had existed for some considerable time. Without power amplification, however, the use of a loudspeaker was pointless, and this was the main reason magnetic recording lagged so much behind disc recording. The output from the mechanical disc gramophone was from a form of (non-electrical) loudspeaker and could be heard over a whole room, whereas the output of the early steel-wire recorders (Poulsen's Telegraphone) could be heard only on earphones.

For each type of microphone, there is a corresponding earphone and loudspeaker type, and we shall look at the various types in detail. Since the basic principles of most of them have been considered under the heading of microphones, these will not be repeated. We shall, therefore, concentrate on the features that are unique to the purpose of transducing electrical signals into audio waves.

Of the two, the task of the earphone is very much simpler, and the construction of an earphone that can provide an acceptable quality of sound is very much simpler (and correspondingly cheaper) than that of a loudspeaker, as any Walkman addict will confirm. The earphone can use a small diaphragm, and ensure that the sound waves from this diaphragm are coupled directly to the ear cavity. The power that is required is in the low milliwatt level, and even a few milliwatts can produce considerable pressure amplitude at the eardrum – often more than is safe for the hearing. The conditions of use are, in other words, strictly defined, and the designer can concentrate on reducing resonances and increasing linearity in the certain knowledge that the effect will be noticeable.

The loudspeaker designer has much more of an uphill task. A loudspeaker is not used pressed against the ear, so that the sound waves will be launched into a space whose properties are unknown. In addition, the loudspeaker cannot be used alone but has to be housed in a cabinet whose resonances, dimensions and shape will considerably modify the performance of

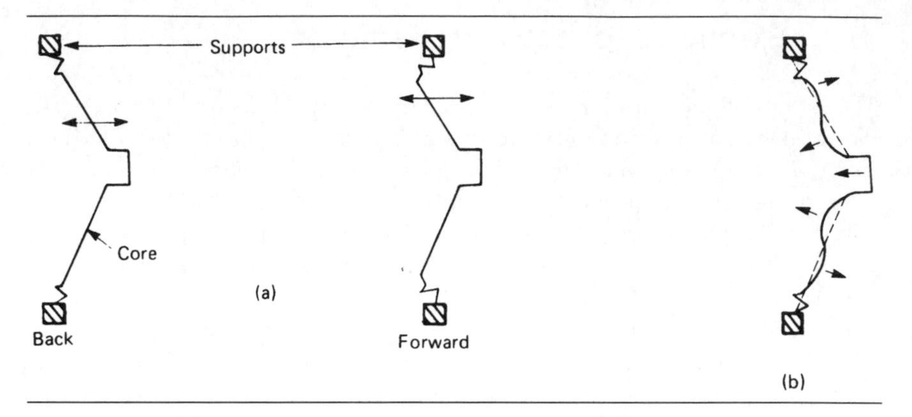

Figure 5.16 The ideal cone movement (a) preserves the shape of the cone so that the whole cone moves in unison. Since a cone cannot be perfectly rigid, however, break-up occurs (b) in which different parts of the cone move independently, with the outer parts of the cone lagging behind the movement of the inner part because the driving force is located at the inner part.

the loudspeaker unit. The assembly of loudspeaker and cabinet will be placed in a room whose dimensions and furnishing are outside the control of the loudspeaker designer, so that a whole new set of resonances and the presence of damping material must be considered. One designer has said that if we had not known loudspeakers before now, it is unlikely that we would even consider trying to make them today, just as we would have banned the use of steam if environmentalists had been around in the 19th century. Fortunately, human beings show a marked reluctance to return to caves after tasting better things.

The transducer of a loudspeaker system is sometimes termed the 'pressure unit', and its task is to transform an electrical wave, which can be of a very complex shape, into an air-pressure wave of the same waveform. To do this, the unit requires a motor unit, transforming electrical waves into vibration, and a diaphragm that will move sufficient air to make the effect audible. The diaphragm is one of the main problems of loudspeaker design, because it must be very stiff, very light and free of resonances – an impossible combination of virtues. Practically every material known has been used for loudspeaker diaphragms at some time, from the classic varnished paper to titanium alloy and carbon fibre, and almost every shape variation on the traditional cone has been used.

The main cone problem is break-up. If the cone is to be able to handle low frequencies, it must have a large area. At high frequencies, however, there will be waves on the cone itself, so that different parts of the cone move in different directions (Figure 5.16). This causes the waves from the different parts of the cone to interfere with each other, considerably modifying the response. The usual solution to this problem is to use more

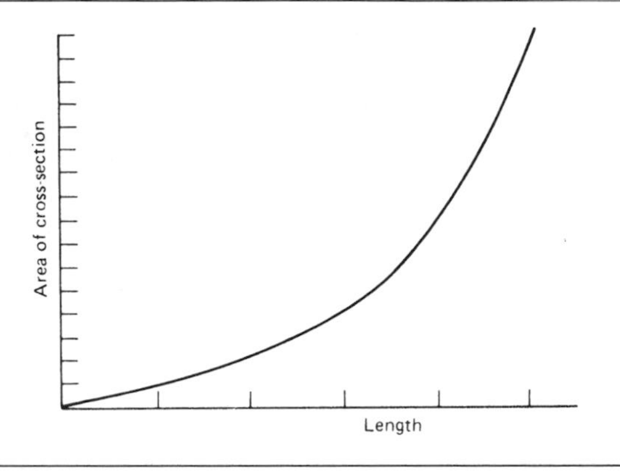

Figure 5.17 The ideal graph of area of cross-section plotted against distance from the driver unit for a horn. A horn of ideal shape needs to be very large in order to make its dimensions comparable with the largest wavelength that it will work with, and compromises such as folding the horn back on itself or driving the horn partway along its length have been used.

than one driver unit and divide the electrical signal into low and high frequency components fed to the appropriate units. This is, after all, reasonably justified because few musical instruments produce a full range of sound frequencies and there is no reason why a single loudspeaker should. The low-frequency units are known as *woofers* and the upper frequency units as *tweeters*, but this attractive terminology is abandoned in the mid range where the units are simply termed mid-frequency units. The few loudspeaker designs that have achieved high-quality results with a single diaphragm, however, are among the best known.

- One solution to the problem of break-up is to use separate loudspeakers for different parts of the sound band, typically one unit for the 20–120 Hz range (a woofer), another for the frequencies above 5 kHz (a tweeter), and a third for the intermediate frequencies.

The efficiency of loudspeakers is notoriously low, around 1%, mainly because of the acoustic impedance matching problem for low frequency signals. In simple terms, most loudspeakers move a small amount of air with comparatively large amplitude, whereas to produce a sound wave effectively they ought to move a very large amount of air at comparatively low amplitude. This mismatch can be remedied to some extent by housing the loudspeaker in a suitable enclosure, but the only type of enclosure that increases the overall efficiency is the exponential horn (Figure 5.17). The sheer size of the horn and the rigid dense construction that is needed make this an unacceptable solution for all but the most avid listeners, one of whom had the foundations of his house cast so that they form an exponential horn shape. Nothing less really suffices, because it is only when the listening end of a horn is about 16 feet square that the benefits of horn-loading become really noticeable. The effect is still noticeable even for small linear horns, however, and the astonishing effect of bringing up the small end of a horn to a Walkman earphone is something that has to be heard to be believed.

- The factor that, more than any other, determines loudspeaker efficiency is the need for a broadband response, and if a loudspeaker is needed to cover only a small band of high frequencies it can be designed to achieve comparatively high efficiency levels. An example is the piezoelectric tweeter used in smoke alarms, which develops a large amplitude of high-pitched sound with only the power of a 9 V battery.

THE MOVING-IRON TRANSDUCER

The first type of earphone, as applied to the early telephones, was a moving-iron type, and this principle has been extensively used ever since. As applied to the telephone, the earphone uses a magnetized metal diaphragm (Figure 5.18) so that the variation of magnetization of the fixed coil will ensure the correct movement of the diaphragm. Without this fixed magnetic polarization, the diaphragm frequency would be twice the frequency of the electrical signal. Earphones of this type are very sensitive, but the sound quality is very poor because of the comparatively stiff diaphragm that causes unavoidable resonances.

Miniature moving-iron earphones are still in use, particularly where sound quality is not of the highest importance, but the principle has died out as far as loudspeakers are concerned. At one time, moving-iron loudspeakers, which were virtually earphone units with a cone attached, were used, but these were soon replaced by the moving-coil type that provided

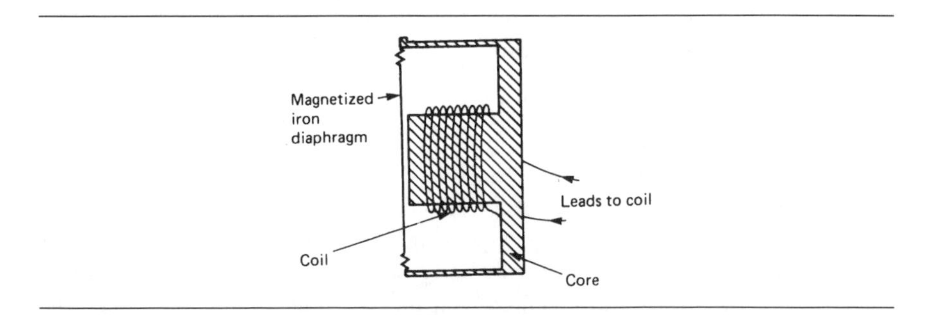

Figure 5.18 The telephone earpiece principle. The magnetized diaphragm is made from iron or a magnetic alloy and is moved by the attraction or repulsion of the core as the signal current flows in the coil. The earpiece is very sensitive, although the sound quality is low.

noticeably better sound quality. The sound of the early loudspeakers has been aptly described as a 'mellow bellow', and this is no longer acceptable except inside cars. Listeners are, thanks to FM radio, gradually learning that the high frequencies, particularly of speech, carry the major amount of the acoustic information and that emphasis of lower frequencies is pointless. Some day, directors of TV dramas will also learn that background noise can easily obscure the dialogue that makes a play intelligible.

THE MOVING-COIL TRANSDUCER

The moving-coil principle as applied to loudspeakers and earphones has been extensively used, and the vast majority of loudspeakers use this principle. The use of moving-coil earphones has been less common in the past, but these are now in widespread use thanks to the miniature cassette player vogue. As applied to earphones, moving-coil construction permits good linearity and controllable resonances, since the amount of vibration is very small and the moving-coil unit is light and can use a diaphragm of almost any suitable material.

The use of moving-coil drivers with cones to form loudspeakers is by no means so simple. One problem of all loudspeakers is that the unit that reproduces the low frequencies needs a very freely suspended cone and must be able to reproduce large amplitudes of movement. This amplitude of movement can be a centimetre or more, and it is very difficult to ensure that the magnetic flux density around the moving coil is uniform over this distance. The design of the magnetic circuit calls for a large magnet and a large core, and a shape that is computed to provide the best available linearity of flux density. In addition, there is a conflict between making the coil and cone very freely suspended and yet maintaining the position of the coil concentric in the core of the magnet.

The magnets of modern moving-coil units are invariably permanent magnets, often using quite exotic alloys. At one time, moving-coil loudspeakers, known as 'dynamic' loudspeakers, used an electromagnet to provide the magnetic field, but this vogue was short-lived because of the demands it made on the power supply of the radio receiver that used it.

At the other end of the frequency scale, there is no requirement for large amplitudes of movement, and the main problems are of resonance and cone break-up. Metal cones are often used with success, but a large number of successful designs of loudspeaker systems use moving coil units only for the lower and mid frequencies, with other types used for the highest frequencies.

A variation of the moving-coil principle that has been successfully used for earphones is the electrodynamic (or orthodynamic) principle. This uses a diaphragm that has a coil built in, using printed circuit board techniques. The coil can be a simple spiral design, or a more complicated shape (for

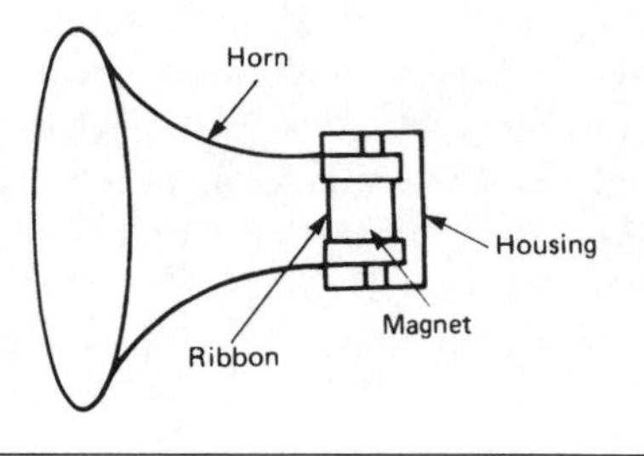

Figure 5.19 The ribbon type of loudspeaker is usually combined with a small horn enclosure because the small size of the ribbon alone is insufficient to produce a large output.

better linearity), and the advantage of the method is that the driving force is more evenly distributed over the surface of the diaphragm. This avoids break-up, and allows the use of a much more flexible diaphragm than would be possible otherwise. Headphones based on this principle have been very successful and of excellent quality.

RIBBON LOUDSPEAKERS

The ribbon principle, already dealt with under the heading of microphones, is also used to provide loudspeaker action. The moving element of a ribbon loudspeaker is necessarily small, and for that reason, the unit is a tweeter rather then a full-range type. The ribbon construction, however, offers a very directional response and can be built into a small horn type of enclosure (Figure 5.19), which makes it a very efficient transducer compared to others. Ribbons of about 5 cm long, corrugated along the length for stiffness, are used, and because the impedance is very low (a fraction of an ohm) matching transformers are usually built into the loudspeaker. With conventional construction, the ribbon tweeter can handle frequencies from 5 kHz upwards, but a more elaborate design can allow this range to be extended down to about 1 kHz.

Wide-range multi-ribbon units are also feasible, but in a very different size (and price) category. The commercially available types use three units, of which the bass unit is very large, and requires its own amplifier to supply about 100 W driving power.

PIEZOELECTRIC LOUDSPEAKERS

The piezoelectric principle has been used in the manufacture of tweeters, for which it is reasonably suited if the problems of resonance can be dealt with. Such tweeters can be of quite high efficiency and are widely used in applications ranging from smoke detectors to computer modems.

The piezoelectric principle has also been used for earphones in the form of piezoelectric (more correctly pyroelectric, since the electrical parameters are temperature sensitive) plastics sheets, which can be formed into very flexible diaphragms. The effect of applying a voltage between the faces of such a diaphragm is to make the dimensions shrink and expand as a waveform is applied. This in turn can be converted into a movement that will move air by shaping the diaphragm as part of the surface of a sphere. This can be done, for example, simply by stretching the material over a piece of spherical plastic foam. The moving mass is very small and sensitivity is high, with no need for a power supply. The linearity that can be obtained depends on the surface shape as well as on the piezoelectric characteristics.

CAPACITOR TRANSDUCERS

The possibility of constructing earphones or loudspeakers along the lines of a capacitor microphone has existed for a long time, but the practical difficulties have been satisfactorily resolved by only two designs: the Quad (Figure 5.20) and the Magna-Planar wide-range electrostatic loudspeakers. The main problem, initially, was that using a single-ended design analogous to a capacitor microphone provides very poor linearity. The breakthrough was the principle, due to Peter Walker, of a diaphragm which had a signal voltage applied to it and which faced a sheet of conducting material at a high voltage on each side. The system has also been used in earphones that, despite the need to provide a high polarizing voltage for the plates, have been very popular and of outstanding audio quality.

The advantage that makes the electrostatic loudspeaker principle so attractive is that the driving effort is not applied at a point in the centre of a cone or diaphragm, but to the whole of a surface that can be large in area. There is, therefore, no break-up problem, since all parts of the diaphragm are driven, and so a single unit can handle the whole of the audio range. Resonances can present a negligible problem, due to the overall drive, and no special enclosure other than is required for electrical safety and mechanical rigidity will be necessary. The original Quad design has now been in production for over 21 years and, although recently redesigned, still follows the original principles and provides outstanding quality, particularly for concert-hall performances.

The most recent design uses the 'point-source' principle, achieved by driving the diaphragm as a set of concentric circles that are not in phase. In this way (Figure 5.21), the sound wave that is created appears to come from a point behind the diaphragm. The practical effect of this is to make

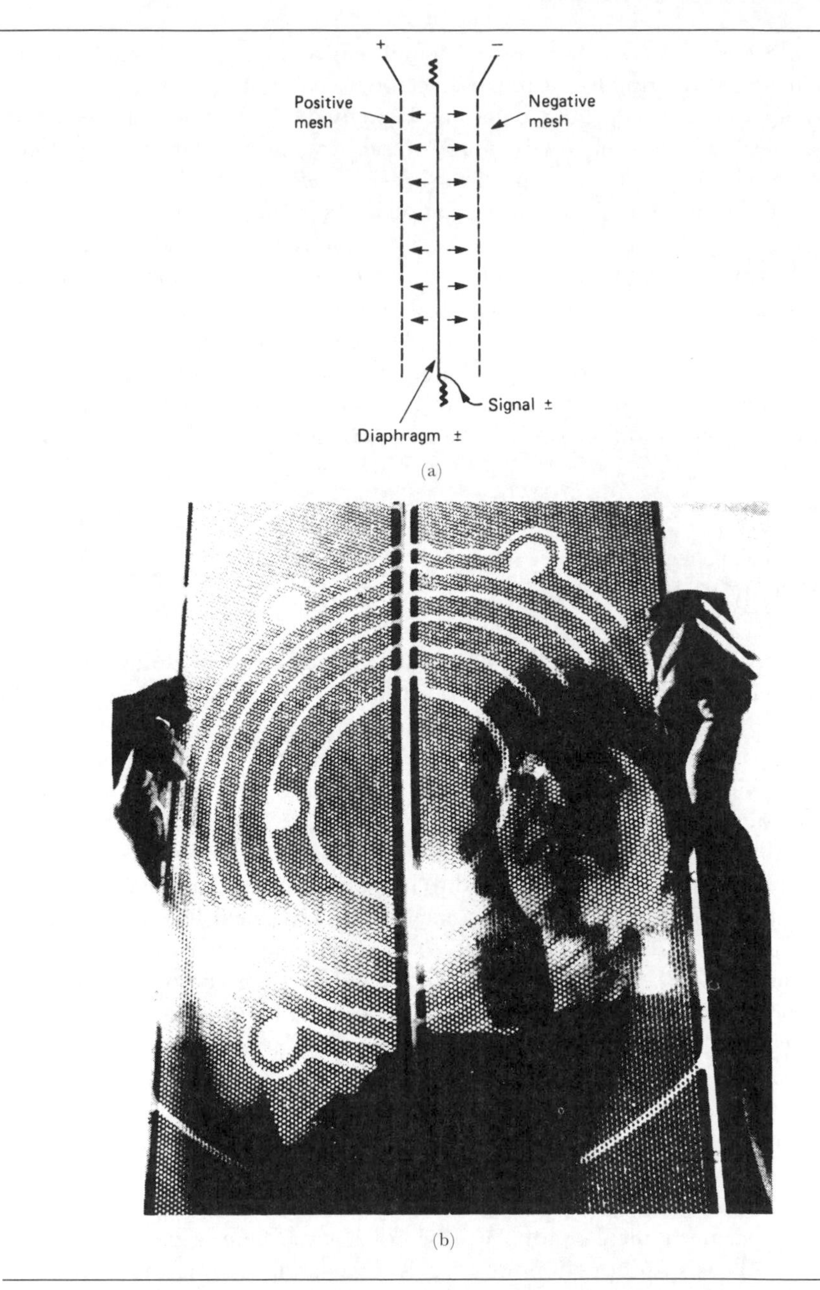

Figure 5.20 (a) The electrostatic loudspeaker principle. The signal voltage applied to the conducting diaphragm will provide charge, and the action of the field provided by the static metal mesh on each side will move the diaphragm. The important feature is that the whole diaphragm, not simply a small portion, is acted on hence there is no break-up effect. (b) The world-famous Quad loudspeakers make use of this type of action.

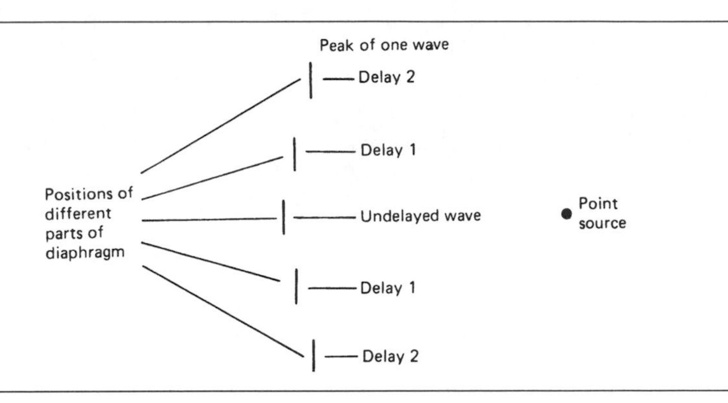

Figure 5.21 The most recent Quad design divides the diaphragm into concentric circles, each driven separately with a small time delay. The effect of this is to make the diaphragm move as if it had been driven from a point source of sound.

the sound appear to be independent of the loudspeaker in a way that is quite remarkable as compared to moving-coil units.

The electret principle has not been used to date for loudspeakers, but earphones have been constructed which follow the basic capacitor type of design but using an electret to provide the fixed charge. These are claimed to give a performance that approaches that of the best electrostatic types, but at a much lower cost and free of the need to supply a polarizing voltage.

ULTRASONIC TRANSDUCERS

Although loudspeakers and microphones may use similar operating principles, the differences between receiving a sound wave and generating it are sufficient to make the actions interchange only very poorly. A loudspeaker can generally be used as a microphone, but as an insensitive and low-quality microphone. A microphone can be used as an earphone with some success, but is not suitable as a loudspeaker because it is not designed to handle the amount of power that is needed.

By contrast, the transducers that we use for ultrasonic waves are usually of a type that is almost totally reversible. The transducers that are used for sending or receiving ultrasonic signals through solids or liquids can operate in either direction if required. For ultrasonic signals sent through the air (or other gases), the transducers are used with diaphragms and in enclosures that can make the application more specialized, so that a transmitter or a receiver unit has to be used for its specific purpose.

The important ultrasonic transducers are all piezoelectric or magnetostrictive, because these types of transducers make use of vibration in the bulk of the material, as distinct from vibrating a motor unit which then has to be coupled to another material. The magnetostrictive principle is not generally applied in the audio frequency range. Magnetostriction is the change of dimensions of a magnetic material as it is magnetized and demagnetized, and the most common experience of the effect is in the high-pitched whistle of a TV receiver, caused by the magnetostriction of the line output transformer. Several types of nickel alloys are strongly magnetostrictive, and have been used in transducers for the lower ultrasonic frequencies, in the range of 30–100 kHz.

A magnetostrictive transducer consists of a magnetostrictive metal core on which is wound a coil. The electrical waveform is applied to the coil, whose inductance is usually fairly high, so restricting the use of the system to the lower ultrasonic frequencies. For a large enough driving current, the core magnetostriction will cause vibration, and this will be considerably intensified if the size of the core is such that mechanical resonance is achieved. The main use of magnetostrictive transducers has been in ultrasonic cleaning baths, as used by watchmakers and in the electronics industry.

The piezoelectric transducers have a much larger range of application although the power output cannot approach that of a magnetostrictive unit. The transducer crystals are barium titanate or quartz, and these are cut so as to produce the maximum vibration output or sensitivity in a given direction. The crystals are metallized on opposite faces to provide the electrical contacts, and can then be used either as transmitters or as receivers of ultrasonic waves. The impedance levels are high, and the signal levels will be millivolts when used as a receiver, or a few volts when used as a transmitter.

- Two important parameters of an ultrasonic transmitter/receiver are beam spread and target angle. The beam spread is the area in which a round wand will be sensed if it is passed through the beam. This defines the maximum spreading of the ultrasonic sound from the transducer. The target angle measures the maximum amount by which the target can be tilted and still be detectable by an ultrasonic transducer.

Although piezoelectric transducers are used in ultrasonic cleaners, their main applications are in security devices and signal processing. An ultrasonic transmitter in air can 'fill' a space (such as a room or a yard) with a standing-wave pattern, and a receiver can detect any change in this pattern that will be caused by any new object. Ultrasonic units cannot necessarily distinguish between a cat and a cat burglar, however. For signal processing, a radio-frequency signal can be converted to an ultrasonic signal by one transducer and converted back by another, using the ultrasonic wave path, usually in a solid such as glass, to create a delay. This use of an ultrasonic delay was at one time essential to the PAL colour TV system that is used in most of Western Europe, but has now been replaced by digital delays.

Ultrasonic transducers are very widely used in security applications for detecting objects, and some systems can detect up to a range of 12 m, and by suitable selection of transducers the angle of coverage can be either wide or narrow. Target material is not critical, so that any surface capable of acoustic reflection can be detected, including liquids and soft or porous materials. Distance measuring can also be achieved with an accuracy of around 0.05% of range.

- Ultrasound waves can also be filtered by mechanically shaping the solids along which they travel, and this is the basis of surface wave filters.

INFRASOUND

The wave frequencies below 20 Hz are not so extensively used as the ultrasonic frequencies, but the sensing of these frequencies is a matter of some importance. The vibrations of the Earth that are accompanied by earthquakes are in the very low-frequency region, and are termed seismic waves. Transducers for seismic waves must be capable of very low-frequency response, which rules out the use of piezoelectric transducers. Most seismic transducers work on the principle of using a suspended mass to operate a transducer of the capacitive or inductive type, such as the example in Figure 5.22. The principle here is that the vibrations of the Earth will move the casing, leaving the suspended mass at rest, and that the relative motion between the supports and the mass will produce the output.

The other natural source of very low-frequency vibrations is the communication of whales, and the signal strength is often enough to permit a piezoelectric type of transducer to be used. A large diaphragm is mechanically coupled to the piezoelectric crystal and the output from the crystal is connected to a MOSFET amplifier, with a chopper stage (see Chapter 12) used for the main amplification. The usual technique with these signals is to tape-record them with a slow-running tape, and replay at a higher speed to make it easier to display (and hear) the waveforms.

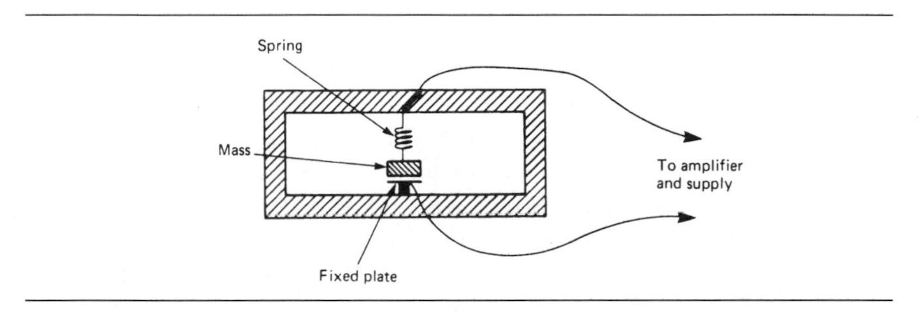

Figure 5.22 Seismic detector principle. The mass remains still while the casing vibrates with the Earth. The relative movement is detected, in this case, by a capacitive transducer. The casing is evacuated to avoid the damping effect of the air.

Chapter 6

Solids, liquids and gases

6.1 Mass and volume

Of all the primary physical quantities, *mass* is the most difficult to define in terms that mean anything to a non-physicist. The official definition is that mass is the quantity of matter in an object, but this is not exactly a step forward in understanding. A more useful, although less fundamental, idea of mass is as the ratio of force to acceleration when a force is applied to a mass. The greater the mass of an object, the less acceleration a given force can produce. Two different masses placed on ice might be equally easy to move, but the larger mass requires more force to accelerate by the same amount.

In everyday life, subject to the almost constant acceleration of gravity, we make no distinction between mass and weight, but weight is a force that is the effect of gravity on a mass. The mass of an object is constant, but its weight on Earth is not quite so constant, because the size of the force of gravity varies slightly from one place to another. In deep space, where gravitation has very little effect, the weight of any object is almost zero, but the mass is unchanged and it determines the amount of acceleration that a force can produce. The traditional method of measuring mass, as distinct from weight, is the *mass balance*, in which an unknown mass is balanced against a set of masses of known values. Because gravity affects both sides of the scales, this is a comparison of masses rather than weights, although it is not effective if the gravity is zero.

Volume is much easier to define as the amount of space that an object takes up. Even this, however, can be deceptive. The volume of a slab of glass is easy to appreciate, but what is the volume of a porous material like a sponge? The volume of a solid non-porous material can be calculated from its dimensions, but for a porous material we have to measure volume by

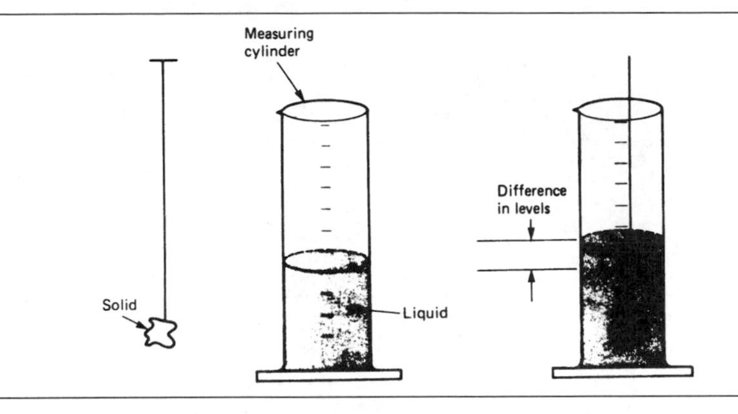

Figure 6.1 Finding the volume of a porous or irregular solid by displacement. The volume of liquid displaced by the solid is equal to the volume of the solid unless the solid dissolves in the liquid.

finding what volume of liquid the object will displace (Figure 6.1). For a gas, the volume is the volume of the container because a gas has no fixed volume or shape.

6.2 Electronic sensors

The mass and volume of solids are quantities that are not easy to sense or measure by electronic methods. The main methods for mass really depend on a weight measurement, but for most purposes this is what is required in any case, since the distinction is academic when you are working on Earth. The available methods make use of strain gauges to measure weight, or position gauges to measure the extension of a spring caused by the weight of a mass.

Figure 6.2 shows the principle of a simple lever-arm weight sensor. The arm is made of an elastic material, in the sense of a material whose change of dimensions will be proportional to the force applied to it. The arm material will usually be metal whose cross-section will be chosen to suit the weights to be measured, which could be in the gram range or in the tonne range, depending on what is to be weighed. A strain gauge, or more likely a pair of strain gauges, is used to obtain an electrical output that is proportional to the amount by which the lever arm is bent. This in turn will be proportional to the weight applied at the end.

Another form of strain-gauge weight sensor is illustrated in Figure 6.3. This consists of a cube of elastic material on which strain gauges have been fastened so as to detect changes in the dimensions. Placing a weight on top of the cube, with the cube resting on a support, will distort the cube, giving

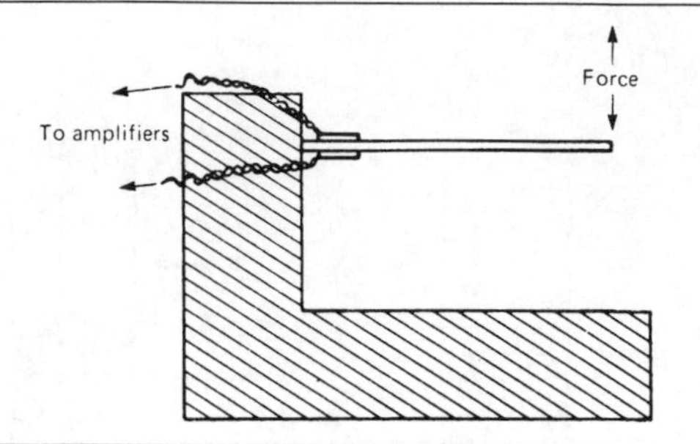

Figure 6.2 The lever-arm weight sensor. The application of a force to the end of the arm causes the arm to bend, and the amount of bending is sensed by the strain gauges. The device requires calibration with known weights.

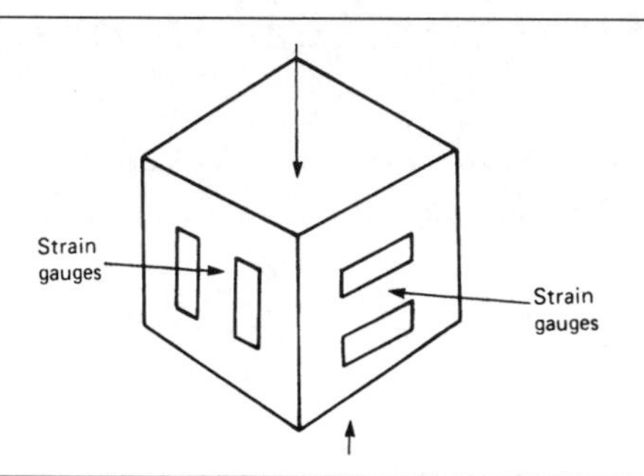

Figure 6.3 A 'force cube' sensor, which makes use of an elastic material in cube form with strain gauges to detect changes of dimension caused by loading. The method is particularly suited to large forces, using a cube of metal rather than a cube of more easily compressed materials.

a reading from the strain gauges that will once again be proportional to the amount of weight or force.

Both types of strain-gauge measurement need to use more than one strain gauge because of the effects of temperature. The effect of a small change of temperature on an elastic material is likely to cause more change of dimension than the effect of a modest force, so that some form of temperature compensation is needed. This usually takes the form of multiple strain gauges, and in the lever type of sensor is best achieved by using one gauge on top of the lever and one underneath. A force will cause one gauge to be lengthened and the other shortened, whereas a change of temperature will cause both gauges to be lengthened, so that if the output of one strain gauge is subtracted from the output of the other, the result will be proportional to force and unaffected by temperature changes.

A third form of mass/weight sensor makes use of a balance principle, although the weight of a mass against a strong spring is usually balanced (Figure 6.4). The effect of the force exerted by the mass is to compress the

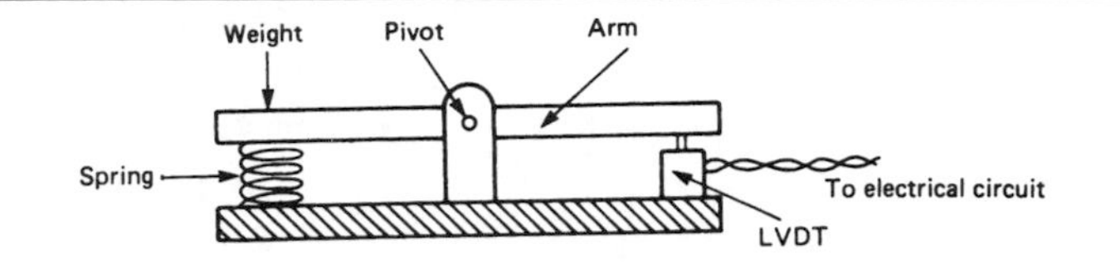

Figure 6.4 A form of lever-arm spring balance in which the force applied to the arm causes compression of the spring and the amount of movement (proportional to force) is sensed by any suitable transducer, such as an LVDT. One advantage of this method is that by changing the position of the spring several ranges of force values can be measured.

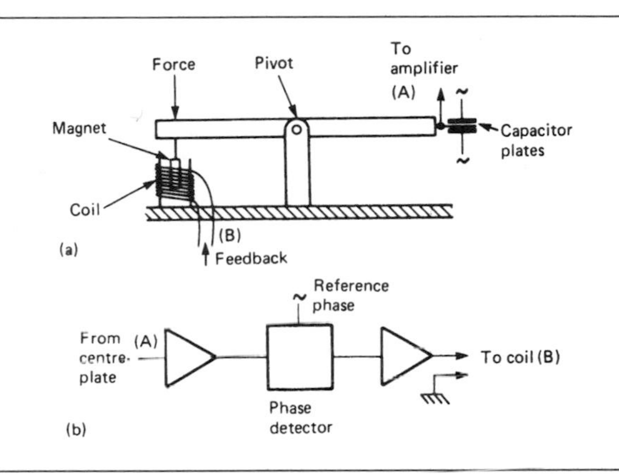

Figure 6.5 (a) The very sensitive feedback balance system in which no springs are used. (b) The electrical circuit shows that the signal from the sensing capacitor is used to restore the balance arm position by applying current to the coil, and the amount of current that needs to be applied in this way is a measure of the force. Calibration is necessary, because the relationship between force and current is not easily calculated.

spring, and this movement can be detected by any of the sensors that were discussed in Chapter 2. A variation on this scheme is illustrated in Figure 6.5. One end of the arm is supported by a light spring, and carries a magnet that is surrounded by a coil. The opposite end of the balance arm forms a capacitive detector, the output from which is amplified and used to provide a current that is circulated through the coil. When the application of a force tries to deflect the arm, the deflection produces a signal at the capacitive sensor, and the output current acts on the magnet so as to restore the original position of the arm. Because of the very large negative feedback, the movement of the balance arm is always very small, so that the linearity of measurement can be good. The force need not be applied directly to the end of the arm, so that a large range of forces can be catered for by making the point of application of the force variable.

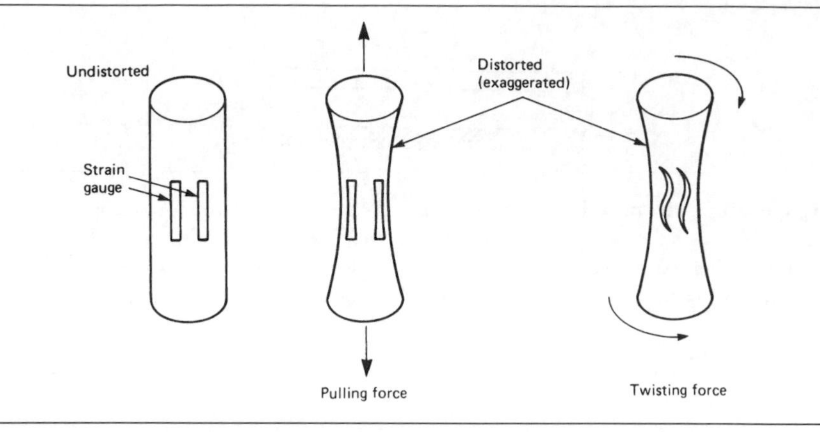

Figure 6.6 The 'elastic cylinder' method of sensing tensile or torsional forces, using a cylinder with strain gauges attached.

For tensile forces (which are pulling forces), an adaptation of the basic strain-gauge design is illustrated in Figure 6.6. This uses an elastic rod or tube that will be distorted when a tensile force is applied, causing an output from the strain gauges. As before, multiple strain gauges are used in order to reduce the effects of temperature changes.

The volume of a solid is not easily amenable to direct electronic measurement. The principle of displacement, however, allows indirect sensing of volume, because when a solid is completely immersed in a liquid it displaces an amount of liquid equal to its own volume. A change of liquid level, which is proportional to a change of liquid volume, can be sensed and this reading used as a measure of the volume of the solid. A displacement method of this type is valid only to the extent to which water, or whatever liquid is used, can completely surround the object, leaving no air-filled cavities. In the case of a porous material, it also depends on the extent to which the liquid can permeate the pores. The liquid must not have any chemical reaction with the solid whose volume is to be measured, and the solid must not be soluble in the liquid. Since the bulk density of all materials is known (or measurable from samples) it is often considerably more accurate to find the mass of a material and calculate its volume as mass divided by density (using units of kg for mass and kg/m^3 for density to get volume in units of m^3). See later in this chapter for a description of liquid level sensors.

6.3 Proximity detectors

The measurement of mass or volume is often much less important than the sensing of the presence of material, known as proximity detection. There

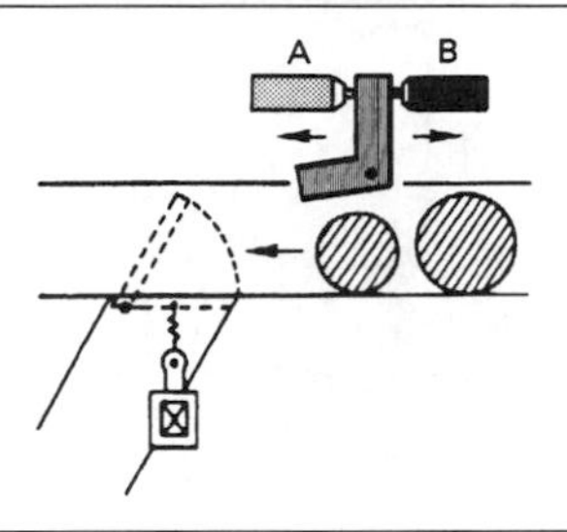

Figure 6.7 Microswitches being used as proximity gauges in a diameter-checking arrangement.

are several forms of proximity detectors, all of which are industrially important and which feature to a considerable extent in all automated equipment. The main uses for proximity detectors range from the simple counting of objects by detecting their presence, to checking for faulty assembly and ensuring that objects are correctly positioned for a subsequent operation.

The simplest form of proximity detection is touch detection, using a microswitch, often connected to a probe, to sense the position of an object. This method has the considerable advantage that it is not affected by the type of material (it is not confined to metal objects, for example), but it does have the drawback of being useful only for objects whose mass is enough to operate the mechanism, and for objects which will not be damaged by being touched.

Figure 6.7 shows a typical application in an automatic sizing gauge for rejecting undersize (or oversize) objects. The lever arm normally touches microswitch A, and will be lifted by each object passing it. If an object is too small to move the arm away from microswitch A, it is too small and can be rejected. If an object is too large, it will press the arm against microswitch B, also actuating the trapdoor rejection mechanism. This very simple example is widely used for the size-grading of objects, and the principle of using two microswitches for maximum and minimum tolerance measurement is one that is found in a large number of applications. The inspection system illustrated in Figure 6.8 depends on the use of a single microswitch to detect warping of a surface – obviously the principle could be adapted to check for convex warping as well as for the concave warping shown here.

Microswitches used in this way are almost universally of the electrical changeover variety, so that either the opening or the closing of a microswitch can be sensed, as in Figure 6.7. Most microswitches allow a maximum travel of the actuating pin which is large in comparison to the range required for detection. In some mechanisms it is necessary to mount the microswitches themselves on a sprung holder to ensure that a grossly oversize object cannot cause damage to the switch itself.

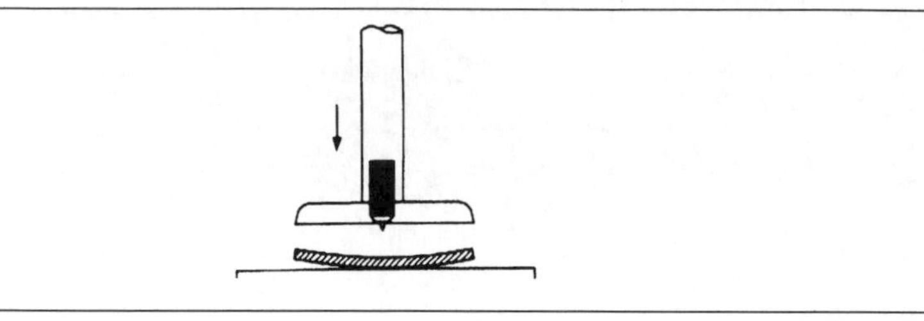

Figure 6.8 An inspection system that uses a single microswitch to check for the warping of an object.

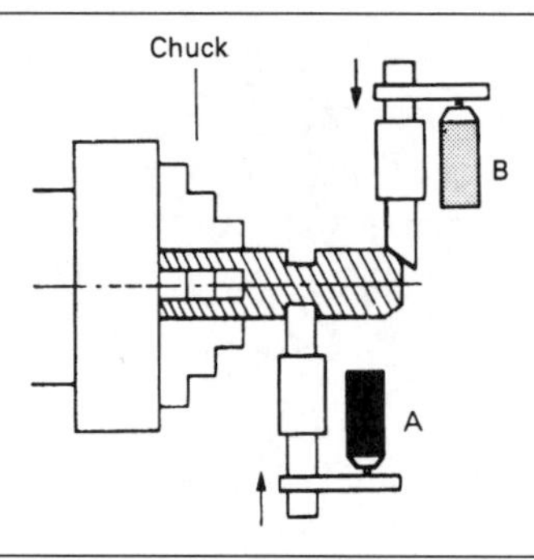

Figure 6.9 Microswitches being used as sensors for work and tool placement in a turning operation.

By far the most common application for microswitch proximity sensing, however, is in tool or workpiece placement on automatic turning, drilling, punching and press equipment. Figure 6.9 shows a typical application in which microswitch A is used to sense that the work is correctly positioned with reference to a flange, and microswitch B is used to position the cutting tool for a chamfer. It is particularly in this type of application that the microswitches must be protected against excessive movement, and since considerable damage can be caused by failure of a microswitch or its associated circuitry, such mechanisms should always use a back-up set of microswitches that will cut off power in the event of an over-travel of either tool or work. Safety considerations will also require microswitches to be used to detect any movement of safety guards or visors.

For some applications, mercury tilt-switches can be used rather than microswitches. Although, as the name suggests, these switches are used to sense an alteration in angular position rather than linear motion, many of the movements that are sensed by microswitches involve the rotation of an arm and are potentially applicable to tilt-switches. Unlike the microswitch, in which spring-loaded contacts are used, there is no switch-bounce effect in a tilt-switch. For some applications in which the number of contacts are digitally counted, the tilt-switch can be preferred, although most counter circuits incorporate anti-bounce circuitry (in which the counter is disabled for a short time after receiving an input so that multiple inputs due to switch-bounce are not counted).

INDUCTIVE PROXIMITY DETECTORS

Inductive proximity detectors operate on the principle that the inductance of a coil is considerably changed in the presence of a metal core. The coil can be part of a bridge circuit or the inductor in a tuned circuit. The presence of metal close to the coil will thus cause a bridge output off-balance condition or a change in the tuned frequency that can be sensed. The change is normally used to operate a switching circuit by way of a transistor or a Triac, so that the output is suitable for controlling a load. The predominant switch action is normally open (NO), although a few types are available in a normally closed (NC) action. The use of changeover (CO) action is unusual.

The overwhelming advantage of inductive proximity switches is that no part of the switch need touch the object being sensed. Their disadvantage is that they sense metallic objects only, and with a sensitivity that depends on the type of metal used. The sensitivity is quoted in terms of average sensing distance for an object of mild steel and, depending on the type of detector, this distance is of the order of 0.8–2 mm for mild steel. Other metals lower the sensitivity, to 70% for stainless steel, 40% for brass, 35% for solid aluminium, and 30% for copper. Most inductive proximity sensors are DC operated, with the choice of current source or current sink action. A current-source type will pass current into a load, a current-sink type will accept (to ground) current from a load. Current ratings are in the 100–200 mA region at DC voltage levels of 5–30 V. A few types allow AC operation, some at mains 240 V level, with current ratings up to 500 mA, allowing light loads to be driven directly.

The housings for inductive proximity switches present a considerable variety, of which two of the smallest styles are illustrated in Figure 6.10.

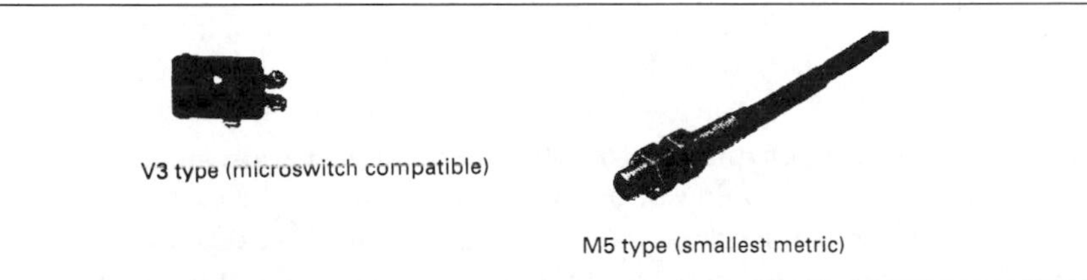

Figure 6.10 Inductive proximity detectors. The diagrams show the smallest units, one cased like the V3 microswitch (to replace an existing microswitch), the other of M5 (smallest metric) size.

Table 6.1 Mechanical details for the popular M-series of proximity switches.

Housing code	Overall length (mm)	Threaded length (mm)	Thread (metric)
M5	27	24	M5 × 0.5
M8	63	40	M8
M12	65	40	M12
M14	66	30	M14
M18	55	50	M18

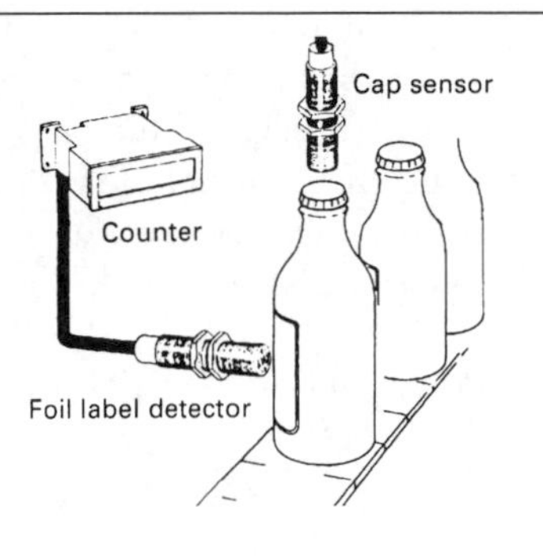

Figure 6.11 Using inductive proximity sensors for bottle tops and foil labels.

Table 6.1 shows the dimensions for some typical units in the metric threaded type. Several of the physically larger types incorporate a LED telltale which will light when the target is sensed. This feature makes setting up much easier, since the load can be disconnected during this time while still allowing the sensor to function.

The applications for inductive proximity switches are more extensive than might be thought of in a device that is primarily a metal sensor. Figure 6.11, for example, shows a counter and sensor for bottles which checks for metal caps and foil labels being correctly fitted and positioned. The use of light metal foils as contacts to a non-metal can make it possible to use inductive sensors in a range of applications for which they would at first sight be rejected, such as the cloth thickness sensors illustrated in Figure 6.12. The comparatively small distance over which a metal object can be sensed can be turned to advantage in using these detectors for sensing incorrect positioning of metal objects, even to the extent of sensing which way round staples have been packed in a cardboard container.

Inductive proximity detectors are also particularly useful in machine tool applications such as injection-moulding presses where they can be used to sense the closure of a mould, and in all of the types of application for which microswitches are also used. Inductive sensors are, however, considerably

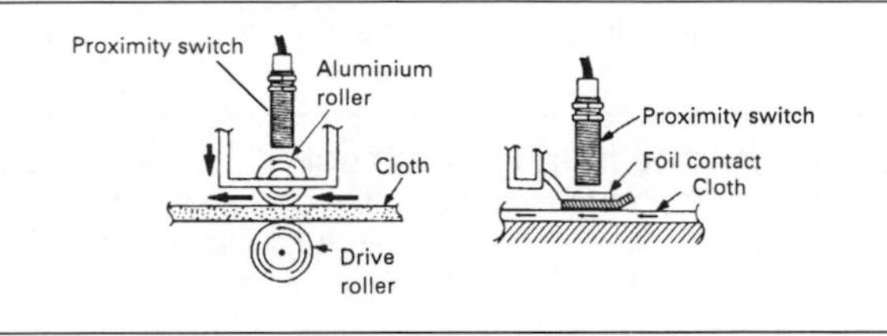

Figure 6.12 Inductive proximity detectors used with metal contacts to sense thickness of a non-metallic material.

more expensive than microswitches, and their use is more likely to be confined to where their advantages are unique in not requiring to make contact with the object that is to be sensed. Other particularly useful features of inductive sensors are the high rates of repetition, ranging from 200 Hz to 2 kHz for the small DC types, to 25 Hz for the AC types, and the working temperature range of, typically, −25°C to +85°C.

CAPACITIVE PROXIMITY SENSORS

Capacitive sensing proximity detectors make use of the stray capacitance that exists between a metal plate and ground, a quantity that can be altered by the presence of non-conducting material as well as by either grounded or isolated conducting material. As before, either a bridge or a resonant circuit detection method can be used, although the resonant system is more common because the stray capacitances are small. A notable feature of capacitive proximity detectors is that the detectable range is much higher than it is for the inductive type. Both AC- and DC-operated detectors are available, and Table 6.2 shows some current versions of these detectors.

In Table 6.3, the ranges are approximate because much depends on the size and shape of the object being sensed as well as on the material. Powders and liquids can be sensed in addition to solids, but any average figure of distance would be misleading for such applications. Grounded

Table 6.2 Characteristics of some current models of capacitive proximity sensors from RS Components and Allen-Bradley.

Manufacturer/model	RS 307–395	RS 341–474	A-B 875C
Supply	80–250 VAC	220/240 VAC	10–60 VDC
I_{max}	200 mA	5 A	200–400 mA
Rate (max)	10 Hz	–	25 Hz
Temp. range	0–60°C	−55°C to +75°C	–
I_{min}	5 mA	–	–

Table 6.3 Typical sensing distances in mm for RS 307–395 sensor, affected by size/shape of object.

Material	Distance in (mm)	Material	Distance in (mm)
Grounded Cu/Al	65	Cu/Al, not grounded	35
Cardboard	25	Wood	25
PVC	20		

metals can be detected at a fairly large range – the figures quoted in the table are for copper (Cu) or aluminium (Al). Repetition rates are in general lower than for the inductive type of sensors. The 200/240 V AC type shown in the table also incorporates a delay feature, hence the omission of any rate or repetition figure. The delay can be varied from 2 s to 10 min to allow for other processes (such as starting a pump or emptying a hopper).

ULTRASONIC PROXIMITY DETECTORS

The ultrasonic type of sensor can make use of diffuse-scan mode (using the same transducer as transmitter and receiver, and detecting a reflected signal) or through-scan mode (using two transducers and detecting when the signal is interrupted). Through-scan mode allows for a much greater range of action, but the diffuse-scan mode is more common for many applications.

The frequency of ultrasound is critical, because a substantial number of industrial machines emit ultrasound in the near-sonic range of 20–50 kHz and sometimes higher, so that the use of such frequencies for proximity sensing is liable to be considerably affected. For that reason, a frequency of 215 kHz is favoured for proximity detection. The output from the transducer is modulated, using frequencies in the range 30–360 Hz, and this modulation acts as another safeguard against interference. The modulation frequency also affects the minimum sensing distance in diffuse-scan mode, since an object cannot be sensed if the transducer has not switched to receiving by the time the echo returns.

Typical ranges for the diffuse-mode types are 100–800 mm for modulation rates of 20–8 Hz; ranges down to 10 mm can be obtained when a higher modulation frequency is used. When through-mode is in use, the ranges can be longer, though 1 m is a typical maximum. When more than one diffuse-mode transducer is being used, care has to be taken so that there is no possibility of one transducer receiving the beam from the other and so working in through-mode.

Ultrasonic proximity detectors are particularly useful in applications that are not well suited to inductive or capacitive types. Figure 6.13 illustrates

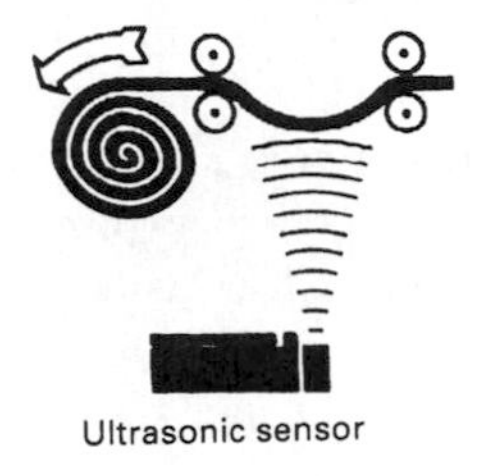

Figure 6.13 An ultrasonic sensor used to detect, without contact, the amount of slack in a ribbon of material.

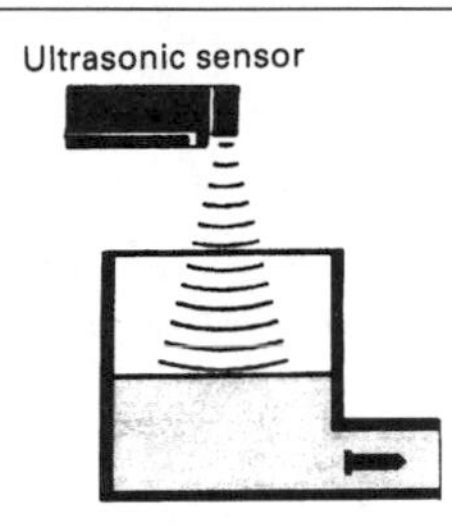

Figure 6.14 Ultrasonic sensing of liquid level. This is a very common application because no contact with the liquid is needed.

an application to maintaining a slack portion of a material that is being wound, preventing stretching due to tensioning of the material, or jamming caused by an excessively slack section. Figure 6.14 shows a liquid level sensor in which the ultrasonic beam is reflected from the liquid surface, detecting (in this example) when the container has been emptied and to what extent it is later filled again.

The use of ultrasonic detection requires some care in the selection, positioning and use of the sensors. The maximum sensing distances that are quoted in data sheets refer to targets with a smooth reflecting surface of at least 30 × 30 mm, arranged orthogonally to the beam (surface plane at 90° to the line of the beam), at 250°C ambient temperature and 60% humidity with negligible air movement. Since the ultrasonic beam is a sound wave in air, it is affected by air movement (which will alter the beam directions of both transmitted and reflected beams) and by the air humidity (which affects the speed of ultrasound). The effect of air temperature is that the speed of ultrasound in air increases by 0.18% for each °C change of air temperature. This is often compensated for within the transducer by altering the clock speed for the time counter as air temperature changes.

A target area smaller than the 30 × 30 mm that is assumed, will reduce the maximum range – as a rough guide, a reflecting surface of 10 × 10 mm will halve the maximum range figure. A target which consists of a smooth flat surface will be undetectable if the surface is angled so as to reflect the return beam away from the transducer – remember that the angle through which the beam is deflected is twice as great as the angle through which the target surface is turned. This effect is more critical at the longer ranges at which the tolerance can be as low as $\pm 2°$, but is not so significant at closer ranges. Nevertheless, ultrasonic proximity detection is much less tolerant of target orientation than other sensing methods.

The types of material that can be detected exclude only a few sound-absorbing substances, such as cotton-wool and rubber foams, but many textile surfaces provide only very weak echoes. In general, very porous materials are not suitable for ultrasonic detection.

PHOTO BEAM METHODS

Optical beam proximity detectors can use *through-scan*, with one source and one receiver aimed at each other, *retroreflective-scan*, with one combined source and receiver used with a mirror, and *diffuse-scan*, making use of a combined source and receiver with light reflected from the object itself. The prices of the respective units are not widely different, so that through-scan methods are very often used where retroreflective methods might be equally applicable. The light source is usually an LED working in either the visible red or the (invisible) infrared range. The use of visible red light makes setting-up easier, particularly over long distances, but the infrared type is less likely to suffer from interference from other light sources.

The range of optical proximity units can be large, up to 15 m for a through-scan type, about half for retroflective use, and typically up to 700 mm for diffuse reflection. The through-scan and retroreflective types depend on the object breaking the beam, but the diffuse-reflection type can be used to sense the approach of an object along the beam as well as cutting the beam. Figure 6.15 illustrates an application of a pair of retroreflective units in maintaining a slack loop in a strip feed, a very common application. The advantage of the optical system is that large loops can be accommodated, much larger than can be sensed by ultrasonic systems, for

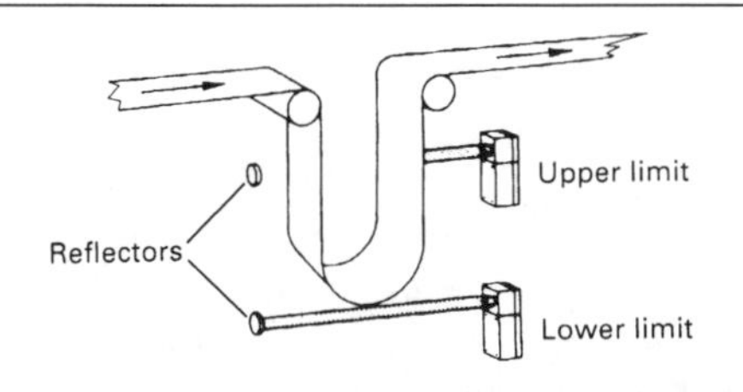

Figure 6.15 Optical proximity detectors using retroreflectors in a system that senses the size of a slack ribbon of material.

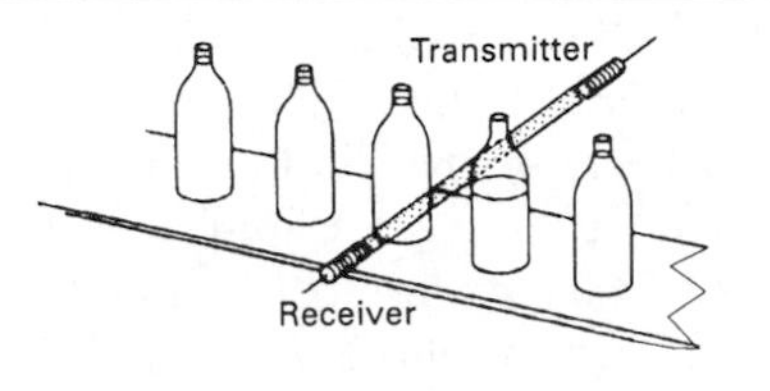

Figure 6.16 Optical detection of the liquid level in bottles – the liquid need not be opaque, but the setting is more critical for clear liquids.

example, making this system applicable to paper, plastic sheet, rubber and other industries. Another problem that is particularly suited to an optical solution is level detection in bottles on a line (Figure 6.16).

Whether visible red light or infrared is used, sensors make use of pulse-modulated signals and polarizing filters to ensure that neither stray light nor multiple reflections cause problems of false triggering. Optical units are available with relay or solid-state switching, DC or AC operation, with or without time-delays, and usually with built-in LED signals so that the sensors can be checked for alignment without the need to actuate other units in an automated system.

Photosensing units can also be used along with fibre optics where the area of use is to be closely confined. The use of 0.5 mm diameter fibres, for example, can allow optical sensing to be used in spaces which are inaccessible to the normal type of unit, with the advantage that the path of the fibre can be bent with negligible loss of light transmission. For building into new equipment, combined LED and photosensor units can be obtained at very low prices, allowing the designer to use whatever amplifying and relay system is preferred.

REED SWITCH/MAGNET AND HALL-EFFECT UNITS

For specialized applications, the use of a reed switch and magnet is a robust and simple proximity detector. The principle is simple – the reed switch is closed (or opened) when the magnet is within range, usually a few millimetres distance, and the reed switches over when the magnet is withdrawn. The system is confined to detecting changes on objects that always mate together, such as doors and doorways, and is widely used in safety interlocking and burglar alarm systems. The reed-switch contacts can carry an acceptable load current, so that no intermediate relay is required for most of the applications of the system. For details of reed-switch mechanisms, see Chapter 12.

Hall-effect units are also magnetically operated, but the magnetic field strength required for a Hall-effect detector can be very low, as low as that of the Earth if needed. Hall-effect proximity detectors are therefore capable of sensing magnets at greater distances than can be used by reed-switch systems and, in addition, because the Hall-effect sensor can detect small changes in magnetic field it can sense interference with the normal magnetic field of the Earth, such as might be caused by steel objects. The switching is bounce free, unlike that of the reed switch, and in some applications the Hall-effect sensor can be used like an optical unit to detect the passing of a ferrous object between a magnet and the sensor. This has the advantage that counting can be carried out in conditions (such as in black liquids) which would be impossible for an optical beam.

DOPPLER DETECTORS

The most exotic of proximity detectors (for moving objects only) is the microwave radar type using the Doppler effect. This operates in X-band in the 10.675–10.7 GHz region, and requires a licence under the Wireless Telegraphy Act of 1949 – the licence is for 5 years and is obtained from a DTI branch in Cardiff. An application form is supplied with each Doppler unit bought from RS Components.

Although the technology may seem exotic, the price of a radar proximity unit is surprisingly low – little more than that of a PIR unit. The unit contains a Gunn diode oscillator with 10 mW output located in a launching horn, and a receiving horn containing a mixer diode. A section of waveguide, between the horns, supplies some of the oscillator output to the mixer for frequency mixing, and the mixed output, a low-frequency signal generated when any object in the field of view moves, is amplified in a separate unit.

The detection ability depends on the size of the target, and a moving human body can be detected reliably at ranges of typically 1–15 m. A moving human generates a Doppler signal whose frequency is centred around 50 Hz, and amplifiers for a Doppler system have their maximum sensitivity at this frequency. This imposes two restrictions on the use of the system. One is that fluorescent lighting in the target area should be switched off when the system is operating. The ionized gas in a fluorescent tube acts as a reflector while the tube is conducting but not when the tube is in its non-conducting state, so that a strong radar return at 100 Hz will be obtained from any fluorescent source. This restriction is not as important as it might seem, because radar detection methods are particularly suited to dark spaces – optical methods can be used in well-lit spaces.

The other restriction is that the abnormally high sensitivity of the amplifier to 50 Hz requires 12 V battery operation rather than a mains supply. The problem is one of hum pick-up from mains wiring rather than from inadequate smoothing. False alarms can be caused by vibration which shakes the alarm module, movement of objects like curtains, doors, banners, etc., and air turbulence in plastic water pipes.

6.4 Liquid levels

A sample of liquid has a mass that is fixed, and a volume that changes slightly with changes of temperature. The shape of a liquid is, in general, the shape of its container, although in zero gravity most liquids take the shape of least surface area, a sphere. The measurement or sensing of the level of a liquid in a container of uniform cross-section is therefore a useful quantity that is proportional to the volume of the liquid.

The simplest form of liquid-level gauge is based on the ancient float-and-arm system, as illustrated in Figure 6.17. Any suitable transducer can be used to detect the angular movement of the arm, and a potentiometer is particularly useful. With a fixed voltage, AC or DC, applied to the fixed ends of the potentiometer, the voltage at the variable tap will be proportional to the position of the float between the low and high extremes.

The float system is less attractive if the liquid is corrosive. Although the float itself, and the arm, can be made from resistant material such as polypropylene, the transducer is still liable to damage from corrosive vapours, and a potentiometer is unlikely to be useful. The LVDT principle can be used, with the metal plunger encased in plastic and the wires also insulated with plastic, but other methods are usually preferable in such circumstances.

One other method is the air-compression sensor (Figure 6.18). A tube is immersed in the liquid so that air (or possibly another liquid) is trapped inside the tube. A change in the liquid level outside the tube will cause the air to be compressed or expanded, and this change of pressure can be sensed. A scheme like this is often used to sense water level in clothes-washers and dishwashers. For switching purposes, a diaphragm and microswitch can be used, but in order to sense the level the transducer should be a manometer or similar pressure detector (see Chapter 1).

A very neat method that can be applied to any liquid, makes use of the permittivity of the liquid that will cause a change of capacitance if the

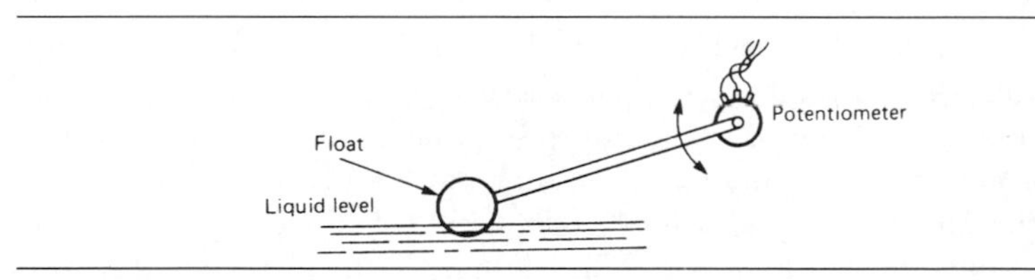

Figure 6.17 The simplest method of liquid-level sensing, using a float arm and a potentiometer.

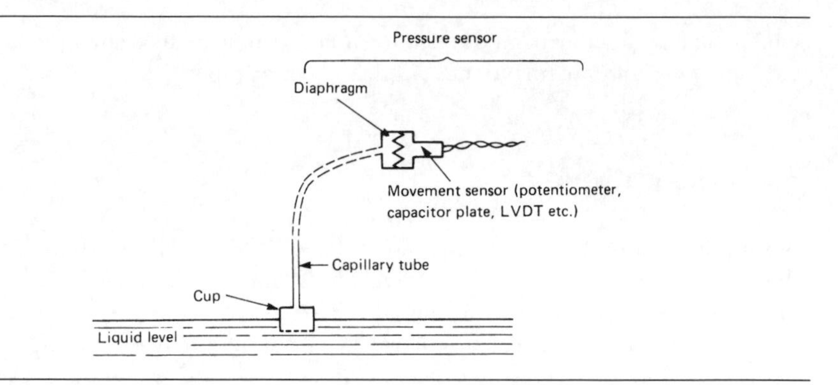

Figure 6.18 A very common form of liquid-level sensor that makes use of the liquid level to compress air. The air pressure is then sensed remotely, with a connection by way of a capillary tube.

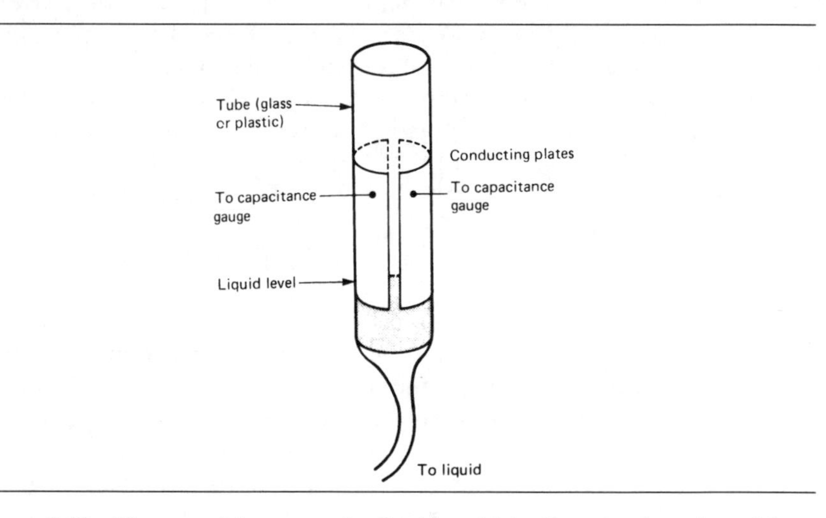

Figure 6.19 The capacitive gauge for liquids, which allows level sensing without any direct contact with the liquid. The permittivity of the liquid is invariably greater than that of air, so if the liquid is a non-conductor, a change in level will cause a change in the capacitance between the plates.

liquid is placed between capacitor plates. The scheme is illustrated in Figure 6.19, showing the transducer as a pair of rectangular foil plates wrapped around a tube so that the edges do not quite touch. These plates now form a capacitor for which the content of the tube acts as a dielectric. Replacing the air in the tube by a liquid will cause a change of capacitance due to the different permittivity of the liquid. The amount of the change will be proportional to the level of liquid in the tube between the limits of the length of the capacitor plates.

The change of capacitance is most easily sensed by making the capacitor part of an oscillating circuit. Some care is needed if the liquid is a partial conductor, as this can damp the oscillation and cause the oscillator to stop. The change of capacitance causes a change of oscillating frequency, with the frequency falling as the liquid rises, and this change of frequency can be converted into a voltage by a frequency discriminator or by a digital–analogue converter.

6.5 Liquid flow sensors

Liquid flow is a very important part of many manufacturing processes, and where the manufacturing process is continuous, rather than a batch process, the sensing and measurement of liquid flow is a vital part of plant control. Flow measurement is, for example, vital in the all-important water industry, both at the supply side for potable water and at the disposal side in sewage treatment works. It is equally important for milk-based industries (e.g. cheese, butter, yoghurt), in chemical works (where relative quantities of liquids entering a reaction vessel may have to be controlled and measured to fairly fine limits), and in printing.

Flow monitoring in one form or another is used in most machines to ensure that lubrication is adequate and that coolant is circulating correctly. As applied to gases, flow monitors are used in healing and air-conditioning plants, in air-cooling of machines and in gas-based chemical reactions. Monitoring is particularly important for both gases and liquids if a stoppage might indicate a blockage or other fault condition, such as pump failure.

The sensing of liquid flow can take three different forms. One is vector flow, in which the speed and direction of the liquid need to be sensed. Another possible requirement is volume flow, measuring the volume per second of liquid passing a point in a pipe. The third possible requirement is mass flow, which is usually calculated from volume flow using the relationship, mass = volume × density. For most automation requirements, with liquids confined in tubes, the volume flow is the important one.

There is no simple and straightforward method of measuring liquid volume flow, although sensing flow is comparatively simple. One of the simplest methods is based on the pressure difference that occurs when the liquid flow is through an orifice, which can be an orifice plate or a nozzle (Figure 6.20). For a liquid whose flow rate is slow enough to be laminar (stream-line), as defined in Table 6.4, the pressure difference across the orifice is linearly proportional to the rate of flow. This pressure difference can be sensed or measured by a manometer or by the difference in level between two tubes. If the flow is not laminar (see Table 6.4) but is

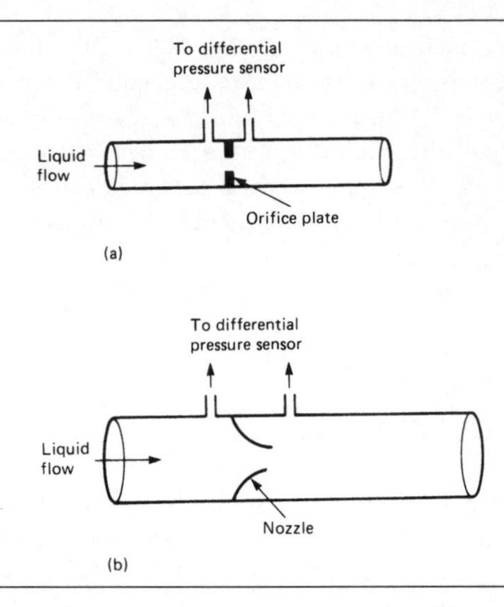

Figure 6.20 Detecting liquid flow rate by means of the pressure difference across (a) an aperture or (b) a nozzle. The pressure difference is proportional to rate of flow only while the rate is low (laminar flow), and the nozzle type usually permits a greater range of flow rates to be measured.

Table 6.4 Laminar and turbulent flow, and how critical velocity of flow is related to liquid viscosity, density and Reynolds number.

Laminar flow means that each layer of a liquid flows in a line that does not intersect with any other layers of a liquid. These lines are called *stream lines*, and laminar flow is sometimes known as stream-line flow. When the rate of flow increases above a critical level, the flow becomes turbulent, with the stream lines breaking up into vortex paths. The resistance to flow then increases greatly.

The critical flow velocity is given by:

$$V_C = (RN)\frac{\eta}{\rho A}$$

where RN = Reynolds number, η = viscosity of liquid, ρ = density of liquid, A = cross-sectional area of tube.

turbulent, the pressure is more likely to be proportional to the square of liquid speed.

If the presence of mechanically moving parts in the liquid stream is permissible, the 'mill-wheel' method can be used. In this method (Figure 6.21), the movement of the liquid turns a turbine wheel which in turn is coupled to a tachometer. This method is quite acceptable for liquids whose flow rate is comparatively slow, and over a fair range of liquid velocity the

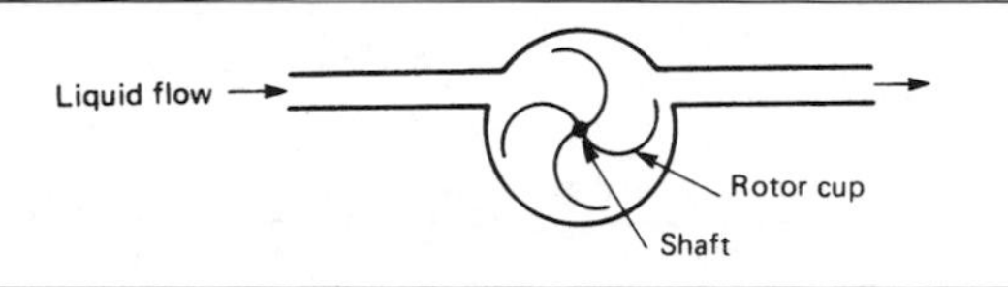

Figure 6.21 The 'mill-wheel' method for liquid velocity sensing. The shaft can be coupled to any angular velocity sensor, such as a tachogenerator or a synchro.

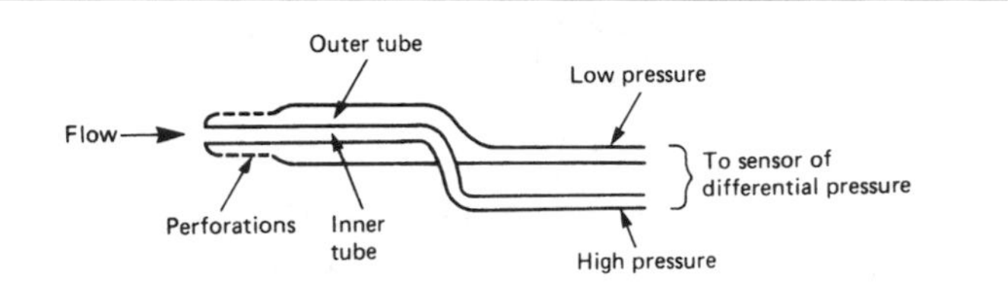

Figure 6.22 The pitot-tube method of measuring gas flow, making use of the pressure difference between gas striking an orifice and gas passing over an orifice.

output is reasonably proportional to the flow rate. The device needs calibration, however, either from another form of gauge or by the ancient method of measuring how much liquid (volume or mass) is delivered in a given time.

Most commercially available methods of flow transducers are based on this principle, although in many processes liquid flow is monitored visually using the old-established tube-and-float method which, because it has no electrical output, is outside the scope of this book. The turbine blades need to be constructed from a material that will be resistant to the type of liquid being measured, and careful selection will be needed, because there are few materials that can be used for a wide range of chemicals. The conversion from turbine speed to electrical output is carried out typically by using a light source and photocell unit, although for some specialized applications (very dark liquids) it may be necessary to use a Hall-effect sensor and a turbine with ferrite magnets included in its vanes. Another, more crude, method uses a vane mounted on a spring-loaded shaft which is placed in the liquid flow. This senses the deflection of the vane by the angle through which the vane is turned. This angle is most simply sensed by a potentiometer connected to the shaft.

For sensing and measuring vector flow, the pitot tube is a method that has been used for some considerable time, particularly for the indication of aircraft air speed. The principle is illustrated in Figure 6.22 in its most elementary form, and consists of a liquid U-tube manometer whose two ends are both fed from nearby points. One of these points is directed at the oncoming airstream, and the other is a perforated tube over which the airstream passes. The pressure difference between these two will depend on

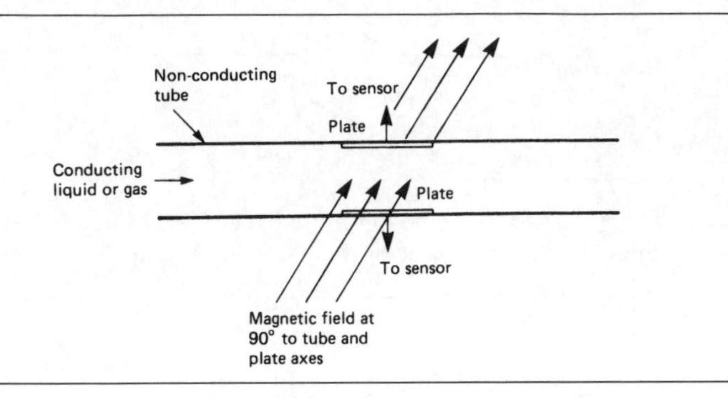

Figure 6.23 The magneto-hydrodynamic principle. The flow of the conducting fluid, so as to cut the magnetic field, causes an EMF to be generated between the plates, and the size of the EMF is proportional to the fluid velocity if the conductivity of the fluid is constant.

the speed and direction of the airstream. The pitot-tube principle can be used with gases and with liquids of low viscosity, and can be adopted for electronic sensing methods by using an electronic manometer.

A method that can be used only for conducting liquids is the magneto-hydrodynamic sensor, whose principle is illustrated in Figure 6.23. Two electrodes are in contact with the moving liquid in a tube, and a strong magnetic field is applied at right angles both to the axis of the tube and to the line joining the electrodes. Motion of the liquid will cause an EMF to be generated between the electrodes, in accordance with Faraday principles. The amplitude of the EMF is proportional to the rate at which the flux of the magnet is cut, so it is proportional to the velocity of the liquid.

The requirement for the liquid to be conducting makes this method of little use for many applications, but it can be used with ionized gases. It has had some limited application as a transducer used to generate electricity from exhaust gases of both gas turbines and coal-fired generating systems.

6.6 Timing

Many sensors of liquid flow are based on the principle that affects waves moving through liquids or gases that may themselves be moving. The equations show how the velocity of a received wave from a stationary source can be affected by movement of the medium that carries the waves, and the relevant equation is illustrated in Figure 6.24. The effect shows that if a wave is detected downstream of the moving liquid that carries the wave, then the time taken to reach the receiver is less than it would be in a liquid at rest. If the wave were moving upstream then the time needed to reach the receiver is greater than it would be in a liquid at rest.

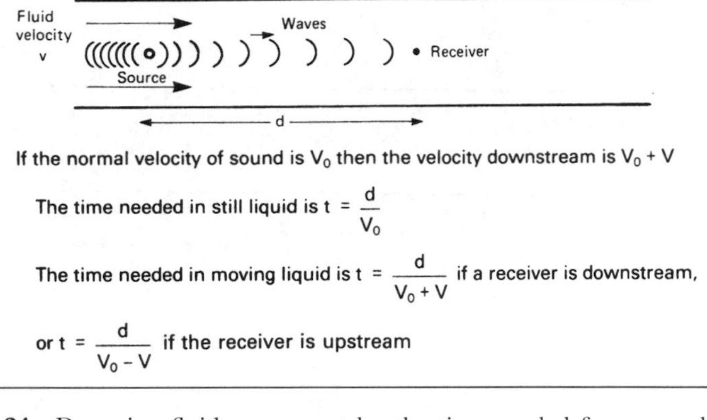

Figure 6.24 Detecting fluid movement by the time needed for a sound wave to travel between two points. This is a form of Doppler effect, but it does not cause any change of received frequency as happens when either source or receiver moves.

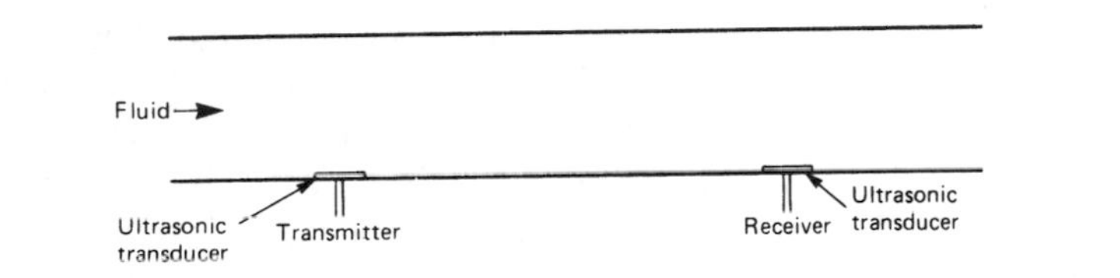

Figure 6.25 Using ultrasonic transducers to measure fluid flow. The method is complicated by the effect of reflections from the pipe, and careful calibration is needed.

The waves that can be used for measurement can be ultrasonic, and ultrasonic flow gauges have been commercially available for a considerable time, ever since small ultrasonic transducers became available. Figure 6.25 shows the principle of the method, using in this example the simpler system of a receiver transducer downstream of the transmitter. This makes the velocity of the received wave equal to the natural velocity (in a still liquid) plus the liquid velocity. The change in wavelength is proportional to liquid velocity, and the absolute value of liquid velocity can be calculated with relative ease.

Ultrasonic methods are most useful when they can be applied in large-diameter tubes, using transducers beaming in the line of the liquid flow. Problems arise when the ultrasonic waves are reflected from the sides of the tube or from bubbles or cavities in the liquid, or when the ultrasonic transducers have to be used beaming across the direction of liquid flow. This latter method is, unfortunately, the most practical from the point of view of measuring the liquid flow with nothing immersed in the liquid. For many types of measurement, however, the crossbeam method is completely satisfactory, depending on the liquid and the shape of the cross-section of the tube.

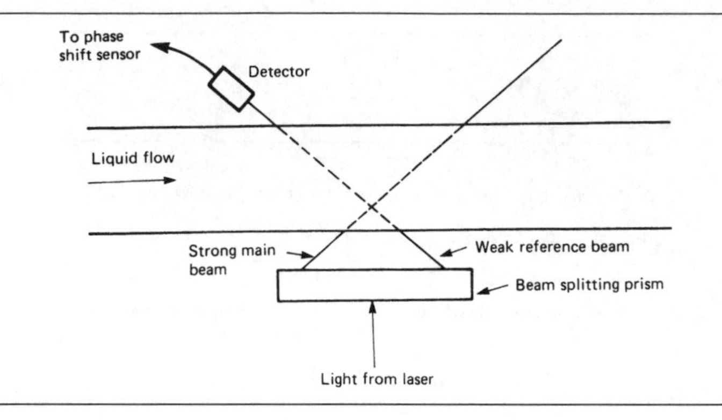

Figure 6.26 Principle of Doppler laser sensor. The light of the main beam is scattered by particles in the liquid, and this scattered light interferes with the weak reference beam, causing a shift in the received interference pattern whose amount depends on the velocity of the liquid.

Laser Doppler methods are also in use now, although they depend to some extent on the use of transparent tubes, or tubes with transparent insets, and on the liquid itself being transparent to the laser frequency. One method is illustrated in Figure 6.26, in which the scattering of the light by tiny discontinuities in the liquid (bubbles, for example) is used to cause interference between two parts of a laser beam, causing interference and a frequency shift of the scattered beam. The Doppler laws as applied to light obey the principles of relativity inasmuch as the velocity of the light is constant throughout a medium, irrespective of the movement of the medium. This means that the frequency shifts that are observed are caused by changes of the light wavelength, never by a change of light velocity. A laser fluid-velocity meter based on these principles can operate with liquid velocities of a few mm/s up to several hundred m/s.

Another method that is considerably less simple is the use of correlation techniques. The sensor system consists of a laser beam that is split so that it can be reflected from two different points in the liquid, one downstream of the other. The received signals that are obtained should be identical if one is delayed relative to the other with a delay equal to the time taken for the liquid to traverse the distance between the beams. Figure 6.27 shows the principle as applied to one reflecting particle in the liquid. The use of the technique depends on electronic methods of correlating the readings, and digital circuits consisting of exclusive-OR gates are used. For calibration, bubbles may have to be injected into the liquid, but most liquids will provide sufficient natural discontinuities to make the system workable.

6.7 Gases

The velocity of gases can be measured in the same ways as are used for

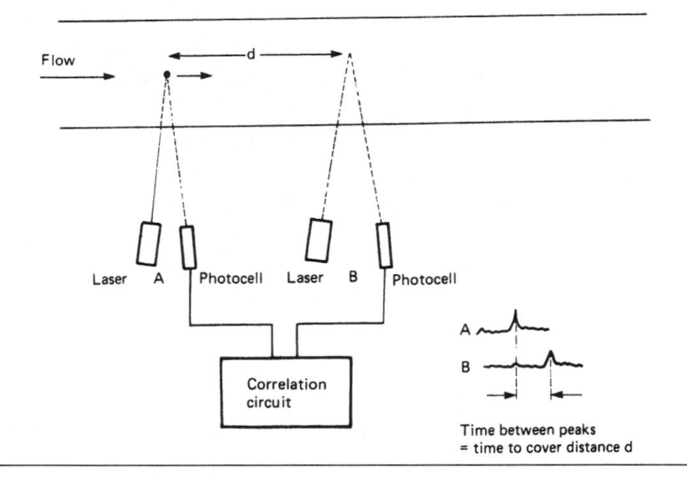

Figure 6.27 The correlation method, which depends on sensing the time difference between identical changes in the reflected beam. Two lasers have been shown for simplicity, but in practice a single laser would be used along with prisms or optical fibres.

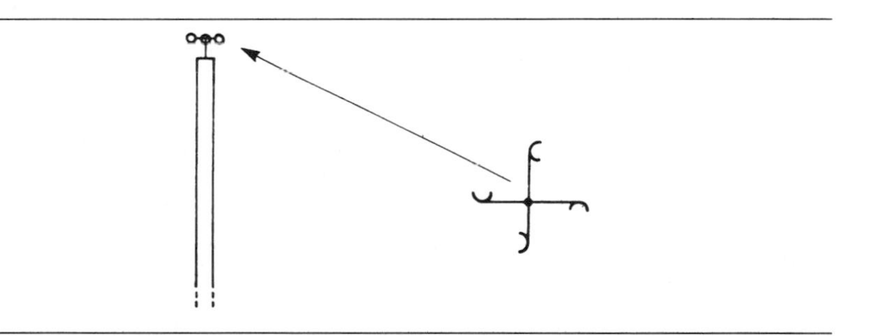

Figure 6.28 The rotating cup sensor for wind velocity. The rotating arms can be mounted on the shaft of a tachogenerator or synchro for a direct electrical output.

liquids, but a few specialized methods are added. The anemometer is used for wind-speed measurements in meteorology, and consists of a set of cup-shaped vanes that rotate in a horizontal plane (Figure 6.28). The speed of rotation is proportional to wind speed, and is most conveniently measured by using a tachometer or a rotary digitizer connected to electronic circuits. The application of electronic methods to anemometers has increased the precision of these sensors in comparison to the older types in which the vanes were coupled to an indicator by a revolving cable.

Hot-wire methods are applied to gas movement as well as to gas pressure (see Pirani gauge, Chapter 1). A wire that is exposed to the moving gas is heated by the passage of a current. The temperature that the wire reaches depends on the cooling effect of the moving gas, so that the resistance of the wire can be used as a measure of its temperature and so of the gas velocity. The relationship between wire resistance and gas flow is not a simple one, and a sensor of this type needs to be calibrated against a more fundamental measurement. In addition, the calibration will need to be repeated if the composition of the gas is changed at any time.

The application of electronics to car engines has created a requirement for a number of gas sensors, particularly flow, temperature and composition sensors. Gas flow can very often be sensed by simple vane-deflection methods, because linearity is seldom an overriding requirement. Gas composition usually means exhaust gas oxygen content, which is a critical factor that measures the efficiency of combustion. The oxygen content of exhaust gas is sensed by using metal oxide 'cells', for which zirconium oxide or titanium oxide are the most suitable types. When either of these oxides is sandwiched between platinum electrodes, an EMF will be generated whose size depends on the availability of oxygen ions surrounding the sensor. By allowing the hot exhaust gas to pass over the cell, then, the amount of oxygen in the exhaust can be sensed, and so the efficiency of combustion assessed. The important feature of such a sensor for use in cars is that the change in combustion conditions from lean (too much oxygen) to rich (too much fuel) causes a very large change of oxygen content (for example, a ratio of 10^{14}) so that the output of the sensor can change very sharply in the range 20 mV to 1 V.

In general, sensors for gas composition rely either on cells whose output is an EMF, or on the effect of the absorption of the gas on a quartz crystal that is part of an oscillating circuit. These types of transducers are very specialized, and many of the cell types make use of the catalytic action of platinum or palladium in thin films. All such sensors are very susceptible to 'poisoning' by unwanted contaminants in the gases, and particularly metallic contaminants such as lead.

A typical gas composition unit for flammable and other gases is the RS Components gas sensor NAP11A. This uses a platinum spiral sensor and a compensating winding in a bridge arrangement, both heated, but only one is exposed to the gas. Following the age-old principle of the Davy lamp, the exposed spiral is enclosed within two layers of wire mesh, preventing explosion of mixtures that contain oxygen mixed with the flammable gases. The outline of the sensor itself is illustrated in Figure 6.29 and its performance with some gases in Figure 6.30. A typical applications circuit, courtesy of RS components, is shown in Figure 6.31. Data sheets can be downloaded from the RS Web site.

Some points to consider in the use of such sensors are that silicone products (greases, oils and polishes) will ruin the performance of the gas detector, and long-term exposure to salty atmospheres will also affect the calibration. These units are intended for detecting small concentrations of gases such as might indicate a dangerous build-up, and must not be used with concentrated gases. Their use in gas alarms must be well thought out,

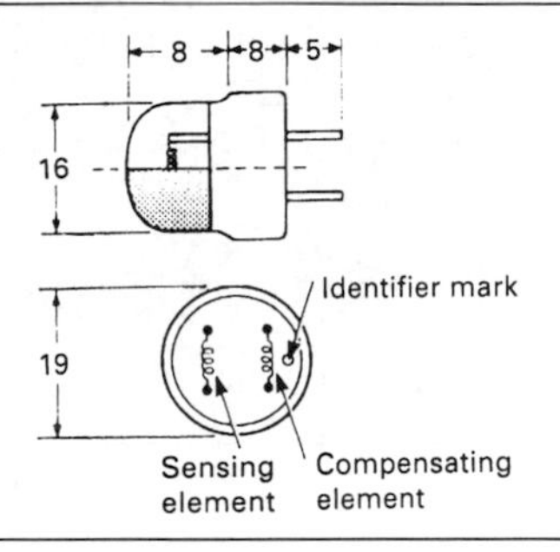

Figure 6.29 The construction of a typical gas sensor, with the sensing winding and the compensating winding (all dimensions in mm).

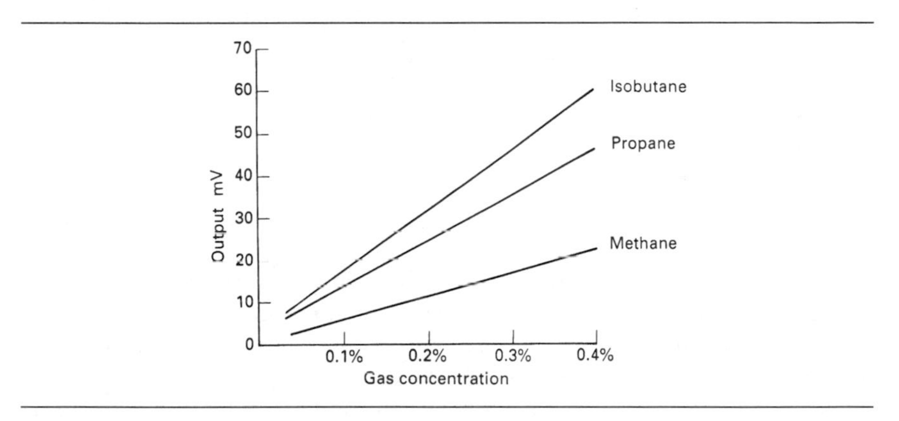

Figure 6.30 The output from a typical gas sensor in terms of mV plotted against gas concentration.

because some flammable gases are less dense than air, and will be found mainly at the top of an enclosed space. Others are denser than air and will be found at the bottom of a space – cellars and ships' bilges are particularly hazardous in this respect.

- A more recent development is the use of metal-oxide gas detectors, particularly the use of tin dioxide. This has led to work on sensors that will detect carbon monoxide in the presence of unburned fuel residues, sensors that will detect unburned fuel despite the presence of carbon monoxide, sensors for hydrogen in the presence of hydrocarbons, and sensors for nitrogen oxides. All of the applications are of major importance in detecting and possibly controlling car exhaust emissions.

6.8 Viscosity

The viscosity of a liquid or gas is the quantity that corresponds to friction

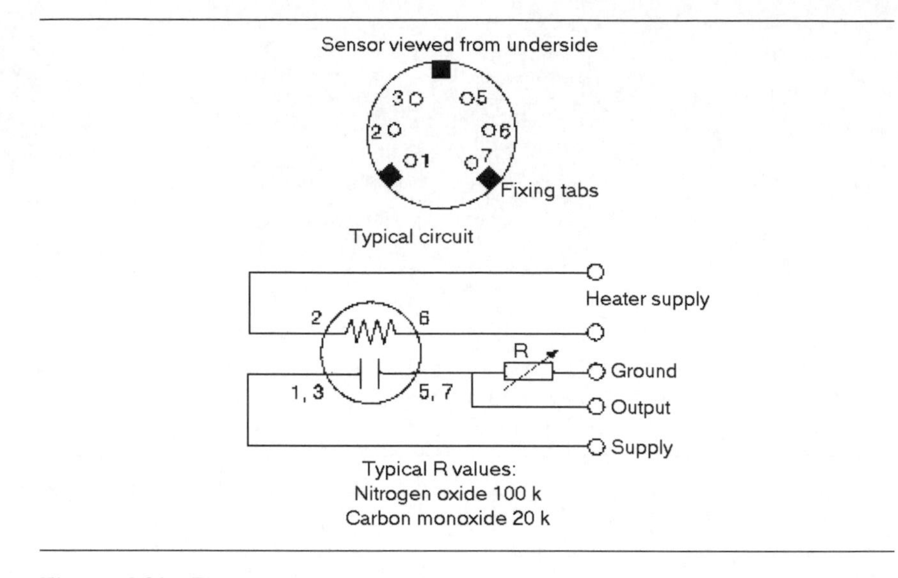

Figure 6.31 Pin layout and suggested circuit for RS Gas detector type NAP11A.

between solids, and which determines how much pressure will be needed to drive the liquid or gas through a pipe. In addition, some chemical reactions result in very marked changes in viscosity, so that sensing methods are needed for liquids that are static and not flowing. For almost all liquids, the viscosity, as measured by a coefficient of viscosity, decreases as the liquid is heated. For gases the reverse is true, and this is one reason why it is so much more difficult to deal with hot gas exhausts as compared to cold gas intakes.

Figure 6.32 shows the definition of the coefficient of viscosity in terms of the force that makes one layer of a liquid move faster than another does. This force cannot be measured directly, but from the definition we can calculate the amount of force needed to spin a solid cylinder in a liquid, the amount of pressure needed to deliver a given volume per second of liquid through a pipe, or the rate at which a dense sphere will fall through the liquid. The traditional methods of measuring viscosity are based on such principles (Figure 6.33), but none of them is appropriate for electronic purposes apart from the pressure difference method that can be important in the control of continuous flow processes.

Note that the methods of sensing liquid flow by means of the pressure difference across a restriction all assume that the viscosity of the liquid is constant – any change of viscosity will be read as a change of velocity. Electronic sensors for viscosity can make use of the damping of mechanical oscillations by a viscous liquid or gas. The principle is illustrated in Figure 6.34 which shows a plunger that can be vibrated by a piezoelectric transducer, and whose amplitude of vibration is sensed by another transducer. If this arrangement is part of an oscillating circuit, then the damping effect of a

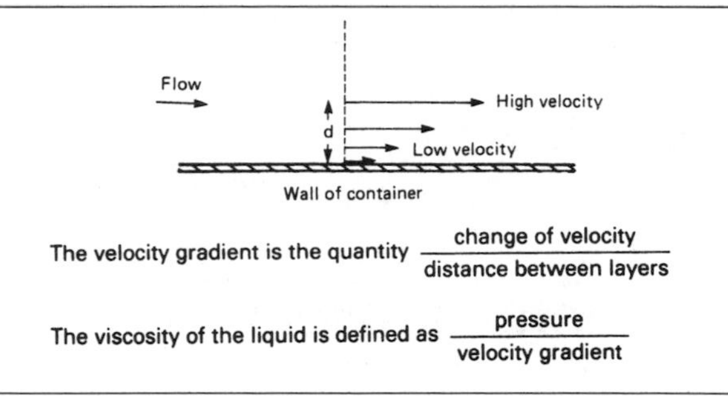

Figure 6.32 The definition of viscosity of a liquid in terms of the pressure and velocity gradient in a moving liquid. The velocity gradient cannot be measured directly.

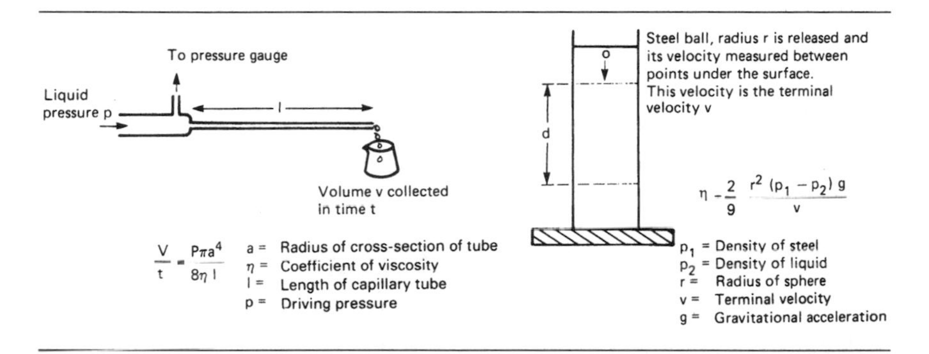

Figure 6.33 Traditional methods for measuring liquid viscosity values.

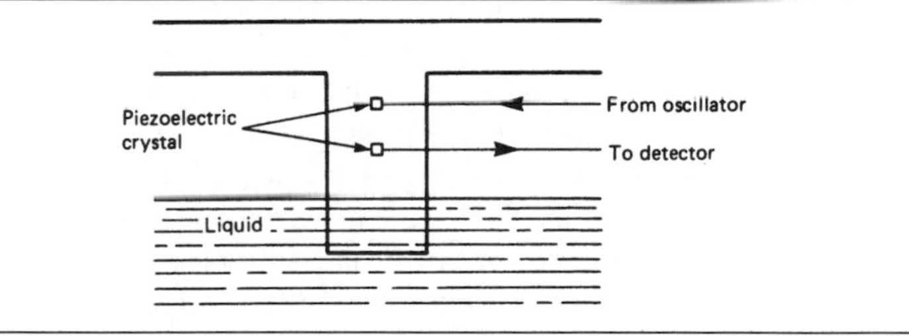

Figure 6.34 The arrangement of piezoelectric crystals on a vibrating rod. One crystal is driven from an oscillator and the frequency is adjusted to make the vibration resonant, as indicated by the peak output from the second transducer. This peak will be considerably damped if the rod is immersed in a viscous fluid, and the change in amplitude of the resonant vibration is a measure of the viscosity if other conditions are equal.

liquid on the amplitude of oscillation will depend on the viscosity of the liquid. This can be sensed electrically as the amplitude of the signal from the receiving transducer.

Chapter 7

Environmental sensors

7.1 Environmental quantities

In the course of the preceding chapters, sensors for many quantities that are of environmental importance, such as wind velocity, have been mentioned. This chapter is concerned with some types of quantities and sensors which have not been covered previously and which are all in some way related to the environment. Some of these quantities need to be sensed in connection with industrial processes, and a few are of considerable importance in all aspects of life, but others are decidedly specialized.

All the devices that are considered in this chapter are sensors or measuring instruments, because none of the quantities is an energy form that permits conversion. There are, of course, natural forms of energy from which efficient transducers would be useful. The low concentration of energy (the amount of energy per m^2 or m^3 of transducer) is usually very much against the use of such devices. It may seem attractive at first sight to generate electricity from wind or waves, but the size of any useful generator is daunting. It is most unlikely that our multitude of amateur environmentalists would ever permit such monstrous contraptions to be built in the huge quantities that would be needed if more than a small fraction of fossil fuel and nuclear sources have to be replaced. Practical generators for large-scale generation must make use of processes such as combustion and nuclear fission which have a high energy concentration. Although the efficiency of conversion by way of steam may seem low at around 40%, it is very much better than is obtainable from other types of transducer systems, few of which ever look like bettering 10% with any degree of reliability.

7.2 Time

The measurement of time is fundamental to many industrial processes as well as to civilized life in general, but it is only comparatively recently that electronic timekeeping has become important. Our units of time are based on the solar year, the average time that the Earth takes for one revolution around the Sun, which is about 365.25 days. Astronomical determination of time is not exactly an everyday practical method, so that clocks and watches have been developed for convenience, and have traditionally relied on mechanical oscillations for a stable standard of frequency. Extraordinary efforts have been made in the past to achieve high precision of timekeeping because of its importance in navigation, but the limitations of a mechanical system have always determined what could be achieved. Really precise timekeeping has only now come about because of the application of electronic methods.

Electronic time measurement makes use of the crystal-controlled oscillator. The piezoelectric properties of the quartz crystal make it an excellent vibration transducer, and if the crystal is cut so that it resonates mechanically, it will act electrically as if it were an electrical resonant circuit with practically no losses. This implies that the frequency is very stable and that the oscillation can be maintained with very little energy input. For domestic electronic clocks and watches, this stability is sufficient, so that a cheap digital watch is a better timekeeper than the most expensive mechanical timepiece of a previous generation. For much more precise timing, the crystal should be kept at a constant temperature, and crystal standard-frequency generators have been built to this specification for some considerable time, since long before digital counting and displays were available. For the utmost stability, *maser* oscillators are now available.

The display of time or the measurement of time intervals can make use of digital methods, counting the waves from the crystal-controlled oscillators. If the system is to be used as a timer, a gate circuit must be added (Figure 7.1) so that the counting can be switched on and off. The digital counter then starts when the gate is switched on and continues until the gate is switched off, with the count number being held (whether it is displayed or not) until another count is needed. If the gating is placed between the crystal oscillator and the counter, then the resolution can be to one oscillation, so that the frequency of the oscillator determines how precise the

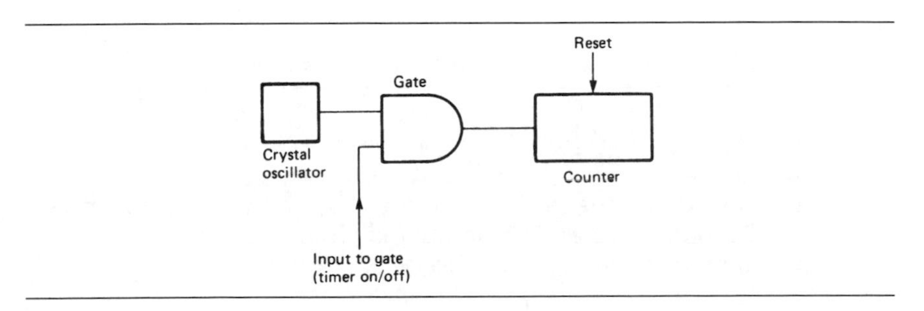

Figure 7.1 The basis of an electronic timer system, using an oscillator, gate and counter/display unit.

timing will be. A frequency of around 4 MHz is commonly used for crystal oscillators, allowing a resolution of around 250 ns. If the gating and counting circuits can use high-speed ECL ICs, then much higher master oscillator frequencies can be used, up to the GHz region.

7.3 Moisture

The presence of moisture in gases or solids often needs to be sensed or measured. A high moisture content in a gas (high humidity) will cause condensation when the gas is cooled, and this can have the effect of depositing liquid in pipes, causing blockages. Conversely, a gas whose humidity is very low can absorb moisture from joints in a pipe, causing the joints to dry up and leak. The moisture content in solids may be desirable, as in the preservation of antique furniture or the working of textiles, or undesirable when mould starts to appear on brickwork. All of this assumes that the moisture is water, but in some specialized applications the presence of other liquids may need to be sensed, and such applications require equipment that is much more specialized.

The presence of moisture (water) in a gas is termed *humidity*, and the *absolute humidity* of a gas (usually air) is the mass of water per unit mass of gas. This absolute humidity is the quantity that needs to be known if the amount of water that can be condensed from a gas has to be determined. The amount of water that can be contained in a gas is limited, and the maximum humidity attainable is called the saturation humidity. This figure depends heavily on the temperature, being very low at low gas temperatures (around 0°C) and very high at temperatures approaching the boiling point of water (100°C).

For many purposes, the *relative humidity* is a more important value than the absolute humidity. The relative humidity at any temperature is the absolute humidity divided by the value of saturated humidity at that temperature, usually expressed as a percentage (Table 7.1). A relative humidity value of 50% at 20°C, for example, means that the air contains

Table 7.1 The relationship between relative humidity and absolute humidity. The absolute humidity is normally measured as kilograms of water per cubic metre of air.

$$\text{Relative humidity \%} = \frac{\text{actual humidity}}{\text{saturated humidity}} \times 100\%$$

Example: One cubic metre of air is dried out and found to contain 0.006 kg of water. The amount of water in saturated air, found from tables, would have been 0.017 kg.

$$\text{Relative humidity is } \frac{0.006}{0.017} \times 100 = 35.3\%$$

Table 7.2 Relating the water content in kg/m^2 to the quoted figures of pressure of water vapour.

Humidity figure in kg of water per m^3 of air

$$\text{Humidity} = \frac{0.00217p}{T}$$

where p = pressure of water vapour (Pascals) and T = Kelvin temperature.

Example: Pressure of water vapour in saturated air at 30°C is 4240 Pa.

$$\text{Humidity is } \frac{0.00217 \times 4240}{303} = 0.030\,\text{kg} \ (= 30\,\text{g})$$

half of the quantity of water that would be needed to saturate it at this temperature. Calculations of absolute humidity from values of relative humidity are never simple because tables of saturated humidity always quote vapour pressures, and Table 7.2 shows how these values can be used to obtain absolute figures.

These calculations depend on using tables of saturated vapour pressure at different temperatures, and such tables have been in use for several centuries now, particularly for steam engineering. Table 7.3 includes the critical values around the boiling point, and the pressures for superheated steam up to 300°C. Tables of vapour pressure for other liquids are not so easy to come by. A few are listed in Table 7.4.

Another measure of relative humidity is the *dew-point*. When a surface is cooled in contact with a gas, a temperature will eventually be reached when the gas deposits water onto the surface by condensation. This surface temperature is called the dew-point temperature, and its significance is that it corresponds to the temperature at which the gas would be saturated. If the dew-point temperature is known, then the vapour-pressure of the water in the gas can be found from tables, and either the relative or the absolute humidity calculated, as illustrated in Table 7.5.

Electronic methods of measuring humidity are based either on dew-point or on the behaviour of moist materials. Of these, methods that use moist materials are by far the simplest to use, although they need to be calibrated at intervals if they are used for measurement as opposed to sensing purposes. The simplest method is a very old one, using the principle of the *hair hygrometer*. A human hair, washed in ether to remove all traces of oil or grease, is very strongly affected by humidity, and its length will shrink in dry conditions and expand in moist conditions.

A very satisfactory relative humidity sensor can therefore be constructed by using a slightly tensioned hair along with any of the standard methods for detecting length changes, such as the linear potentiometer, LVDT, capacitive (with oscillator) and so on. As usual, the use of a capacitive

Table 7.3 The saturated vapour pressure for water. The temperature intervals have been reduced for critical regions.

Temp. (°C)	Pressure (kPa)	Temp. (°C)	Pressure (kPa)
−20	0.10	30	4.24
−10	0.26	40	7.38
−8	0.31	50	12.33
−6	0.37	60	19.92
−4	0.44	70	31.16
−2	0.52	80	47.36
0	0.61	90	70.11
1	0.66	95	84.53
2	0.71	96	87.67
3	0.76	97	90.94
4	0.81	98	94.3
5	0.87	98.5	96.00
6	0.93	99.0	97.75
7	1.00	99.2	98.45
8	1.07	99.4	99.16
9	1.15	99.6	99.88
10	1.23	99.8	100.60
11	1.31	100.0	101.32
12	1.40	100.2	102.04
13	1.50	100.4	102.78
14	1.60	100.6	103.52
15	1.71	100.8	104.26
16	1.82	101	105.00
17	1.94	102	108.78
18	2.06	103	112.67
19	2.20	104	116.67
20	2.34	105	120.8
21	2.49	110	143.2
22	2.64	120	198.5
23	2.81	130	270.1
24	2.98	140	361.4
25	3.17	150	476.0
26	3.36	170	792.0
27	3.56	200	1555
28	3.78	250	3978
29	4.00	300	8592

Note: The boiling point of a liquid is the temperature at which the saturated vapour pressure of the liquid equals the atmospheric pressure, so that boiling point temperatures fall when the atmospheric pressure is low, and are raised at higher pressure.

gauge allows the reading to be made in terms of change of frequency and makes it easy to treat the information by digital methods. The amount of force on the hair must be very small, so that whatever method of sensing length changes is used must not place an excessive load on the hair. The

Table 7.4 Saturated vapour pressures (in kPa) for benzene, ethanol, ether and mercury. The mercury vapour pressures were once important for vacuum pump engineering, but lower pressures are obtainable using silicone oils.

Temp. (°C)	Benzene	Ethanol	Ether	Mercury
0	3.47	1.67	24.7	0.0000247
20	10.0	5.88	58.7	0.000160
40	24.3	17.8	123	0.000765
60	51.9	46.8	232	0.00328
80	100	108	400	0.0118
100	179	225	654	0.0374

Table 7.5 How the relative humidity is found from dew-point readings. The method depends on having tables of saturated vapour pressure.

Temperature of air = 20°C
Dew-point = 9°C

The air is therefore saturated at 9°C. The pressure of water vapour is therefore 1150 Pa (from tables of saturated vapour pressures). The saturated vapour pressure at 20°C is 2340 Pa (from tables).

$$\text{Relative humidity} = \frac{1150}{2340} \times 100 = 49\%$$

system can be calibrated either against another indicating hygrometer, or against a standard chemical humidity measurement (Figure 7.2).

Although the hair type of hygrometer is capable of surprisingly good results, the lithium chloride type is more directly suitable for electronic purposes. Lithium chloride has a very high resistance in its dry state, but the resistance drops considerably in the presence of water, and over a reasonable range the resistance reading can be used to measure relative humidity. The lithium chloride cell is made part of a measuring bridge, and the output is used to drive a meter movement, trigger a switch action at some critical value, or is converted to digital form for display.

Other effects can be used to determine the moisture content in absolute terms. The presence of moisture in air alters its value of permittivity, so that the capacitance between two fixed metal plates in air will change very slightly as the moisture content of the air changes. This, as usual, can be sensed in the form of a change of frequency of an oscillator. The readings have to be corrected for changes of temperature, since the dimensions of the plates will change as the temperature around them changes. Another effect that can be used is the heat conductivity of the air that will alter as the moisture content is altered.

The most useful effect, particularly for remote measurement, is

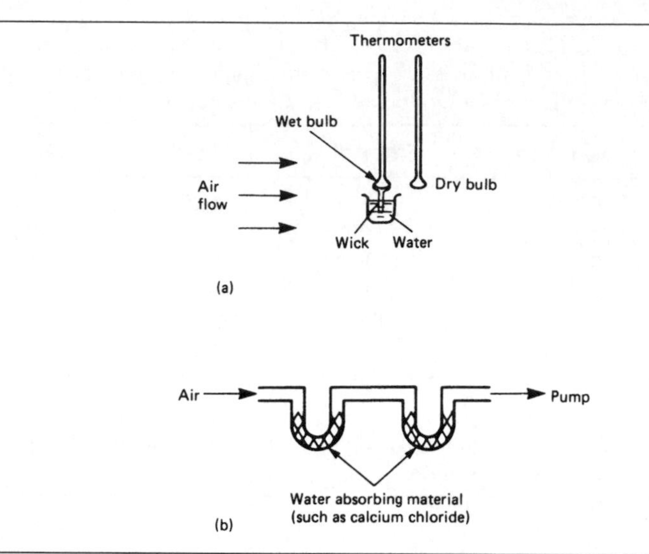

Figure 7.2 Standard methods for humidity. (a) The wet-and-dry thermometer method is not very precise, and depends on the use of tables. Its merit is simplicity. (b) The chemical absorption method draws a metered amount of air past a desiccating material that is weighed before and after the process. The weight difference is the weight of water present in the air. This latter method is precise but slow and difficult to set up.

microwave measurement. Microwaves at a frequency of 2.45 GHz are absorbed strongly by water (the principle of microwave cooking). If microwave signals at this frequency are beamed across air and returned from a fixed reflector, the amount of returned signal will depend very considerably on the water content in the intervening air. This method has the considerable advantage that it senses the average water content along a path in air, which can be quite large. The system needs to be calibrated, but once calibrated, the amplitude of the returning signal can be used as a measure of humidity (the greater the returned signal, the lower the humidity). Dew-point methods can provide fairly precise measurements of either absolute or relative humidity, but require the addition of a microprocessor system if direct readings are needed.

An outline of such a system is illustrated in Figure 7.3. This uses a Peltier junction as a cooling element, using the well-known principle that a pair of junctions between metals, which provides an EMF for an applied temperature difference will also operate in reverse, providing a temperature difference, when connected to a supply. When semiconductor junctions are used, the change in temperature when current is passed can be large enough to allow one junction to attain temperatures lower than 0°C while the other junction is held at room temperature. Since the effect is current-controlled, the rate of fall of temperature can be controlled electronically.

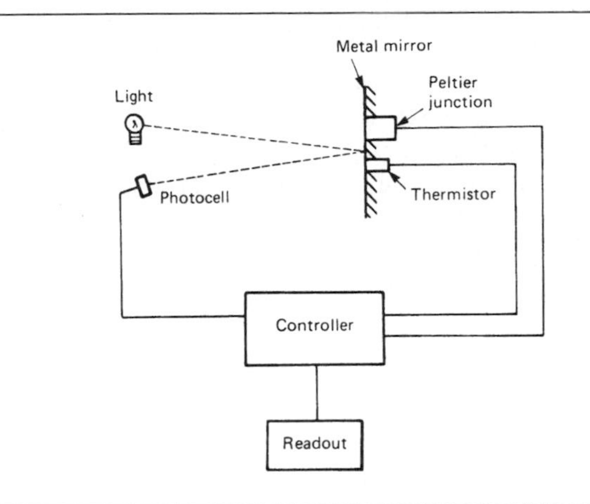

Figure 7.3 An outline of an electronic relative humidity detector system using a Peltier junction for cooling, a thermistor to detect temperature, and a photocell to detect the formation of dew. It is easier to use two thermistors in a-wet-and-dry system, but the results are much less reliable.

The Peltier junction is therefore used to lower the temperature, and a thermistor measures this temperature. A beam of light is reflected from the mirror-surface of the metal attached to the junction, and a photocell senses the reflected light. At the dew-point, the mist on the surface causes the amount of reflected light to drop sharply. The amplifier connected to the photocell operates the reading gate of the system, so that the output from the thermistor temperature gauge is converted to digital form and stored. The microprocessor system also stores the output from another thermistor gauge which senses room temperature. Its ROM contains a set of table values for the temperature readings and for relative or absolute humidity readings (the absolute humidity can be found only if the volume of the space is known). The method of obtaining relative humidity has been outlined in Table 7.5.

As an example of a room humidistat, the RS type 331–118 uses a hair element which permits a setting range from 30–90% relative humidity (RH), with a switching differential of 1.5% based on 45% RH. The calibration holds good up to an air temperature of 40°C and maximum air speed of 1.5 m/s. The switch is rated at 480 V, 10 A AC or 250 V, 0.5 A DC. A more specialized unit for air-conditioned spaces uses a probe to measure both temperature and humidity, and the sensing principle for humidity is the capacitor type. The humidity range is 10–90% RH, with a tolerance of ±5%, and the unit is available in four versions, with low-voltage supplies of 12–30 V required. A more specialized unit for high-pressure gases uses a remote probe for both temperature and RH readings at pressures up to 10 MPa (1450 psi), with an RH range of 0–100%.

MOISTURE IN SOLIDS

Moisture in solids can be sensed in terms either of the conductivity of the materials, changes in permittivity, or of the absorption of microwaves. For sensing the presence of moisture in materials of comparatively fixed composition (even where some variation occurs, as in masonry), a simple resistance reading between connectors set at a fixed distance is often all that is required. For measurement purposes, readings have to be calibrated, and calibration is a long and tedious procedure that requires samples with various moisture levels to be checked for resistance, then weighed, baked to remove moisture and weighed again. The difference in weights shows the moisture content which can be expressed as a percentage of the original weight of material.

A very common requirement is for moisture content of soil. This is very seldom required to be precise, which is just as well, because the relationship between the resistance of the soil and its moisture content is not a simple one. A simple resistance indication, however, is enough to tell whether a plant needs watering or not, which is the main reason for using moisture indicators. For civil engineering purposes, the use of a soil-resistance moisture meter is only a preliminary indicator, and a complete soil sample analysis would be needed before any decisions on the suitability of soil for foundations were made.

7.4 Acidity/alkalinity

The acidity or alkalinity of water is an important factor for water suppliers, and also for all users of water such as chemical plants, generating stations, agriculture and horticulture. The acidity or alkalinity of water is measured on the pH scale, on which perfectly neutral water has a pH value of 7, fairly strong acid solutions have a pH of 2 or lower and fairly strong alkali solutions have a pH of 12 or more. The basis of this scale is the relative amount of free hydrogen ions in the water, and is outlined in Table 7.6.

The sensing of pH can be fairly simple if all that is needed is an indication of a change in the ionization. Perfectly neutral water, with a pH of 7.00, has a very high resistivity value, but any trace of ionization will cause the resistivity to drop very sharply. Some care is needed in the measurement of resistivity, however, because most metals dissolve to some extent in water, creating ions and causing a resistivity change. Metals such as platinum or palladium are best suited to this type of use, but this type of indication does not show whether the conductivity is due to hydrogen ions (acidity), hydroxyl ions (alkalinity) or metallic ions (contamination).

The standard electrical method for reading pH depends on the glass-electrode system, illustrated in Figure 7.4. The glass bulb is very thin and contains a mildly acidic solution which is a good conductor. A platinum

Table 7.6 The basis of the pH scale for acidity and alkalinity of water solutions.

Pure water	Two hydrogen atoms are attached to each oxygen atom: 1 in 10^7 is dissociated to ions H^+ and OH^-. This corresponds to pH = 7
Acid	Hydrochloric acid is almost completely dissociated into H^+ and Cl^- ions. If the H^+ is present to the extent of 1 in 10^3, this corresponds to pH = 3
Alkali	Sodium hydroxide dissociates into Na^+ and OH^- ions. The OH^- ions suppress the dissociation of water molecules, so that 1 in 10^{12} of water molecules might be dissociated. This corresponds to pH = 12.

The definition of pH is $-\log_{10}$ (hydrogen ion concentration)

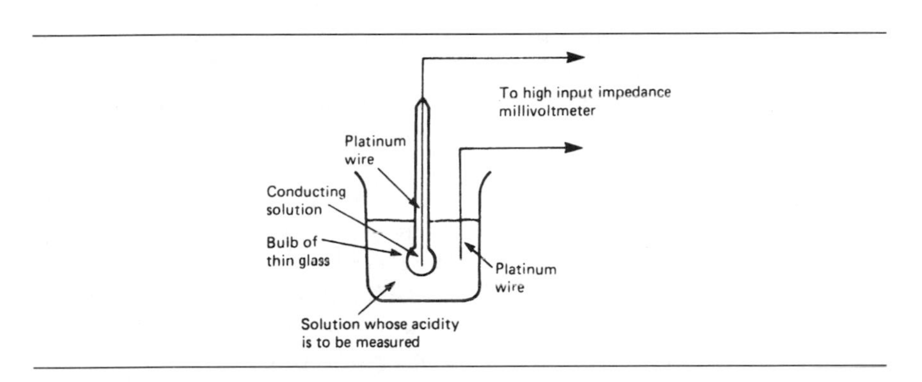

Figure 7.4 The glass electrode method of measuring pH. The output is a voltage which is proportional to pH, but the very high impedance of the glass requires the use of an FET (or valve) input stage for the millivoltmeter.

contact is inside the bulb, and another platinum contact is immersed in the water whose pH is being measured. The EMF that exists between the glass electrode and the external platinum wire is then a measure of the pH value in the water. The resistance of this cell is high, of the order of 100 MΩ, so that high input-impedance MOS DC amplifiers are needed to read the few mV of output. The glass-electrode pH measuring system is fragile, and the readings are easily upset if the glass is allowed to dry out or become stained, but of the available methods it is the most reliable for any kind of use outside laboratory conditions. Calibration is easily carried out, because solutions of standard and constant pH value (buffered solutions) are easily obtained.

7.5 Wind-chill

The effect of wind on heat-loss has been known for a long time, but wind-

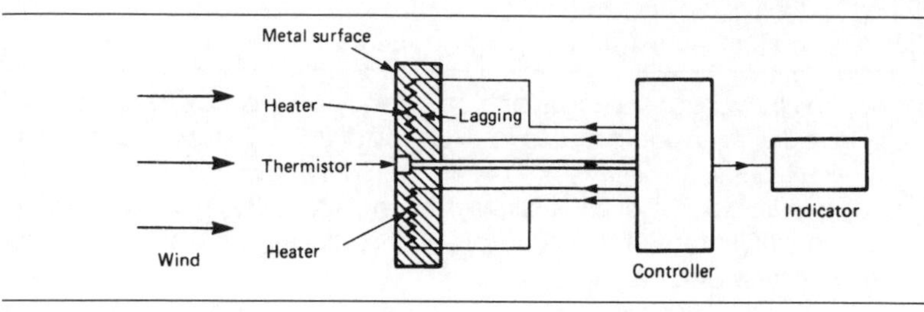

Figure 7.5 Principle of a wind-chill indicator system.

chill factors have only recently been quoted in weather forecasts to the general public. The principle is that an object located in still air will lose heat comparatively slowly, because the air itself acts as a thermal insulator. Even for a comparatively large temperature difference, then, the rate of loss of heat can be low. When the air is moving, however, its cooling effect is much greater. The layer of air that is in contact with the warm surface is constantly being removed and renewed, taking its heat with it, so that the effect of moving the air is the same as the effect of being immersed in much colder, still air. The wind-chill temperature expresses the temperature of still air that will provide the same rate of cooling as the moving air at a higher temperature. The factor is often misunderstood – if the air temperature is 8°C and the wind-chill temperature is 2°C, then the temperature of an object in the air will drop only to 8°C, but the rate at which the temperature drops will be faster, as if the heat were being lost to a 2°C air temperature.

Wind-chill is easily amenable to electronic measurement, and Figure 7.5 shows one system. The thermistor measures the temperature of the sensor, which is maintained at a constant temperature by a heating element whose current is measured. In still air, the amount of current required to keep the sensor at a constant temperature will be low, because the loss is comparatively low. In moving air, considerably more current is required to maintain the temperature constant, and this change of current is a measure of the wind-chill factor. A more practical arrangement uses one sensor in the moving air and one kept in still air at the same air temperature, comparing the current readings so as to indicate the wind-chill. Calibration is needed, and this can be done by taking readings for several values of temperature difference between the sensors.

7.6 Radioactive count rate

An environmental factor that has sprung into prominence in the last few decades is the radioactive count level, and this figure has been used in such a way as to prove that a little learning can be a remarkably dangerous

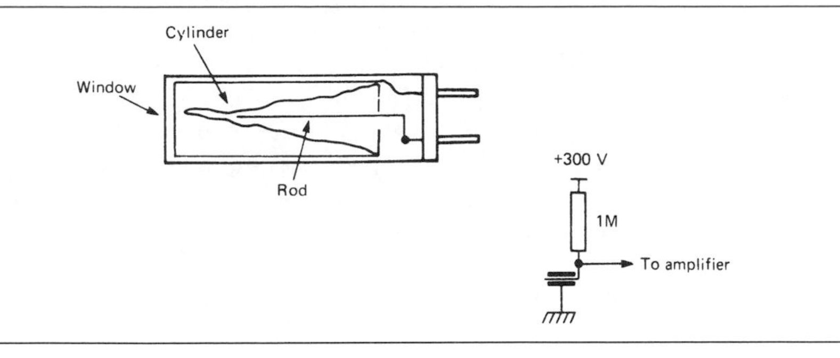

Figure 7.6 The construction of one form of Geiger-Muller tube and the electrical circuit. The presence of an ionizing particle or ray in the tube ionizes the gas and causes conduction until the ions are absorbed. The output is one brief pulse for each ionizing event.

thing. The quantity that is measured in these readings is the number of ions produced per second by all types of radiation and by particles from radioactive materials.

Since the whole Earth consists of materials that are to some extent radioactive, and is constantly bombarded with radiation from the Sun and other stars, there is no place that is free of radiation. In addition, the normal level of radioactivity varies very considerably from one place to another, and is particularly high where there are old granite rocks, or in the presence of deep-mined materials. One very revealing test is to measure the radiation count downwind of a coal-fired power station or on moors of granite rock. Most detectors can be driven completely off-scale by the radiation from any pre-war luminous watch, because these used radium for the luminescence.

The basis of many counters of ionizing radiation or particles is the Geiger-Muller tube (see Figure 7.6). This consists of a tube that contains a mixture of gases (usually the inert gases of the air, such as krypton) at a low pressure, about 1/80 of atmospheric pressure. The electrodes in the tube are maintained with a voltage difference of about 400 V, and in the presence of radiation or ionizing particles, the gas in the tube is itself ionized, allowing a brief pulse of current to pass. The current is brief because the gas mixture contains traces of bromine or iodine that have the effect of neutralizing the ions rapidly so that the gas does not continue to conduct after the cause of the ionization has passed.

The output from the tube is taken across a load resistor and, when the particles do not arrive in great profusion, consists of a pulse for each particle. The pulses can be amplified and used to operate a counter, a ratemeter, or a loudspeaker to give the clicking noise beloved of films on radioactivity. The count rate is obtained by passing the pulses into an integrating circuit whose DC voltage is then proportional to the rate at which the pulses arrive. The GM counter, however, detects only the ionization caused by radiation or particles – it does not indicate the cause of the ionization, nor is it equally sensitive to all causes of ionization.

Natural sources of radioactivity can produce ionization from three causes. The first cause is the alpha particle, which is an ionized nucleus of the gas helium. This has a comparatively large mass and very strong ionizing effect (and a correspondingly large effect on living cells), but is absorbed strongly in all materials, including air, so that its range from its source is usually only a fraction of a millimetre. Only GM tubes with a very thin entry window (usually mica) can detect alpha particles, and only when the window end of the tube is held against the source of the particles.

The second type of particle is the beta particle, which is the familiar electron. This particle ionizes materials quite efficiently but is very much smaller than the alpha particle so that its effects on living cells are much less. The range in air can be several cm, but a sheet of paper is enough to stop electrons from all but a very active source. Electrons are very efficiently detected by the GM counter.

The third type of radiation is the short-wavelength ray, such as the gamma rays from radioactive materials and the cosmic rays from outer space. These are only weakly ionizing, but are immensely penetrating and require several feet of concrete or several inches of lead for screening. Because these rays have a rather weak ionizing effect, they are not efficiently detected by the GM counter. These rays, however, are the most dangerous to life.

As well as these natural sources of radiation from the Earth itself and from outer space, the use of natural radioactive materials in concentrated form (as in nuclear power sources) gives rise to other particles. The main particle of concern is the neutron, which has virtually no ionizing effect, since it is electrically uncharged, and which therefore does not affect the GM counter unless the bombardment is dense. The alternative is to use a GM tube that contains a vapour that will absorb neutrons and give out electrons. Neutrons are very penetrating and harmful, but difficult to detect. Shielding can make use of water, paraffin wax or other low-density materials.

The main alternative to the GM type of counter is the scintillation counter. The basis of this is a crystal that will give off a faint flash of light when affected by an ionizing radiation. The crystal is kept in a dark space and the faint light is detected by a photomultiplier (see Chapter 3) so that the pulses from the photomultiplier can be counted. Several types of sensitive crystals can be used, so that one type of radiation can be detected rather than another. This makes it possible to estimate the relative contribution that different types of radiation make to the whole.

One of the main problems of using detectors like the GM counter is to determine what amount of the reading is due to the presence of an unwanted radioactive material. The continual bombardment of ionizing particles and radiation from the Earth and from space produces a background count, and this count has to be determined, often over a long period, and subtracted from the count obtained in the presence of suspected radioactive material. Low-level radioactive material, such as laboratory glassware and clothing used in the radiological industries, have a count level so close to the natural background level that it is difficult to establish the level of contamination. Given the choice of having a nuclear dump or a heap of manure next to your house, it's better to plump for the nuclear waste!

7.7 Surveying and security

The application of electronic methods to security systems has considerably improved the chances of detecting and avoiding the hazards of intrusion, fire, smoke and theft. Earlier methods, such as the use of microswitches on doors and windows, and conductive films to detect broken windows, are still applicable, but this chapter deals mainly with the sensors and transducers in more recent electronic systems. For details of security systems, consult the books by Gerard Honey (Newnes).

CAPACITANCE DETECTORS

Capacitance detectors are proximity detectors that operate when an earthed object is taken close to the detector. The detecting surface is a conductor, so that this form of detection is particularly useful for metal objects like safes. Non-metallic objects can be protected using metal strips (preferably not visible from outside). A single cable can be used to link the detecting metal object back to the sensing electronics portion of the alarm, but unless this cable is shielded it will form part of the sensing system and may cause false alarms.

The associated electronics panel is typically an oscillator that uses the capacitance to earth of the sending panel as part of the tuning capacitor in an oscillator circuit. When the capacitance changes due to the presence of an intruder close to the sensing panel, the frequency of oscillation will also change and a phase-sensitive detector can be used to make this change of frequency cause a change of output voltage that will trigger a thyristor, so sounding an alarm.

PIEZOELECTRIC SENSORS

The piezoelectric effect has been noted in Chapter 5. Piezoelectric crystals are very efficient detectors of vibration, particularly at ultrasonic frequencies, so that these detectors are very useful in sensing breaking glass, for which a peak sensitivity at 150 kHz is usually specified. The sensor, a piezo-

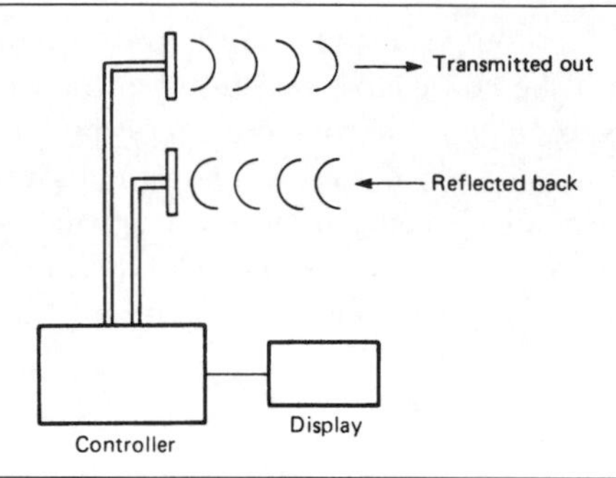

Figure 7.7 The sonar distance gauge. Pulses are generated in the controller and used to send waves from the transmitter transducer. The received echo signals are amplified, and the time delay between these and the transmitted pulses are used to compute distance. When two transducers are used, as illustrated, the signals can be continuous rather than pulsed, and phase change is measured rather than time. Pulsed systems can use a single transducer, switched between transmitting and receiving.

electric crystal microphone, need not be in contact with the window or even close to the window, and several windows in a room can be monitored by one detector.

The same type of device can also be used in vibration detectors to sense when a door or other surface is being attacked. In this role, however, false triggering can be a problem because vibration can be caused by traffic or aircraft noise.

Piezoelectric devices are also used in ultrasonic detection systems. In such systems, one transducer is used to radiate ultrasonic energy, and another to detect it at a distance. The reflection of the radiated acoustic waves forms a standing-wave pattern, and any movement in this area will cause phase shifts and amplitude changes that can be detected by the receiving transducer. False triggering of ultrasonic systems can be caused by air currents or natural noises that contain an ultrasonic component.

The sonar distance gauge principle is illustrated in Figure 7.7. A beam of ultrasound, usually in the 40 kHz frequency range, is sent out from a transducer, usually as short pulses. At this frequency, the wavelength in air of the ultrasound is around 7.5 mm, so that the resolution of the system is of this order. The reflected beam is picked up, in this example by another transducer, and the time interval between sending and receiving is converted into a reading of distance, using either analogue or digital methods. The same transducer can be used both for transmitting and for receiving if the time interval between pulses is large enough to allow for the maximum delay of the beam in the largest space that will have to be measured.

The system requires some care if it is to be used correctly, because the

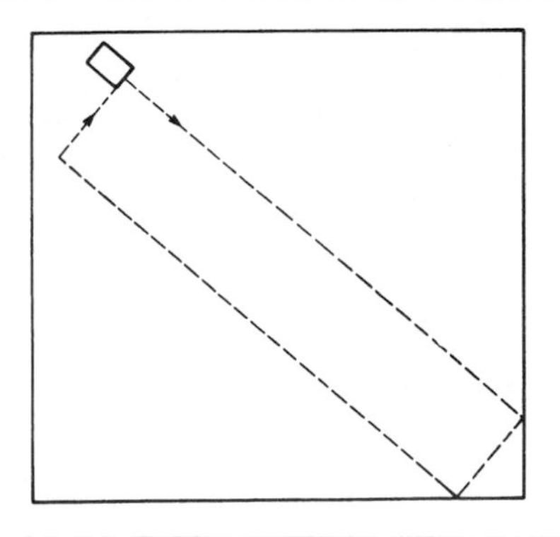

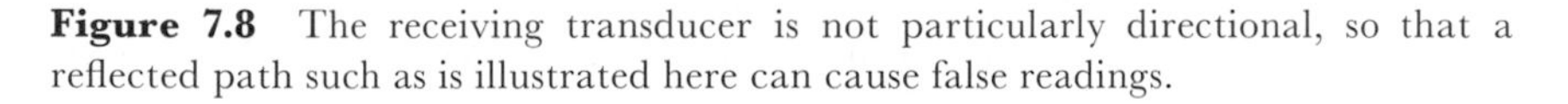

Figure 7.8 The receiving transducer is not particularly directional, so that a reflected path such as is illustrated here can cause false readings.

beam is invisible, making it impossible to be sure where the main reflecting point is located. False readings can be obtained if the beam is angled in a room, so that the distance that is measured includes more than one reflection (Figure 7.8).

For security systems, the ultrasonic beam is usually continuous, with a separate receiver transducer. Inside a room or other closed space, a complex pattern of standing waves is set up and the signal from the receiver transducer is compared with a phase-shifted signal from the transmitter, and the amount of phase shift and amplitude adjusted until the signals are of equal amplitude and phase. If these signals are then applied to a phase-sensitive comparator, then any change in the phase of the received signal will cause an output from the comparator that can be used to trigger an alarm. Such a phase change can be caused by an intruder, but the sensitivity of the system is often such that the alarm can be triggered by cats, mice or pieces of paper moving in a breeze. The more elaborate systems, therefore, use amplitude discrimination as well as phase shift to trigger the alarm.

TEMPERATURE SENSORS

Temperature sensors can be of either the bimetallic or thermistor (see Chapter 4) types. Bimetallic sensors are cheap and robust, and less liable to false triggering caused by brief exposure to heat. They can also be wired directly to the alarm device because their terminals are capable of handling larger amounts of current than the semiconductor type of sensor. In addition, the bimetal thermostat has a built-in hysteresis, meaning that once it has triggered it will stay in that state until the temperature is reduced to below the triggering temperature.

Thermistor sensors have the advantages that they can be much more sensitive, and can be microprocessor controlled so as to provide any desired hysteresis. This allows the response to short-term high temperature to be adjusted so as to eliminate most of the conditions that cause false triggering.

SMOKE SENSORS

Smoke sensors are either of the photoelectric or the ionization type, and because the characteristics are so different, both types should be used in critical situations. The photoelectric type uses a light source, usually an LED, along with a photocell, and responds to any obstruction of the light beam by smoke. This type of sensing is particularly suitable to detect smouldering (for example, from bed linen that has been in contact with a lighted cigarette) that is likely to lead to a fire, and is not affected by strong air currents.

The ionization type uses a weak radioactive source in a space (the ionization chamber) that contains metal surfaces (electrodes) with a voltage difference. The radiating charged particles ionize the air in this space and so permit a small current to pass between the electrodes. This current is in the picoamp region, so that a DC amplifier is needed to sense it. Smoke particles, which may be so small as to be invisible, and fumes from a wide range of chemicals will cause the ionization current to drop, triggering the alarm. This type of detector is more suitable for detecting a fire that has broken out, before smoke has had time to trigger the photoelectric type. A snag is false triggering caused by humidity changes and vapours that are not a fire hazard. This can be overcome to some extent by suitable design of the ionization chamber.

INFRARED

Infrared sensors can be active or passive. The passive infrared sensors operate by detecting the thermal radiation from a body and the active systems use an infrared source to flood an area so that phase/amplitude changes can be sensed, or infrared cameras used to produce an image. The PIR systems generally use a pyroelectric sensor (see Chapter 4). Precautions are needed against false triggering produced by moving reflecting surfaces (such as pools of water, vibrating metal panels, etc.).

MICROWAVE

Microwave detectors operate by flooding an area, which can be a large area, with microwaves so that standing-wave patterns are established. A detector that is phase sensitive will respond to any disturbance in this pattern caused by any object moving within the range of the unit.

NIGHTFALL DETECTORS

Nightfall detectors use light sensors that switch the lighting system on when the incident light falls to a low level. These provide permanent floodlighting in the hours of darkness, and are useful when allied to CCTV. The position and sensitivity of the sensor has to be chosen so that it will not be affected by the lighting itself, nor by other stray light from surrounding houses, cars, the Moon, or other sources. This is not always easy, and a system that behaves perfectly in clear weather may cause problems in foggy conditions because of light scattering from the small water droplets in fog.

In addition, problems will arise if the sensor becomes obstructed. Because the sensor must not respond to the security floodlighting, it is often aimed at the sky, and is subject to obstruction by frost, rain or leaves. Unless regular cleaning is carried out, the glass cover of the sensor can become obscured by green slime or other algae.

TIMED SYSTEMS

Timed systems of lighting can be systematic or random. A timed system can be set, for example, to switch lights on at 10.00 p.m. and off again at 6.00 a.m., ensuring that the lights are on during hours when the premises are empty. This is another form of permanent lighting, and the regularity of the pattern is a disadvantage for any intruder who has observed the lighting for a few nights.

Random timed switching can be more effective. This switches the lighting on and off at random intervals during the hours of darkness, so presenting an element of surprise. Once again this is at its most effective when used along with CCTV.

SOUND ACTIVATION

Sound activated lighting systems are an effective method of deterring intrusion, making use of acoustic sensors operating some or all of a set of security lights. The sensing level can be chosen to avoid triggering from distant or faint sounds, although false triggering is not such a problem for security lights as it is for alarm systems. Miniature sound activating sensors can be incorporated in bulb-holders for domestic use, and can be combined with random-triggering circuits.

PIR SYSTEMS

PIR lighting is a very popular form of security system, particularly in low-cost form for domestic premises. The sensor may be part of the lighting unit, or a separate component, and is adjustable for aiming, sensitivity, and ambient light sensing. The larger units use one or more separate sensors, switching a number of lights.

STANDALONE SMOKE DETECTORS

See later in this chapter for details of smoke detector principles. Standalone smoke detectors are battery-operated and intended for domestic premises. They are predominantly of the ionization type, and should carry a warning about radiation hazards (the main factor being safe disposal of the radioisotope when the alarm is scrapped). The signalling unit is a piezoelectric tweeter that emits a volume of sound that is out of all proportion to its size and to the battery power. The battery is usually a 9 V alkaline type that provides a life of a year or more. End of battery life is signalled by a beep at a slow rate, typically one each 3 minutes. A test button is provided to check that the sounder is still operative. The whole device is packaged in a casing that is typically a shallow box about 13 cm square.

These alarms should be located in critical areas, particularly sleeping areas and in places where fire can originate such as kitchens or boiler rooms. In a kitchen, false triggering is likely from fan ovens and from grills, so that some care is needed over positioning.

7.8 Animal fat thickness

One of the more specialized uses of sonar distance meters has been in measuring the thickness of the layer of fat under the skin of animals, pigs in particular. The speed of sound in a fatty layer is different from the speed in the more dense meat, so that an ultrasonic beam will be reflected from the place where fat ends and meat starts. By making the usual measurement of time elapsed between sending the ultrasonic pulse and receiving the reflection, the thickness of the fat layer can be measured painlessly.

7.9 Water purity

There is no single instrument that can assess all aspects of water purity, because this would require the assessment of both ionizing and non-ionizing contaminants, bacterial and viral presence, dissolved gases, colour and taste; hardly a task for which any one instrument is suited. One reading that can be a very useful guide, however, is electrical conductivity. This does not need to be carried out on an absolute basis, measuring the conductivity in units of Siemens per metre, but can be comparative, measuring the current flow in an electrometer cell at a fixed voltage, as compared to a sample of demineralized or distilled water under the same conditions. So many common impurities such as nitrates, nitrites and other salts will increase the conductivity of water, that this figure can be used as one which will indicate when a more complete analysis should be undertaken.

Readings of pH are also useful as they can help to establish what type of contaminants are likely to be present, but since perfectly non-conducting water must also be perfectly neutral, the pH reading is of rather less importance compared to conductivity. A useful secondary test is clarity, using an optical gauge to assess the transparency of a long column of water. Many contaminants that do not affect the conductivity of water have a noticeable effect on its transparency.

The most difficult contamination problems, from the point of view of continuous monitoring, are the biological and non-ionizing types. Although there is a possibility of manufacturing biological transducers which will provide an electromotive force in the presence of specific bacteria, we are a very long way from any device that will provide an electrical readout for the whole range of bacteria and virus contaminations that can be found in water. Solvents present another type of problem, because being electrically non-conducting they do not alter the conductivity of water and are very difficult to detect by other than chemical methods.

7.10 Air purity

We once took air-conditioning for granted, but the incidence of Legionnaire's disease has concentrated attention on the problems that can arise if air-conditioning is fitted and forgotten. In addition, there is an energy wastage aspect to consider when air-conditioning is run continuously, and the problem of the noise which, even at a low level, is always present and which can be heard even in some concert halls.

Many buildings that use air-conditioning do not need the full flow to be operating all the time, because the extent of occupancy can be very variable, particularly in cinemas, restaurants, kitchens and other spaces. The answer is an air-quality sensor, using a semiconductor surface whose conductivity will alter when contaminants are present in the air. The types of contamination that can be sensed include perspiration, tobacco smoke, oil and grease vapours (from cooking) and similar noxious fumes. The semiconductor surface is used at an elevated temperature so that there is no tendency for contaminants to condense on the surface.

The calibration of an air-quality sensor is fairly arbitrary, and its positioning is important. Doors, windows, fireplaces, corners and recesses need to be avoided because the air samples will not be typical, and the preferred height for fitting the sensor is 1.5 m above floor level. When the sensor is first installed, its output should not be used, because it will take up to 2 days of operation to stabilize the unit. Once this has been done, the sensitivity has to be set with reference to clean air, so that the output switches when the air becomes noticeably contaminated. Once the sensitivity is set, the sensor can be coupled to the air-conditioning system to trigger the fans when the contamination level is reached.

7.11 Smoke and fire detectors

Smoke detectors for domestic use may become compulsory in the lifetime of this book, so that the methods used are now of considerable importance in domestic situations as well as industrially. In domestic premises, smoke detection is considered more important than the detection of fire by temperature change, because the asphyxiating effect of smoke from foam plastics is the prime cause of death in domestic fires. Industrial fires are more complex, and often involve very fast rates of temperature rise, sometimes with very little smoke.

Smoke detectors operate on two main principles: ionization detectors which will detect the ionized air from a fire even when there is little or no visible smoke, and the optical type, which is designed to detect smoke which is present even when there is little or no rise in temperature. In general, the ionization types are used industrially and the optical types in domestic uses.

The ionization type of detector uses a radioactive source, usually americium-241 ($^{241}Am_{95}$), with a low activity level, typically 0.8 μCi. The arrangement of the ionization chamber is as shown in Figure 7.9. The gap between the source plate and the electrode is normally conducting due to the emission of alpha particles (helium ions) by the americium, which has a half-life of 458 years. The ion current is very small, of the order of 10 pA ($1\ pA = 10^{-12}\ A$), so that any additional insulation leakage would be sub-

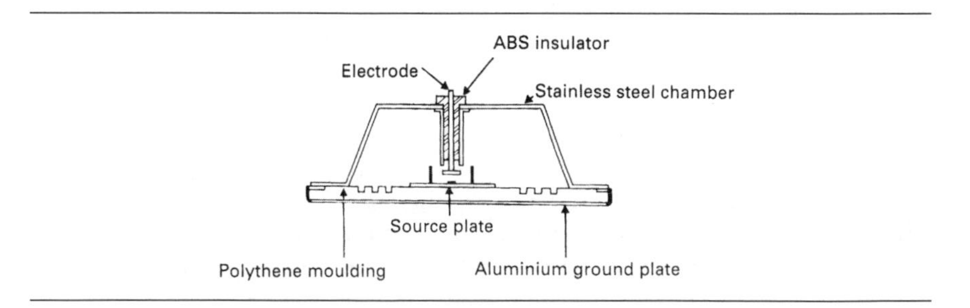

Figure 7.9 The construction of the ionization chamber of a radioisotope type of fire and smoke detector. When such a unit is scrapped, the isotope source should be returned for recycling.

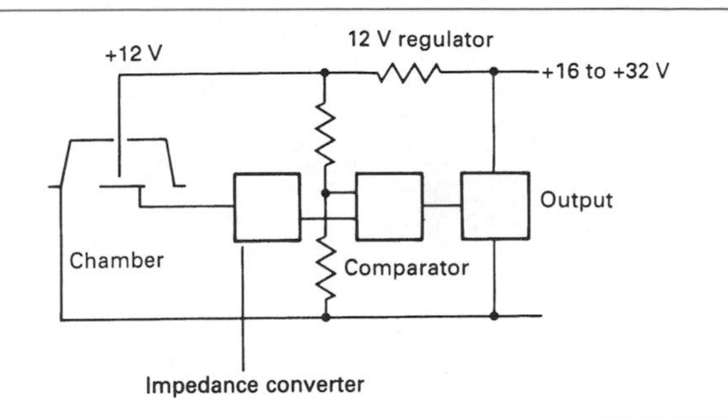

Figure 7.10 Block diagram of the electronic part of a radioisotope fire and smoke detector.

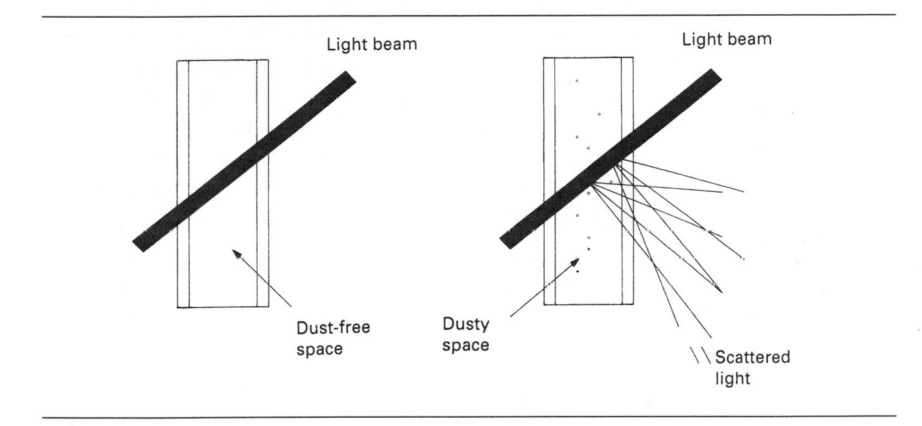

Figure 7.11 The Tyndall effect: a light beam through clean air causes no detectable scattering, but the presence of smoke or dust will result in a noticeable amount of light outside the beam area.

stantial in comparison. The acceptable level of insulation leakage is of the order of 0.5 pA. This implies that the insulators must not be touched, and if the ionization chamber has to be replaced, utmost care must be taken to avoid any contamination of the insulation by, for example, solder flux.

In the presence of smoke from a fire, particles entering the ionization chamber will be struck by the alpha particles, and the alpha particles will cling to the much larger particles of smoke. Because the charged units are now much larger, they cannot travel so quickly in air, and the current is reduced. By detecting the reduction in current, the detector can be made to activate an alarm. A typical circuit arrangement is illustrated in block form in Figure 7.10, using a MOSFET impedance converter (source–follower), comparator and output stage.

The optical type of detector operates on the well-known Tyndall effect of scatter (Figure 7.11). When a beam of light passes through clean air, the

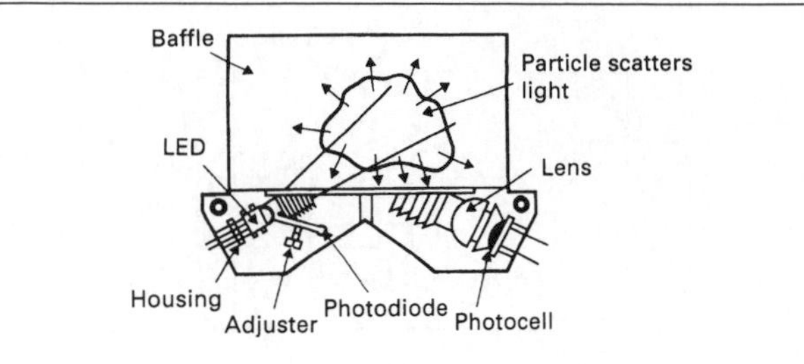

Figure 7.12 The arrangement of an optical smoke detector, using two photodiodes.

beam is invisible. No light can be detected looking along an axis at right angles to the beam. In the presence of smoke, however, light is scattered (randomly reflected) from the tiny particles that make up smoke, so that there is an appreciable amount of light visible on the transverse axis.

A typical optical scattering detector chamber is illustrated in Figure 7.12 with the size of a smoke particle shown in a magnified view. In the absence of smoke, the light-beam from the LED does not illuminate the photocell. There is a balancing photodiode provided so that any light reaching the photocell from internal reflection can be balanced out – this should be almost negligible because the interior of the detector is coated in matt black. In the presence of smoke, the scattered light from the smoke particles will cause an increase in the intensity of light reaching the main photocell with very little change at the compensating photodiode.

A typical circuit is illustrated in Figure 7.13, in block diagram form. The pulsing circuit for the LED is operated from a constant-current source

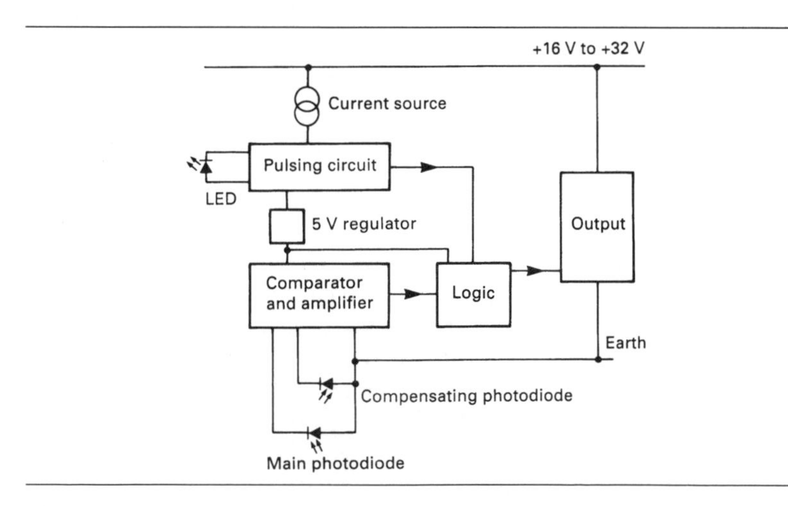

Figure 7.13 Block diagram for the circuit of the photoelectric smoke detector.

using a high peak current to provide a concentrated light beam. The inputs from the main photosensor and the auxiliary photodiode are compared, and the output of the comparator is used to drive the logic circuit which suppresses any output until three successive pulses of light all provide a smoke warning output. If this happens, the output is triggered, operating the alarm.

SERVICING

Both the ionization and the optical types of detector can be housed in similar casings that incorporate an insect screen and cover. Despite the covering, smoke detectors of both types are subject to a build-up of dust and cleaning is required at intervals. The insect screen must be cleaned using methylated spirit if any tarry or sticky deposits have formed. The cleaning procedures for the two types of detector then diverge.

The ionization chamber type of detector must have its central electrode removed, taking care not to damage the connecting lead. This exposes the low-level radioactive source, which must be treated with some care, because even though its intensity is lower than that of some granite rocks, you may be working with several units each day. You should therefore take the simple and sensible precautions that apply to any low-activity radioactive source, although the inhalation of fumes from methylated spirit is more of a hazard.

Radioactivity is dangerous only if you remain close to an intense source for some time, or swallow or breathe in particles of low-level material. The obvious steps are not to touch the radioactive source (which is sealed in any case) and not to place your fingers close to it. The recommendations are that you do not put your fingers closer than 50 mm (2 in) from the source and do not place your eye closer than 150 mm (6 in) from the source. The unit can be cleaned using a brush dipped in methylated spirit, and drained (to avoid leaving drops). Use the same brush to clean the source, source plate, mounting and baffle to remove any traces of dirt.

When the methylated spirit has evaporated, use another clean dry brush to remove any residues. When cleaning of the unit is complete, clean the first brush in fresh methylated spirit and the drying brush in soap and water, and leave both to dry thoroughly before using again. The used methylated spirit should be kept in a clearly labelled bottle until the bottle becomes full, when it should be disposed of in the same way as any other solvent (there will be no measurable radiation from the methylated spirit). The detector components can then be reassembled.

The optical type of detector can be cleaned when the gauze and the baffle have been removed. A soft brush moistened with methylated spirit is used, taking care to brush the interior of the chamber to remove any traces of light-coloured dust which will cause excessive reflection. A dry dusting brush should be used to remove dust from the LED and the photocells, lenses and housings. Brush the interior of the chamber only very lightly, because hard brushing can polish the surface, causing reflections. Following cleaning, the unit is reassembled.

TEMPERATURE TYPE

The fixed temperature type of fire detector is used extensively in industrial surroundings, and consists of a thermistor wired in a bridge circuit. The usual operating temperature is 57°C – a compromise between giving reasonably early warning of a real fire and possible false alarms in very hot weather. The rate of rise system uses two thermistors in a bridge, with one thermistor embedded in a heat sink. Slow changes in the ambient temperature will affect both thermistors equally, so that no alarm is actuated even if there is a substantial change of temperature. In fire conditions, the thermistor that is in free air will heat much more rapidly than the thermistor in the heat sink, unbalancing the bridge and actuating the alarm. The rate of rise type is particularly effective in detecting fires in spaces which are normally held at a low temperature (such as cold rooms), since a normal temperature detector will not register until a fire is well advanced, because the initial temperature is so low.

7.12 Building acoustics

The acoustics of buildings have always been of interest in connection with concert halls and opera houses, theatres and debating chambers, but it has become even more important in the 20th century because of the use of broadcasting studios and public address systems. In addition, the sound quality obtainable from recording media has considerably improved. For example, the 50–60 dB range of the old vinyl disc has been superseded by the 90–100 dB range of the compact disc, although later developments such as Minidisc™ and MP3 have not necessarily been forward steps in range or other aspects of sound quality.

This has made the acoustics of studios much more important, so that defects that once were tolerated can no longer be allowed. In particular, ambient noise levels need to be reduced. This is a process that is particularly difficult at a time when external noise from traffic (particularly air traffic) has greatly increased. At the same time, there has been progress in improving the acoustics of public buildings, including concert halls and theatres, although there is still a tendency to consider the appearance of the architecture long before considering the acoustics. This has the usual effect that a new concert hall may need several years of painstaking work before its acoustics are acceptable. All of this effort might have been saved

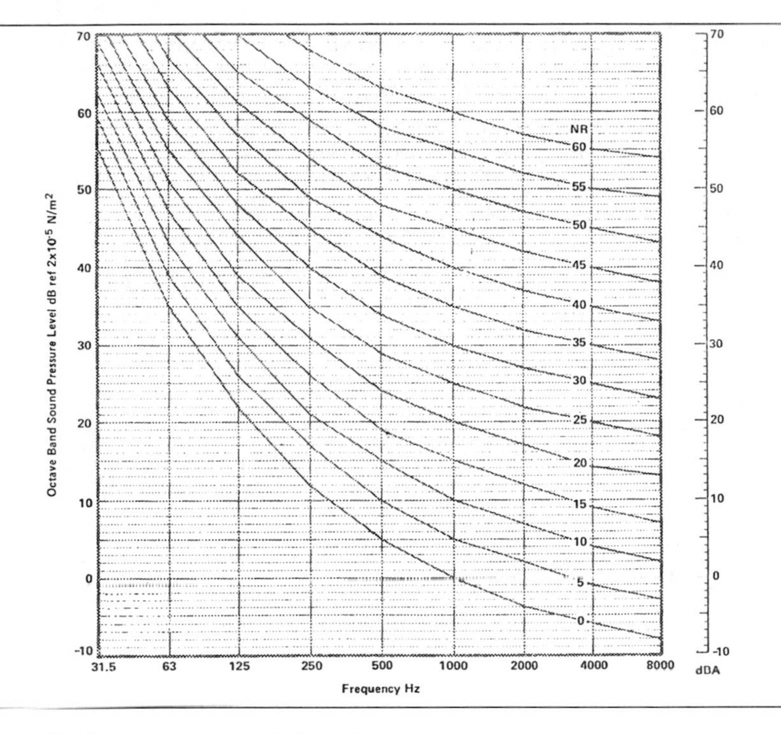

Figure 7.14 A typical set of NR rating curves.

if the architect had been aware that fashionable design does not necessarily lead to good acoustics.

Consider one important factor, background noise. At one time a background noise measurement would be used to produce a single figure, but this is no longer sufficient. A background noise assessment now is derived from a set of measurements that take account of the spectrum (the frequency spread) of the noise. This prevents noise at frequencies too low to hear from affecting the noise rating. The systems used for rating noise are NC, used mainly for air-conditioning noise, and NR used, particularly in Europe, for other forms of background noise. Figure 7.14 illustrates a set of NR curves.

REVERBERATION TIME

The reverberation time of an enclosed space is an important factor in determining the quality of sound, and is particularly important for the intelligibility of speech in buildings. The highly reverberant 'cathedral sound' is appropriate for some types of music, but disastrous if you want to announce that the train on Platform 7 will be delayed for a couple of days.

Cathedrals and similar large spaces can have reverberation times of 4–8 s. Studios designed for orchestral or music recording work need reverberation

Table 7.7 Quantities used to determine % Alcons intelligibility.

D	Distance from sound source to most distant listener
RT	Reverberation time in seconds
V	Volume of enclosed space in cubic metres
e	Directivity ration of sound
M	Acoustic modifier, usually unity

times in the range 1.2–2 s, but for pop sounds a lower figure, typically 0.4–0.7 s is used. Where speech intelligibility is important the reverberation time must be less than 1.5 s, and another factor, the percentage loss of consonants (% Alcons), should be measured. This is calculated from measurements of the acoustic quantities D, RT, V, e and M, as defined in Table 7.7.

In the past, reverberation time was estimated by methods such as firing a gun and timing the echo return, but much more precise methods are now used, with the sound source and the timing all electronically controlled. Specialist acoustics firms can supply equipment for any measurements ranging from simple reverberation tests to complete analysis of acoustic characteristics.

Chapter 8

Other sensing methods

8.1 Unusual measurements

Laboratory work, whether comparatively routine in nature or involving new ideas, often requires sensing and measuring systems that have not been covered by any of the examples up to this point. In many cases, this does not imply that completely new sensors have to be developed, because an unusual application can often be catered for by using a combination of existing sensors. Research laboratories, however, generally have to construct their own sensing and measuring systems because the nature of the work implies that nothing is available. In this chapter, we shall consider some types of sensors and measuring methods that are used in laboratory work, in engineering development and in some types of geological work, but which find little or no application outside these uses.

8.2 Permittivity

The *permittivity* of an insulating material determines its effect on the capacitance of a pair of conducting plates that sandwich the material. The usual measurement of permittivity is as *relative permittivity*, symbol ε_r, meaning the permittivity of the materials divided by the permittivity of space (a vacuum). This quantity was formerly known as *dielectric constant*. The permittivity of a vacuum is close enough to the permittivity of air to allow this amount to be used except for the most precise measurements. The measurement can be for the purpose of determining how effective a new material (usually a plastic or ceramic material) would be when used as a dielectric for capacitors, but it can also be used in assisting to determine the structure of a material. Table 8.1 shows permittivity values for some common materials.

Table 8.1 Relative permittivity values for various selected materials. The values of relative permittivity are seldom large.

Material	Value of ε_r
Vacuum	1.000000
Air	1.00004
Aluminium oxide	8.8
Araldite resin	3.7
Bakelite	4.6
Barium titanate	600–1200
Magnesium silicate	5.6
Nylon	3.1
Polystyrene	2.5
Polythene	2.3
PTFE (Teflon)	2.1
Porcelain	5.0
Quartz	3.8
Titanium dioxide	100

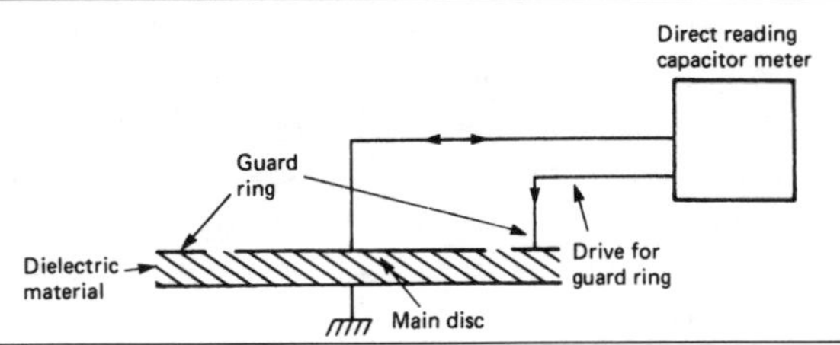

Figure 8.1 Measurement of relative permittivity. The system uses a circular disc capacitor with a guard ring whose action is to suppress stray capacitance. The voltage on the guard ring is identical to the voltage on the main plate, but no measurements of capacitance are taken from the guard ring.

The outline of a system for measuring relative permittivity is shown in Figure 8.1. The parallel-plate capacitor is constructed with circular plates and a 'guard-ring' to offset the effects of stray capacitance at the edge. The value of capacitance is measured (using a capacitance meter) with, and then without, the dielectric, maintaining the same spacing between the plates. The ratio of capacitance values is taken as the relative permittivity of the material.

The principle of the capacitance meter is shown in Figure 8.2. The capacitor is alternately charged from a constant-voltage source and discharged through a meter or into an integrating counter. The switching is carried out by a MOS array, and for high sensitivity has to be done at a high frequency, e.g. several MHz. The capacitance is obtained as shown in the example from a reading of current, and the whole system can be made direct-reading.

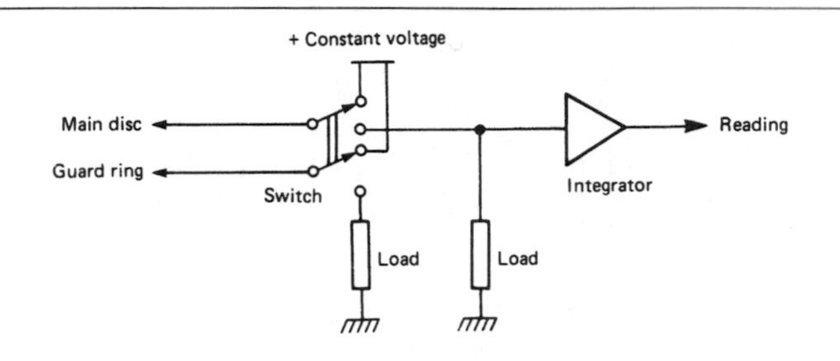

Example: **Switch operates with 50% duty cycle at 1 MHz. For the capacitance this means that the charging and discharging is each completed in 0.5 μs. If the voltage is 10 V and the capacitance is 50 pF then the charge is 500 pC, which is discharged 10^6 times per second. This corresponds to a current average of $10^6 \times 500 \times 10^{-12}$A = 5×10^{-4}A or 0.5 mA, an easily measurable quantity.**

Figure 8.2 The principle of the direct-reading capacitance meter and how the output current is related to the capacitance value. The switching device would normally be a MOSFET bridge, although reed relays (see Chapter 12) can be used for large capacitance values.

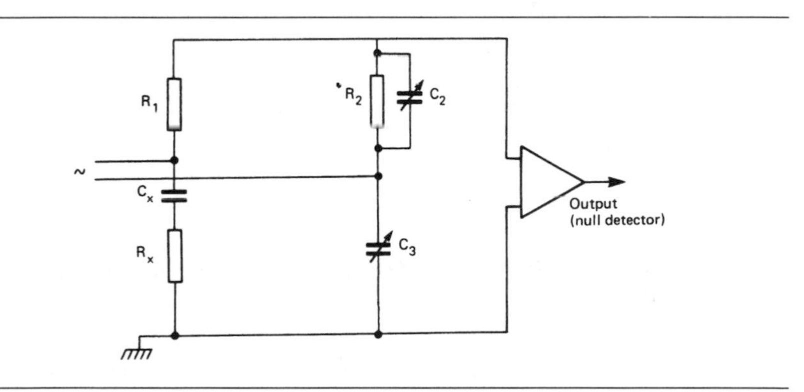

Figure 8.3 The classic Schering bridge method for measuring capacitance and leakage resistance. Modern methods for leakage resistance charge the capacitor and measure the time constant from the rate of fall of voltage.

In addition to the relative permittivity of a capacitor dielectric, its *loss factor* may have to be assessed. This is a very much more difficult measurement unless the loss factor is large, although for some purposes (such as soil sample analysis) large loss factors are normal. The measuring method that is used is a bridge in which the capacitor with the dielectric under test is balanced against a standard low-loss capacitor and a phase-shift resistor, using the circuit of Figure 8.3, the Schering bridge, or a more specialized type.

Electrostatic charge sometimes has to be measured by electronic methods, and the simplest method is to measure the potential at various distances

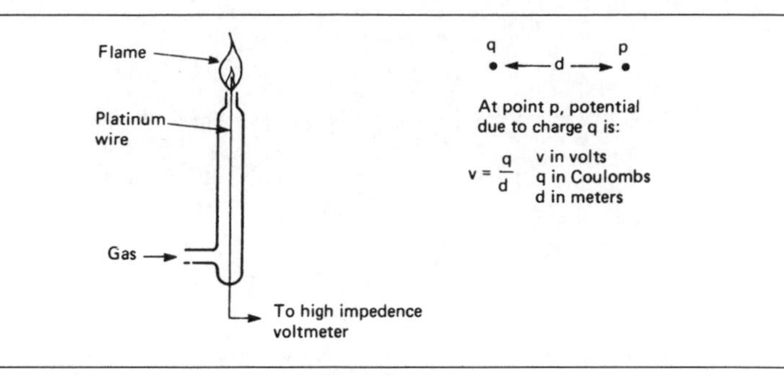

Figure 8.4 The flame electrode method of measuring the potential at any point in the air. Obviously this is unsuitable if the air contains flammable vapours. The relationship between charge, distance and potential is also shown.

from the charge – the relationship between potential and charge is shown in Figure 8.4. The 'connection' between the voltmeter and the air is established by using a flame to ionize the air around the detector. The voltmeter must have a very high input impedance, and a MOS type is normally used to replace the older valve voltmeters that were once common in measuring laboratories. Two determinations of potential at different distances are usually needed to establish the amount of charge, although this method will not be suitable if the charge leaks away rapidly. Note that there is no electrical connection between the charged object and the voltmeter; the potential is due to the field set up by the charge.

8.3 Permeability and magnetic measurements

Permeability is analogous to permittivity, and its effect on the inductance of a coil is illustrated in Figure 8.5. The important difference, however, between these types of quantities is that permeability and relative permeability are not constants for a given material. We can quote a single figure of relative permittivity for a sample of mica, but for a sample of steel the

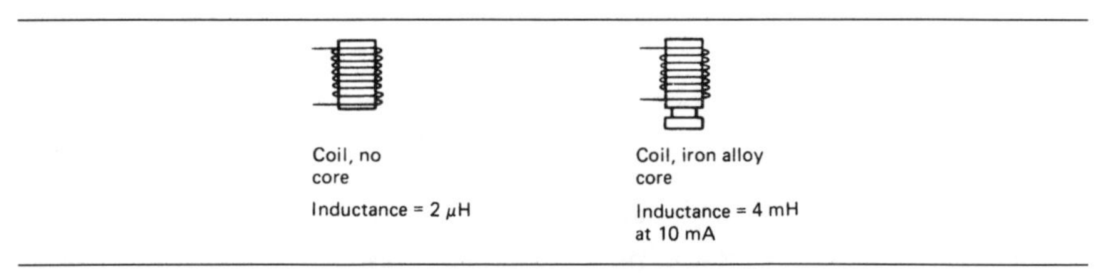

Figure 8.5 The effect of relative permeability on the inductance of two otherwise identical coils. Whereas values of relative permittivity of 2 or above are comparatively rare, most ferromagnetic materials have relative permeability values that can be measured in thousands or tens of thousands.

Table 8.2 Typical maximum relative permeability (μ_r) values for important types of magnetic materials.

Material	Max. μ_r value
Silicon-iron	7000
Cobalt-iron	10 000
Permalloy 45	23 000
Permalloy 65	600 000
Mumetal	100 000
Supermalloy	1 000 000
Typical dust core	10–100
Ferrite core	100–2000

value of relative permeability is not a constant, although the maximum possible value may be fairly constant. Relative permeability values depend on the previous history of the material, in the sense that if the material has been in a magnetic field, then its value of relative permeability will be affected by the fact that it has been magnetized (see Table 8.2).

Therefore, rather than quote a single figure to describe the behaviour of a material in a magnetic field, we have to quote a number of figures, and these are illustrated in Figure 8.6 with reference to the magnetic hysteresis curve of a material. The graph shows the magnetic flux density (B) in the material plotted against the ampere-turns of magnetizing force in a coil sur-

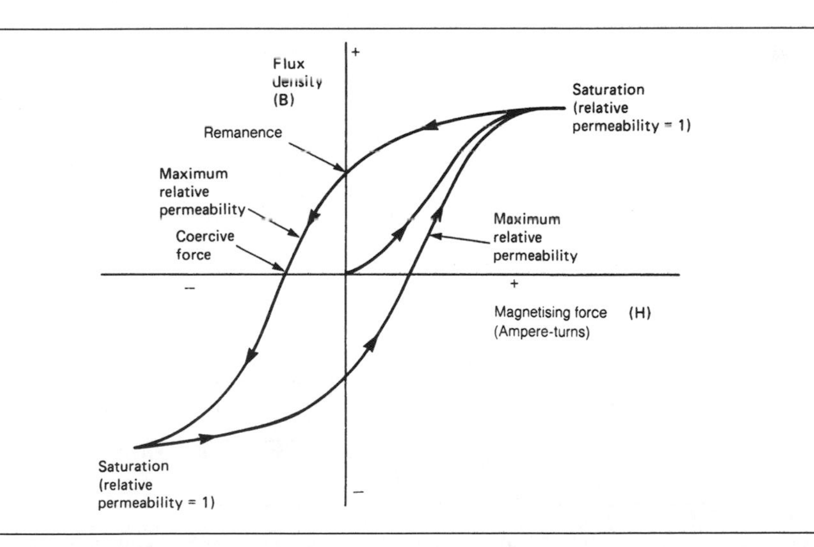

Figure 8.6 The B–H hysteresis curve for a magnetic material. The value of relative permeability is obtained from the slope of the curve, and is unity at saturation. The graph also shows the value of remanence and coercive force.

rounding the material. We assume that the material starts unmagnetized, so that with no magnetizing force applied, there is no flux density. As the material is magnetized, then, the flux density value rises in a non-linear way.

At some value of magnetizing force the flux density reaches a peak and will from that point show very little increase, increasing only at the rate of increase of the magnetizing force itself. The peak value of flux density is called the *saturation flux*, and it determines how effective a material can be when used as the core of an electromagnet. If from this point the current through the magnetizing coil is reduced, the graph does not follow the same path. When the magnetizing force is zero, the flux density in the magnetic material will have some value called the *remanence*, or remanent flux density. A high remanence is a necessary quantity for materials intended as permanent magnets, but is highly undesirable for cores of electromagnets. If the current through the magnetizing coil is now reversed and increased, a point will be reached when the flux density in the sample of material is zero. The amount of magnetizing force needed to achieve this is called the coercive force, and a high value of coercive force means that a material will be difficult to demagnetize, a desirable feature of, for example, material used for recording tape.

The use of Hall-effect sensors allows flux density to be measured even for small samples of material, so that evaluation of these magnetic quantities is much easier now than it was previously. The type of set-up illustrated in Figure 8.7 can be used to display a hysteresis curve on an oscilloscope, and the same technique of cycling the material from one saturation level to the other and back can be used to allow digital presentation. A completely automated system is microprocessor controlled, and can be made auto-ranging to allow for the enormous differences between different magnetic materials.

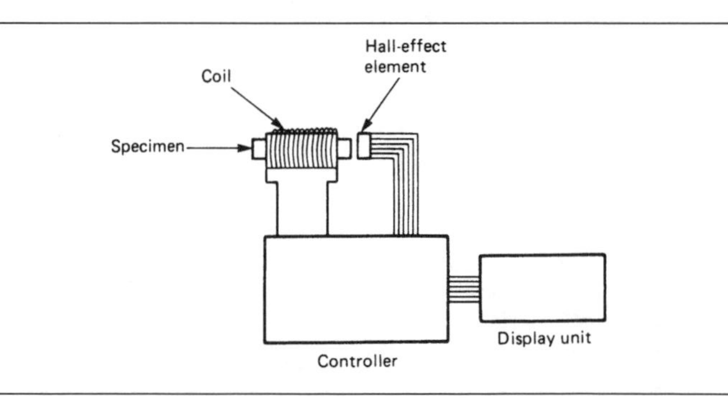

Figure 8.7 Outline of an automatic hysteresis curve tracer. A timebase generates the rising and falling currents through the coil, and this same waveform is applied to the X-plates of the cathode ray tube display. The output of the Hall-effect sensor is amplified and used to provide the Y-plate signal.

8.4 Nuclear magnetic resonance

The use of nuclear magnetic resonance has become increasingly important in the last few years, both as a major method of non-destructive testing and also as a method of medical investigation. The principle, simplified, is that the combination of a very strong magnetic field and a radio frequency signal can cause the nuclei of atoms to absorb and emit radio-frequency signals at certain values of the magnetic field. The importance of the system is that the received signals can be processed by a computer into the form of an image which can show a remarkable amount of detail, so that cracks in a weld or defects in a heart are visible on the image.

The very powerful magnetic field that is required can be provided by conventional electromagnets when the specimens are small, but for purposes such as head and body scanning, a superconducting magnet is required with a maximum field strength of about 0.5 T. For medical purposes, the RF power is about 100–200 W, and the frequency range of signals is 1–100 MHz.

8.5 Gravitational sensing

For mineral surveying, the tiny changes in the Earth's gravitational field caused by dense deposits of minerals can be detected and used to locate the deposits. Gravity meters of remarkable sensitivity have been available for a long time, but in recent years the addition of electronic sensors has improved both the sensitivity and the portability of gravitational meters.

The basic methods are the swinging pendulum and the spring balance. The time per swing of a pendulum depends on the gravitational acceleration, as shown in Figure 8.8, and can be measured very precisely by

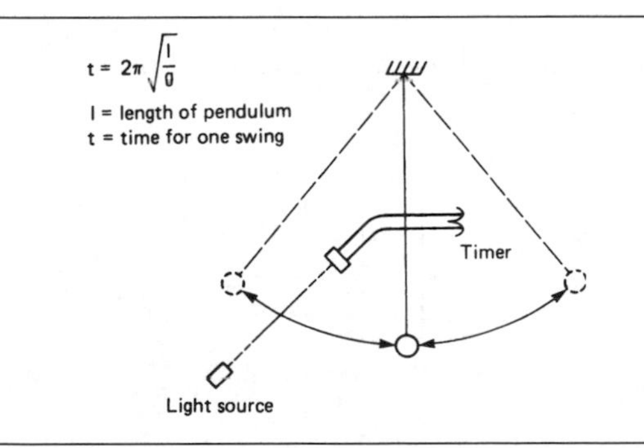

Figure 8.8 Measuring the time of a swinging pendulum with a quartz crystal clock. For a pendulum of fixed length in an evacuated enclosure, the time for one small-amplitude swing depends almost only on the value of the gravitational acceleration, g.

allowing the pendulum to interrupt a light beam and sensing the interruption with a photocell. By the use of a quartz crystal timer, a very small variation in the swing time can be detected and read as a variation in the gravitational constant. The pendulum should be mounted in an evacuated enclosure, and should be swinging freely when the measurement is taken.

The alternative is to measure the changes in the deflection of a spring, loaded by a mass. This calls for the sensing of very small changes in length, and a laser interferometer is the best method available. Once again, the whole balance should be mounted in a vacuum to avoid the effect of air damping and air currents.

8.6 Spectrum analysers

Spectrum analysis means obtaining information on the amplitude of waves for each frequency of waves in a set, the spectrum of the title. Originally, the phrase applied only to the visible spectrum, but developments in wave detection led to infrared and ultraviolet spectrum analysers being developed, and the title is now also applied to instruments that analyse the radio-frequency spectrum in a similar way. The principle of all types of spectrum analysers (Figure 8.9) is that each frequency in the spectrum is selected by a filter and its amplitude measured. The display of the information can be on a cathode ray tube, or as a graph or a printout from an attached computer.

For the radio-frequency spectrum, the filtering action is carried out by tuned circuits, using LC circuits for the lower frequencies and cavities for the microwave region. The range of frequencies requires any radio-frequency spectrum analyser to operate in switched bands, because no single tuning circuit can operate over more than a small part of the useable range. For the visible spectrum of light, the filtering action is obtained by using a glass prism, using the principle that the angle through which a light beam is bent will depend on the frequency of the light.

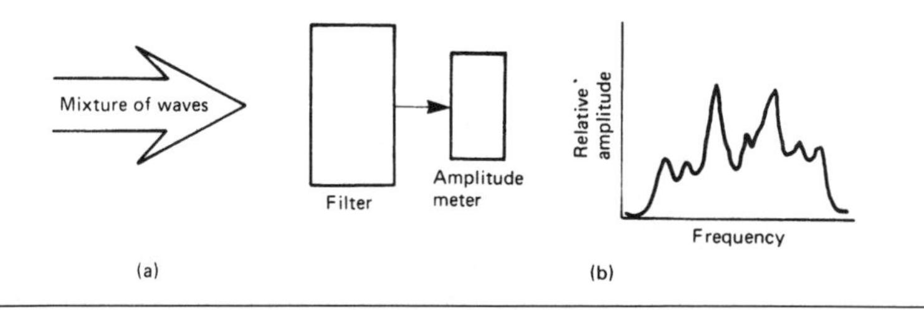

Figure 8.9 (a) The operating principle of a spectrum analyser, and (b) a typical output. For radio waves, the filters and amplitude detectors are the familiar radio circuits, but for optical frequencies prisms are used for filtering and photocells for detection.

Infrared and ultraviolet spectrum analysers cannot make use of glass prisms apart from the small part of the spectrum that lies next to the visible range. Materials such as quartz for ultraviolet and rock salt for infrared can be used, and in the region where the longer-wave infrared meets the short-wave microwave range, prisms of paraffin wax are effective. The shorter the wavelength of the radiation, the more difficult it is to find a material for a prism, because the action of a prism depends on dispersion, meaning that the speed of the waves in a material depends on their frequency. The shorter the wavelength, the less is the interaction with the atoms of materials and the lower the dispersion. As we approach the X-ray frequencies, nothing is suitable as a prism and practically all materials are transparent.

Spectrum analysers in all classes are usually operated by electronic methods, and make use of electronic detection. For the radio-frequency ranges, the filters can be electrically controllable and the filtered frequencies detected by a diode, so that the plotting of received amplitude for each frequency can be controlled by a microprocessor. The higher frequencies, of which the visible range is at the higher end, are dealt with using a servo-motor to rotate the prism and a detector such as a photocell to measure the amplitude of the received signal.

Chapter 9

Instrumentation techniques

In the course of this book several circuit methods have been described, and this chapter is a summary of these methods that apply to instrumentation together with a longer description of the methods that are used for analogue to digital (A–D) and digital to analogue (D–A) conversion. Of these, the A–D and D–A methods are the more important because the rate of development of these methods has been rapid and many readers will not be familiar with the current state of the methods that are used. For that reason, the more familiar methods are presented in summary form, but the D–A and A–D conversions are described in detail from first principles. In particular, although many texts describe methods that are applicable to DC or low-frequency analogue conversion, few deal with the methods that have to be used for wide amplitude and frequency ranges.

The main non-digital instrumentation methods can be grouped as impedance converters, bridge circuits and operational amplifiers. Of these, the impedance converters are mainly the familiar emitter–follower and source–follower circuits. The use of a MOSFET as a source–follower has largely replaced the older techniques using electrometer valves for very high input impedance applications, although electrometer valves are still used in some equipment for insulation testing. Bridge circuits have been covered as required in this book, and the only linear method that requires more detailed description is the use of operational IC amplifiers.

9.1 Op-amps

For most purposes, an IC of the type called the operational amplifier (usually shortened to op-amp) will be used in place of a multi-stage amplifier constructed from individual transistors. The internal circuit of

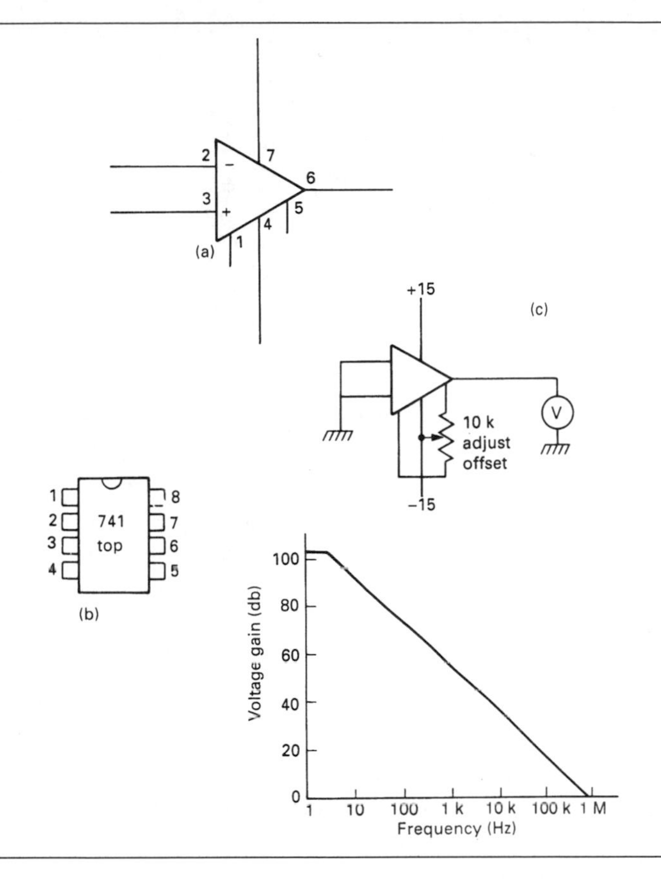

Figure 9.1 The 741 op-amp, typical of a range of amplifiers used extensively in instrumentation circuits: (a) symbol; (b) pin layout; (c) offset adjustment; (d) gain–frequency characteristic.

such an IC is not necessarily known to the user, and the important point to recognize is that the gain and other features of the op-amp will be set by external components such as resistors. This is true also of most amplifiers that have been constructed by using individual transistors, but since the op-amp consists of one single component, it is very much easier to see which components control its action, since only these components are visible on the circuit board. In many applications, only one or two resistors will be required to be of close tolerance.

The 'typical' op-amp is the type 741, which is a very old design of bipolar op-amp but one whose basic principles are followed by most of the modern types also. Figure 9.1 shows the connections to, and the symbol for, an op-amp of this type – the triangular symbol is used for any op-amp, but some of the connections are not used for modern op-amp types. The important point to note is that the op-amp has two signal inputs, marked with + and

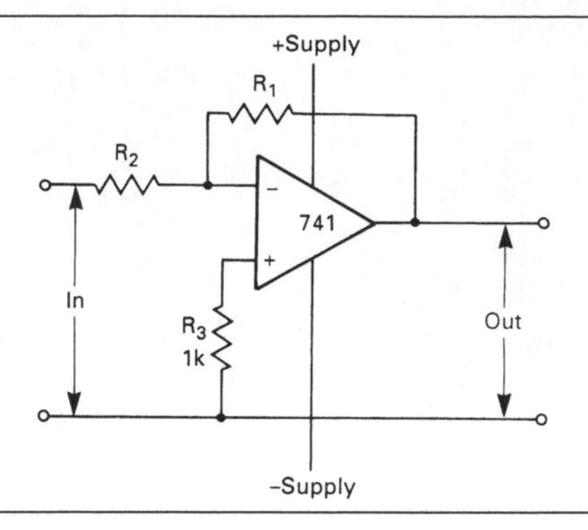

Figure 9.2 The basic op-amp inverting circuit, in which the gain is calculated from the resistor values.

– signs. A signal at the + input will appear amplified at the output and in phase. A signal at the – input will also appear amplified at the output, with the same figure of gain, but inverted. If the same signal is applied to each input the output should be zero, because the action of the op-amp is to amplify the difference between the signals at its two inputs. If both inputs are used in this way, the amplifier is acting as a differential amplifier.

For many purposes, however, the op-amp is used single-ended, meaning that only one input signal is supplied consisting of a voltage whose amplitude varies with respect to ground. For such a signal, the op-amp can be used in two ways – as an inverting amplifier or as a non-inverting amplifier. The gain of the op-amp in its natural state is very high, of the order of 100 dB, equivalent to a voltage gain of 100 000 times, so that for almost all practical uses this gain has to be reduced by using negative feedback. The negative feedback connection is from the output terminal to the (–) input. If we simply connect a resistor from the output to the (–) input, the negative feedback is 100% – all of the output voltage signal is connected back to the input, and the gain will be unity. We need to be able to take a signal to the input, however, and if we want the output signal to be inverted relative to the input, the signal input has to be to the (–) terminal. Connecting the input signal by way of a resistor (Figure 9.2) will reduce the amount of negative feedback, and so permit the gain to be greater than unity if required. In the circuit of Figure 9.2, the voltage gain is given by R_1/R_2 (in decibels, $20 \log [R_1/R_2]$). To achieve unity gain, we can make these resistor values equal, and by using different values we can achieve various figures of voltage gain. The value of R_2 includes any circuit resistance of the source of the signal, and more easily predictable results are obtained if this resistor is of several kΩ in value, swamping source resistance values.

The DC bias of the op-amp is achieved most simply by using two supplies of equal positive and negative voltages, typically +12 V and –12 V. When

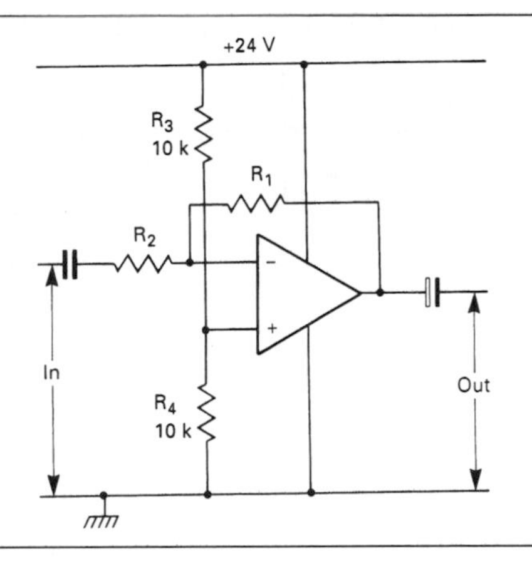

Figure 9.3 Using an op-amp from a single-ended supply. This is suitable for AC signals only; a balanced power supply must be used for DC amplifiers.

this is done, provided that there is a resistor of not too high a value connecting the output with the (−) input, the op-amp will be correctly biased. If it is inconvenient to have to supply voltages like this, the circuit can be rearranged to the type shown in Figure 9.3, using a single voltage supply. In this case, however, capacitors are needed to isolate the DC voltages. The DC voltage on the (+) input is set to half of the supply voltage by using resistors R_3 and R_4 of equal value, but the remainder of the circuit is constructed in the same way as before, with capacitors now used to feed signals in and out. The DC level of both input and output will be approximately half of the supply voltage.

The inverting amplifier connection of the op-amp is used to a large extent, but for some purposes a non-inverting connection can be useful. The amplifier is non-inverting if the signal is taken to the (+) input, but the feedback must still be connected to the (−) input, and the amount of feedback will decide how much gain is obtained. Figure 9.4 illustrates a typical non-inverting circuit, using balanced + and − power supplies. The voltage gain is now equal to $(R_1 + R_2)/R_2$, so that if unity gain is needed, the resistor R_2 should be omitted. The input of signal is taken to the (+) input terminal, with a resistor connected to ground to ensure correct DC conditions as usual. The input resistance of the op-amp is very high, but the use of R_3 makes the value equal approximately to the size of this resistor. The output resistance, as always, is low – a few ohms. The version of this circuit for a single-ended power supply is illustrated in Figure 9.5.

SLEW AND BANDWIDTH

The use of op-amps does not solve all the problems of providing gain,

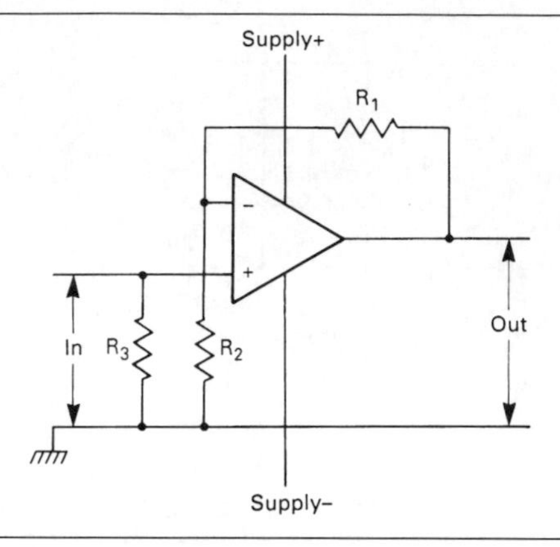

Figure 9.4 A non-inverting op-amp circuit using balanced power supplies, ±24 V.

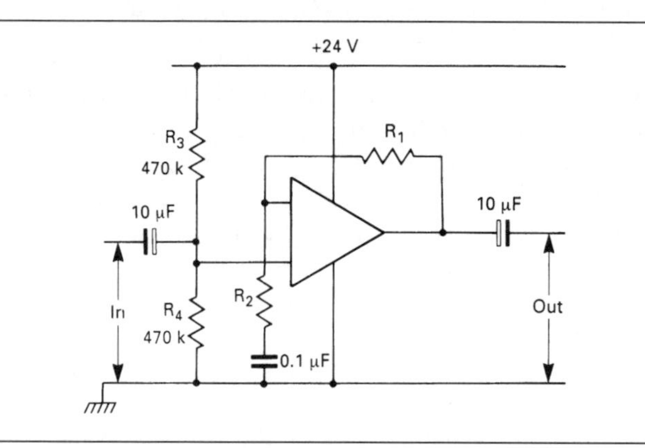

Figure 9.5 The non-inverting op-amp circuit adapted for use with a single-ended power supply and AC signals.

although for low-frequency signals there are few problems to overcome. Because the op-amp is a very complicated circuit that makes use of many transistors in common–emitter circuits, its gain for high-frequency signals will be lower than its gain for low-frequency signals. More seriously, its gain for large-amplitude fast-changing signals will be lower than for low-amplitude signals of the same frequency. This latter problem is one of slew rate.

The slew rate of a signal is the rate of change of voltage in the signal, in terms of V/s. In practice, this gives figures that are much too large, and the more practical unit of V/μs is used. The importance of slew rate is that two waves of identical frequency and waveshape can have very different slew rates, depending on their peak amplitudes. Suppose, for example that we have a wave that is of sawtooth shape and which repeats at 100 kHz. It is

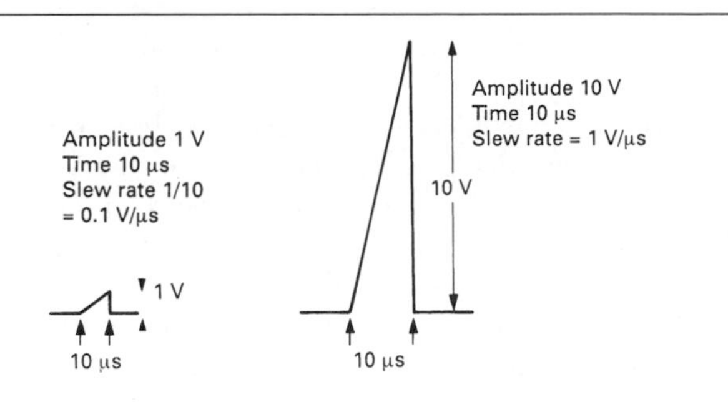

Figure 9.6 Illustrating the differences in slew rate for two signals of the same frequency.

easy to design an op-amp circuit that will provide a large figure of voltage gain for a wave with this frequency, but the slew rate also has to be considered, as Figure 9.6 shows. If the amplitude of the wave is 1 V, then for a time of 10 μs (corresponding to a frequency of 100 kHz), the slew rate is 1 V in 10 μs, equal to 0.1 V/μs. For a 10 V amplitude, however, the slew rate is 10 V/10 μs, equal to 1 V/μs. The amplifier must cope with the slew rate of the highest amplitude of signal at the output. As it happens, many op-amps will cope with a slew rate of 1 V/μs, but the 741 type is limited to a slew rate of 0.5 V/μs. For higher slew rates, however, it may be necessary to specify different op-amps.

The maximum slew rate cannot be altered by circuit tricks such as negative feedback; it is fixed by the internal design of the op-amp itself. Unlike the bandwidth for small sine wave signals, which can be increased by using more negative feedback and sacrificing gain, slew rate is a fundamental figure for the op-amp. High slew-rate op-amps can be obtained for applications in which this figure must be large. To take an example, the RS 5539 op-amp has a quoted slew rate of 600 V/μs, making this op-amp suitable for very fast-changing signals, including sine waves approaching the 1 GHz range. This, however, is exceptional, and slew rate figures in the range 0.5–10 V/μs are more common.

The slew rate is a very useful figure to use when an op-amp is employed in circuits that use sawtooth, pulse and other non-sine waveshapes. For sine wave use, an alternative figure – the power bandwidth – is more useful. The power bandwidth is the maximum frequency of signal that can be amplified at full power. If higher frequencies are used, lower gains must be used so that the product of gain in decibels and frequency in (usually) MHz or kHz is the same. In other words, if the gain of an op-amp is quoted as 100 dB (gain of 100 000 times) and its power bandwidth as 10 kHz, you can expect to be able to use the full 100 dB of gain on signals up to 10 kHz, and if you make the gain 10 000 (80 dB) you can use signal frequencies up to 100 kHz, and so on. To put it another way, each 20 dB reduction of gain (by using negative feedback) will give a 10-fold rise in maximum operating frequency.

This, however, is still subject to the slew rate limitation, so that the slew rate should always be calculated first. For a sine wave, the slew rate in V/μs is given by $6.3 \times V_o \times f$, where V_o is the peak amplitude, and f is the frequency in MHz. For example, a sine wave of 5 V peak and 0.5 MHz frequency would have a slew rate of $6.3 \times 5 \times 0.5 = 15.75$ V/μs. Only a limited number of op-amp types can cope with this slew rate, although many could give the required bandwidth for signals of lower amplitudes.

CDA AND TRANSCONDUCTANCE AMPLIFIERS

Another type of op-amp that is used for some purposes is the current-difference amplifier (CDA), otherwise known as a Norton amplifier. In an amplifier of this type, the output voltage is proportional to the difference between the currents at the two inputs. This leads to circuits that look very odd when compared with the usual type of op-amp, because the input currents are set with the aid of large-value resistors, as Figure 9.7 illustrates. The current to the '+' input is set with a 2M2 resistor, and since the voltage at the output is normally required to be about half of the supply voltage, a 1M resistor from this point to the '−' input will set the bias current correctly. The advantage of using a CDA is that the voltage swing at the output can be very close to the limits of the supply voltage, making this type of amplifier useful in interfaces to digital equipment.

Another comparatively rare type of op-amp is the transconductance amplifier, in which the transconductance (milliamps of current output per voltage difference between the inputs) can be chosen by setting the bias current. Such amplifiers are used in gain-controlled circuits.

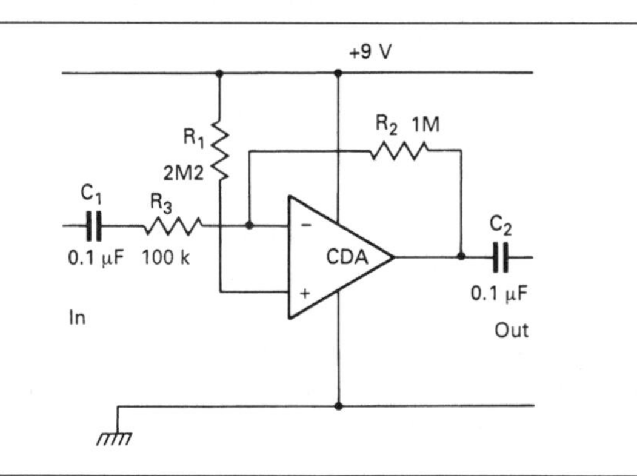

Figure 9.7 A typical transconductance amplifier circuit, illustrating the large resistor values.

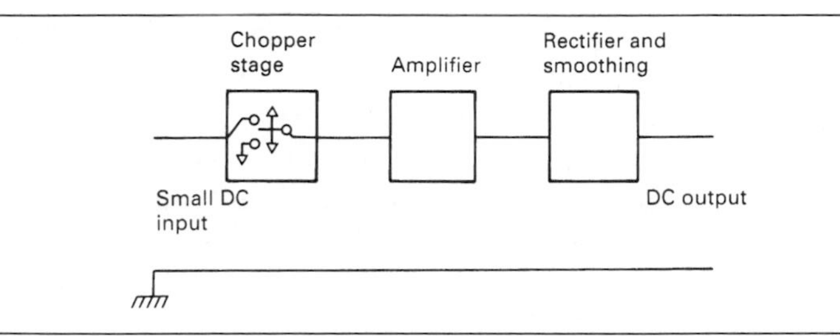

Figure 9.8 Outline of a chopper amplifier. These are used to a lesser extent now that digital techniques are available.

An alternative to the use of operational amplifiers is the chopper amplifier. The principle is illustrated in Figure 9.8, and is used to attain very high-gain DC amplification. A low-amplitude DC signal is converted into a square wave by 'chopping' the signal on and off, using FET switches, and using this square wave as the input to a pulse amplifier. Very high gain of such a waveform can be obtained, and the output square wave can be rectified and smoothed to obtain a high-level DC signal whose amplitude will be very precisely proportional to the amplitude of the original signal. Chopper amplifiers are less liable to drift than direct-coupled DC amplifiers and chopping techniques can also be applied to more conventional op-amps in order to reduce offset. See also Chapter 12 for chopper principles.

9.2 Analogue to digital conversions

The conversion of analogue signals into digital form is the essential first step in any system that will use digital methods for counting, display or logic actions. For some purposes it is necessary to distinguish between conversion and modulation in this context. Conversion means the processing of an analogue signal into a set of digital signals, and modulation means the change from the original digital signal into a type of digital signal that can be stored or transmitted by an error-free method. The two are, however, bound up with each other because many forms of conversion are also forms of modulation. The more modern methods of analogue to digital conversion are mainly based on a modulation system called pulse code modulation, in which the digital pulses represent in binary or other coded form the amplitude of a sample of the analogue signal. This has replaced older methods based on pulse amplitude or frequency, because its output is a stream of digital numbers that can be processed using familiar computing techniques.

The first point to settle about an AD conversion is how many bits should be used for a number. The use of eight bits permits the system to distinguish 256 different amplitude levels, and this may be quite sufficient for many applications – the fewer the number of bits in the conversion, the faster and more easily the signals can be handled. Much depends on the range of signal amplitudes that need to be coded. If, for example, the signal amplitude range is 90 dB, this corresponds to a signal amplitude range of about 32 000 : 1. Using just 256 steps of signal amplitude would make the size of each step about 123 units, too much of a change. This size of step is sometimes referred to as the quantum, and the process as quantization. In these terms, the use of 256 steps is too coarse a quantization for such an amplitude range. If we move to the use of 16 bits, allowing 32 767 steps, we can see that this allows a number of steps that is well matched to the amplitude range of 90 dB or 32 000:1. This is the level of quantization that is used in CD systems.

The choice of the number of bits has wider implications, however. Ideally, the amplitude of a signal at each sample would be proportional to a number in the 16-bit range. Inevitably, this will not be exact, and the difference between the actual amplitude and the amplitude that we can encode as a 16-bit number (or whatever is used) represents an error, the quantization noise. The greater the number of bits that are used to encode an amplitude, the lower the quantization noise will be. What is less obvious is the effect on low-amplitude signals. For low-amplitude signals, the amount of quantization noise is virtually proportional to the amplitude of signal, so that when the digital signal is re-converted to analogue the low-amplitude signals appear to be distorted.

The greater the number of digital bits used for encoding, the worse this effect gets, and the cure, by a strange paradox, is to add noise. Adding *white noise* – noise whose amplitude range is fairly constant over a large frequency range – to very low-amplitude signals helps to break the connection between the quantization noise and the signal amplitude and so greatly reduces the effects that sound so like distortion. This added noise is called *dither*, and is another very important part of a conversion process for signals with a wide range of amplitude. For many A–D conversions of industrial significance such refinements will be completely unnecessary. The noise level is very low, corresponding to a one-digit number.

SAMPLING RATE

All A–D conversion starts with *sampling*. Sampling means that the amplitude of the audio analogue signal is measured and stored in a short interval of time, and if sampling is to be used as a part of the process of converting from analogue to digital signals, it has to be repeated at regular intervals. The principles of the process are illustrated in Figure 9.9, from which you can see that if an analogue waveform is to be converted into a digital form with any degree of fidelity, a large number of samples must be taken in the course of one cycle. If too few samples are taken, the digital

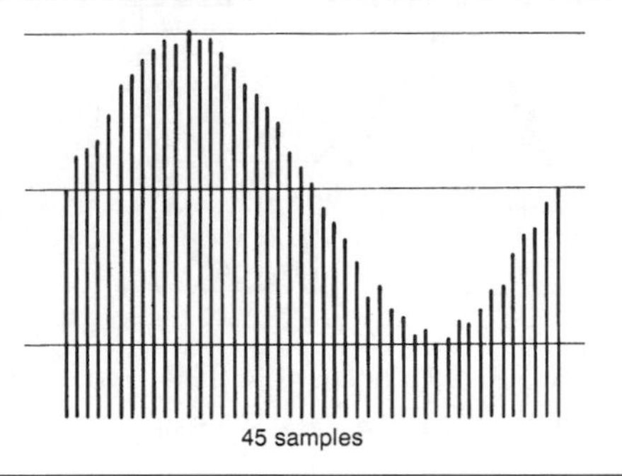

Figure 9.9 Sampling a waveform to obtain pulses whose amplitude is that of the waveform at the instant of sampling.

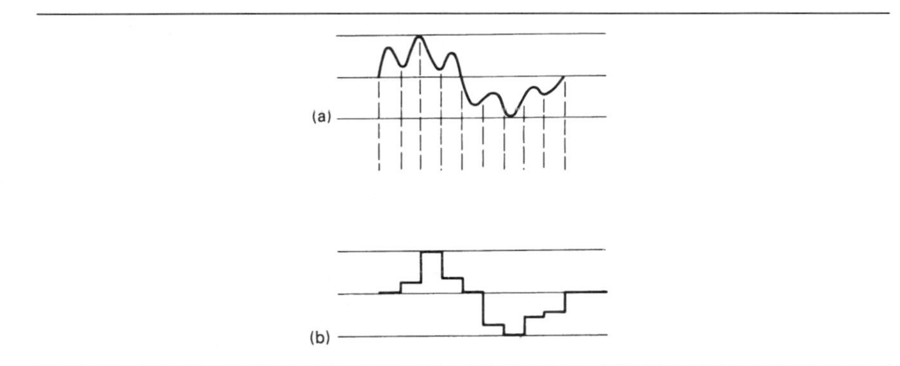

Figure 9.10 (a) An analogue signal that has been sampled at too low a rate to provide (b) a pulse output that is not truly representative.

version of the signal (Figure 9.10) will look quite unlike the analogue version. On the other hand, if too many samples are taken per cycle, the system will be working with a lot of redundant information, wasting processing time and memory space. Whatever sampling rate is used must be a reasonable compromise between efficiency and precision and, as it happens, the theory of sampling is by no means new.

In 1948, C. E. Shannon published his classic paper *A mathematical theory of communications* on which the whole of digital conversion is based. The essence of Shannon's work is that if the sampling rate is twice the highest frequency component in an audio signal, the balance between precision and excessive bandwidth is correctly struck. Note that this pivots around the highest frequency component in a mixture of frequencies – it does not imply that for a 1 kHz sine wave a sampling rate of 2 kHz would be adequate unless your reverse conversion was arranged always to regenerate a sine wave output. What Shannon's theory is about is non-sinusoidal waves which can be analysed into a fundamental frequency and a set of

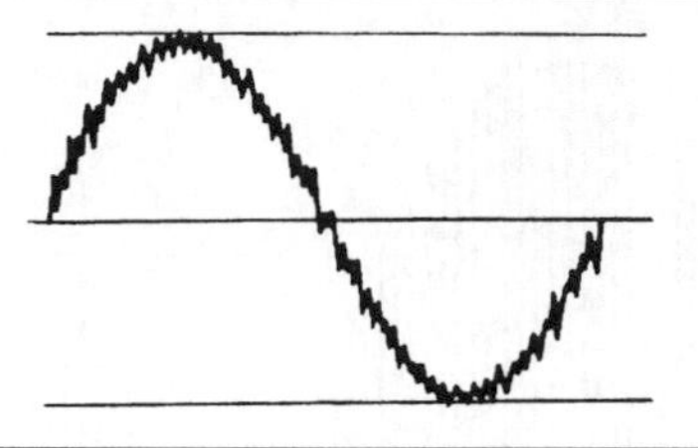

Figure 9.11 In a waveform which contains high-frequency components, these components represent only a very small part of the amplitude.

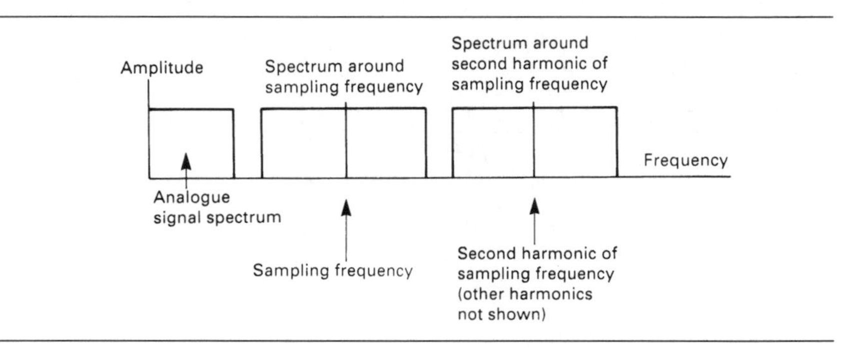

Figure 9.12 The spectrum of a sampled signal, which contains the frequencies of the analogue signal along with sidebands of the sampling frequency and its harmonics.

harmonics. What this boils down to is that if we look at a typical waveshape (Figure 9.11), the highest frequency component is responsible for a small part of the waveform, whose shape looks nothing like a sine wave and which could equally well be represented by a sawtooth.

Sampling the highest harmonic, then, at twice the frequency of that harmonic will provide a good digital representation of the overall shape of the complete wave, all other things being equal. Sampling under these conditions provides a set of pulses whose amplitude is proportional to the amplitude of the wave at each sampled point. A spectrum analyser will then reveal something like the illustration of Figure 9.12. This consists of the range of frequencies that were present in the analogue signal (the *fundamentals*) plus a set of *harmonics* centred around the sampling frequency and its harmonics. This is not a problem when the pulse amplitudes are being converted into digital form, but when the pulses are recovered a filter will be needed to separate out the wanted part, which is the lowest range of frequencies – the frequencies of the original signal.

The existence of these harmonics makes it important to ensure that the sampling rate is high enough. If the analogue signal had an 18 kHz bandwidth and a sampling rate of 30 kHz were used, the harmonics

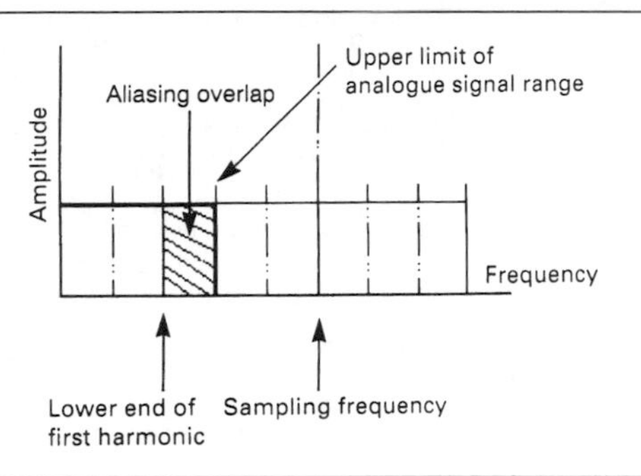

Figure 9. 13 The result of using too low a sampling frequency. The lower sideband of the signals around the sampling frequency overlap the upper frequencies of the analogue signal.

around the 30 kHz sampling frequency would extend down to $30 - 18 = 12$ kHz and up to $30 + 18 = 48$ kHz, but the lower sideband of this set will overlap the 18 kHz of the original sound (Figure 9.13). This is an effect called *aliasing*, meaning that over a range of frequencies in the original range there will be a set of 'aliases' from the lower sidebands of the sampling frequency.

Even if the sampling frequency is made twice the highest audio frequency, difficulties still arise because there may be harmonics in the audio signal that extend to higher than half of the sampling frequency. This can be dealt with by using an anti-aliasing filter that is a steep-cut filter that will remove frequencies above the upper limit of the audio range. If the sampling frequency were too low, this filter would need to have an impossibly perfect performance. The sampling frequency must therefore be high enough to permit an effective anti-aliasing filter to be constructed. As before, such methods are needed only for signals whose amplitude and frequency ranges are large, such as audio signals, and in many instrumentation systems, using signals at DC or low-frequency AC, such methods are unnecessary.

SAMPLING AND CONVERSION METHODS

The process of sampling involves the use of a sample-and-hold circuit. As the name suggests, this is a circuit in which the amplitude of a waveform is sampled and held in memory while the amplitude size can be converted into digital form. The outline of a sample-and-hold circuit is shown in Figure 9.14, with a capacitor representing the holding part of the process. While the switch is closed, the voltage across the capacitor is the analogue

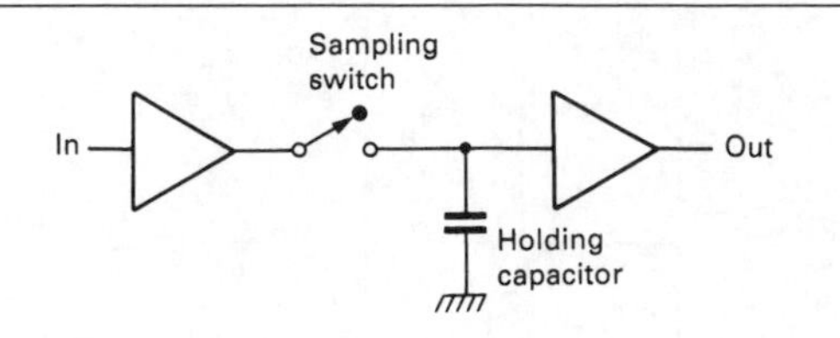

Figure 9.14 The basis of the sample-and-hold system.

voltage, maintained by the buffer amplifier that has a low output impedance. No conversion into digital form takes place in this interval.

When the switch is opened, the amplitude of the analogue signal at the instant of opening is the voltage across the capacitor, and this amplitude controls the output of the second buffer stage. This, in turn, is the signal that will be converted to digital form. The instant of sampling can be very short, but the time that is available for conversion to digital form is the time between sampling pulses. The 'switch' that is shown will invariably be a semiconductor switch, and the capacitor can be a semiconductor memory, although for a fast sampling rate a capacitor is perfectly acceptable when its only loading is the input impedance of a MOS buffer stage.

The effect of sampling is only to quantize the signal. The signal is still an analogue signal in which the variation of amplitude with time carries the information of the signal, and the change that has come about as a result of sampling is the substitution of a set of pulses for the original varying signal. The signal is now an amplitude-modulated set of pulses at the sampling frequency, but this is not a digital signal. The actual conversion from analogue to digital form is the crucial part of the whole encoding process. There is more than one method of achieving this conversion, and not all methods are equally applicable to all uses. There are two main methods and the following is an outline of the problems that each of them presents.

The integrator type of A–D converter is used widely in digital voltmeters. The principle is simple enough (Figure 9.15). The central part of the circuit is a comparator, which has two inputs and one output. While one input remains below the level of the other, the output remains at one logic level, but the logic level at the output switches over when the input levels are reversed. The change at the inputs that is needed to achieve this can be very small, a matter of millivolts, so that the action is that the output switches over when the input levels are equal. If one input is the signal being converted, this signal will be held at a steady level (by the sample-and-hold circuit) during the time that is needed for the conversion. A series of clock pulses are applied both to a counter and to an integrating circuit. The output of the integrating circuit is a series of equal steps of voltage, rising by one step at each clock pulse, and this waveform is applied to the other input of the comparator.

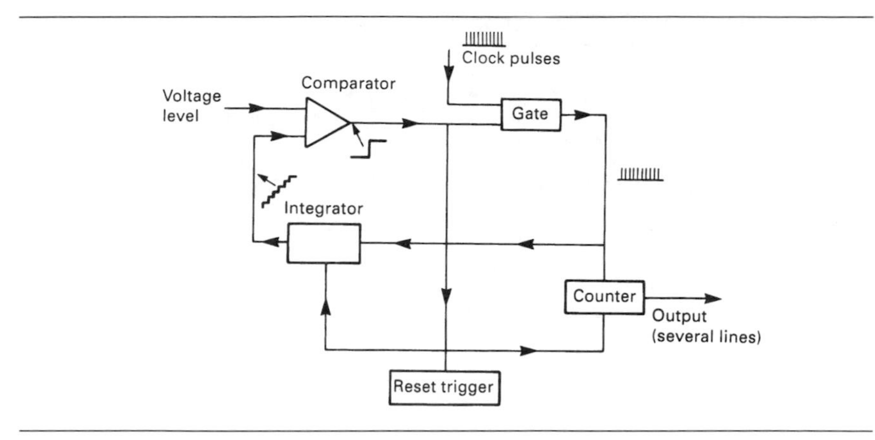

Figure 9.15 The block diagram of the integrator type of A–D converter.

When the two inputs are at the same level (or the step waveform input just exceeds the sample input), the comparator switches over and this switch-over action can be used to interrupt the clock pulses, leaving the counter storing the number of pulses that arrived. Suppose, for example, that the steps of voltage were 1 mV, and that the sample voltage was 3.145 V. With the output of the integrator rising by 1 mV in each step, 3145 steps would be needed to achieve equality and so stop the count, and the count would be of the number 3145 in digital form, a digital number that represents the amplitude of the voltage at the sampled input. The switch-over of the comparator can be used to store this number into a register as well as stopping the clock pulses. A master clock pulse can then reset the integrator output to zero and terminate the conversion ready to start again when a new sample has been taken.

These numbers are, of course, for illustrative purposes only, but they demonstrate well how a signal level can be converted into a digital number by this method. Note that even with this comparatively crude example, the number of steps is well in excess of the 256 that would be available using an 8-bit digital number. Note also that the time that is needed to achieve the conversion depends on how fast a count rate is used. This will obviously need to be much faster than the sampling rate, since the conversion of each pulse should be completed by the time the next pulse is sampled.

The quality of conversion by this method depends very heavily on how well the integrator performs. An integrator is a form of digital to analogue (D–A) converter, so that this presents the paradox of using a D–A converter as an essential part of the A–D conversion process, rather like the problem of the chicken and the egg. The very simple forms of integrator, like charging a capacitor through a resistor, are unsuitable because their linearity is not good enough. The height of each step must be equal, and in a capacitor charging system the height of each step is less than that of the

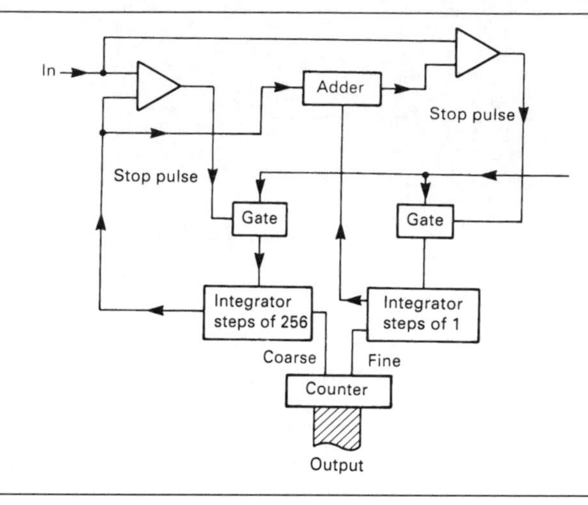

Figure 9.16 The use of twin integrators to allow faster processing of analogue signals into digital.

step previous to it. The integrator is therefore very often a full-blown D–A converter in integrated form.

For low-frequency analogue signals there are few problems attached to this method, but for wide ranges of amplitude and frequency the circuits may not be able to cope with the speed that will have to be used. This speed value depends on the time that is available between samples and the number of steps in the conversion. For example, if a time of 20 μs is available to deal with a maximum level of 65 536 steps, then the clock rate for the step pulses must be:

$$\frac{65\,536}{20 \times 10^{-6}}$$

which gives a frequency of 3.8 GHz (*not* MHz), well above the limits of conventional digital equipment. This makes the simple form of integrator conversion impossible for the high sampling rates and large numbers of bits, such as are used for audio signals. Many of the recent advances in D–A and A–D conversion are due to the research that has been triggered by the use of CD technology and digital tape equipment.

One solution, retaining the integrator type of converter, is to split the action between two converters, each working with part of the voltage. The idea here is that one counter works in the range 0–255, and the other in units that are 256 times the steps of the first. The voltage is therefore measured as two 8-bit numbers, each of which requires only 255 steps, so that the counting rate can be considerably lower. Since the counters work in succession, with the smaller range of counter operating only after the coarser step counter has finished (Figure 9.16), the total number of steps is $2 \times 255 = 510$, and the step rate is now, still assuming a 20 μs time interval:

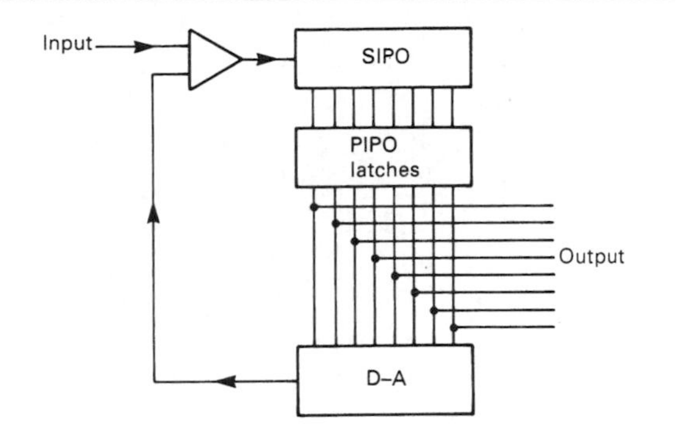

Figure 9.17 Outline of the successive approximation method of A–D conversion.

$$\frac{510}{20 \times 0^{-6}}$$

– a rate of 25.5 MHz. This is well within the range of modern digital circuitry in IC form.

Another form of A–D converter is known as the successive approximation type. The outline of the method is shown in Figure 9.17, consisting of a serial input parallel output (SIPO) register, a set of latches or PIPO register, a D–A converter and a comparator whose output is used to operate the register. To understand the action, imagine that the output from the D–A converter is zero and therefore less than the input signal at the time of the sample. The resultant of these two inputs to the comparator is to make the output of the comparator high, and the first clock pulse arriving at the SIPO register will switch on the first flip-flop of the PIPO register, making its output high to match the high input from the comparator. This first PIPO output is connected to the highest bit input of the D–A converter, which for a 16-bit register corresponds to 32 767 steps of amplitude.

Now what happens next depends on whether the input signal level is more than or less than the level of output from the D–A converter for this input number. If the input signal is less than this, the output of the comparator changes to zero, and this will in turn make the output of the SIPO register zero and the output of the PIPO register zero for this bit. If the input signal is greater than the output from the D–A circuit, the 1 bit remains in the first place of the register. The clock pulse will then pass this pulse down to the next PIPO input – this does not, however, affect the first PIPO input which will remain set at the level it had attained.

Another comparison is now made, this time between the input signal and the output of the D–A converter with another input bit. The D–A output will be either greater than or less than the input signal level, and as a result, the second bit in the PIPO register will be set to 1 or reset to 0. This second bit represents 16 384 steps if its level is 1, zero if its level is 0. The process is repeated for all 16 stages in the register until the digital number that is connected from the PIPO outputs to the D–A inputs makes the output of the D–A circuit equal to the level of the input signal. Using this method, only 16 comparisons have to be made in the sampling period, giving a maximum time of 1.25 μs for each operation. This time, however, includes the time needed for shifting bits along the registers, and it requires a fast performance from the D–A converter, faster than is easily obtained from most designs. Speed is the problem with most digital circuits, which is why there is a constant effort made to improve methods of manufacturing ICs, even to the use of alternative semiconductor materials (such as gallium arsenide) that could allow faster operation.

9.3 Digital to analogue conversions

Conversion from digital to analogue signals must use methods that are suited to the type of digital signal that is being used. If, for example, a simple digital amplitude modulation were being used, or even a system in which the number of 1s were proportional to the amplitude of the signal, the conversion of a digital signal to an analogue signal would amount to little more than smoothing. As it happens, it is possible to convert a digital signal into a form that can be smoothed simply, and this is the basis of bitstream methods, noted later in this chapter. For the moment, however, it is more useful to concentrate on the earlier methods.

Smoothing always plays some part in a digital to analogue conversion, because a converted signal will consist of a set of steps, as Figure 9.18 shows, rather than a smooth wave. With a fast sampling rate, however, the steps are close spaced, and very little smoothing is needed. Even a crude resistor–capacitor smoothing circuit can work wonders with such a wave, and the more complex integrators can reproduce the analogue signal very faithfully, often closer to the original than is possible when a purely analogue method has been used. The faster the pulse-rate of the output of the D–A converter, the easier it is to smooth into an acceptable analogue signal.

The pulse code type of digital system does not allow a set of digital signals

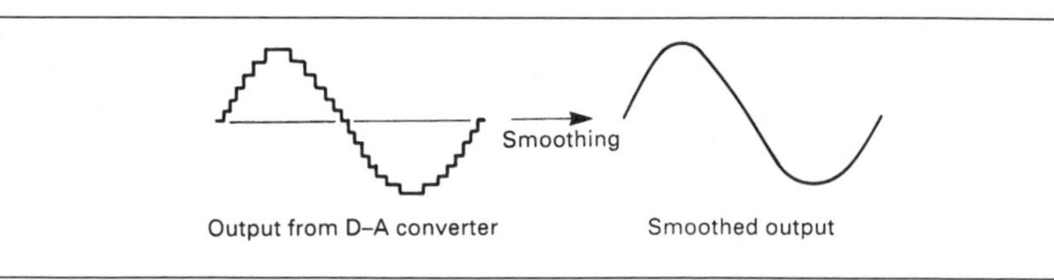

Figure 9.18 The use of smoothing following D–A conversion.

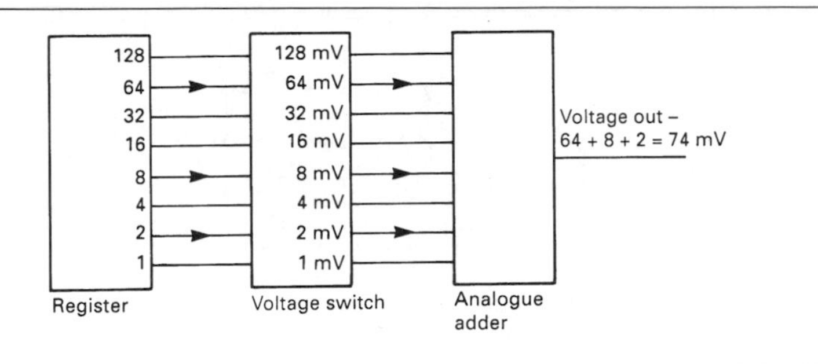

Figure 9.19 Outline of a voltage-adding type of D–A converter.

to be converted into an analogue signal by any simple method, however. Remember that the digital signal consists of a set of digitally coded numbers that represent the amplitude of the analogue signal at each sampling point. The circuits that are used for conversion to analogue will convert each number into a voltage amplitude that is proportional to that number. The nature of the conventional conversion requirement is more easily seen if it is illustrated using a 4-bit number. In such a number (like any other binary number), each 1 bit has a different weighting according to its position in the number, so that 0001 represents one, 0010 is two, 0100 is four and 1000 is eight.

The progression is always in steps of two, so that if a register were connected to a voltage generator in the way illustrated for an 8-bit converter in Figure 9.19, each 1 bit would generate a voltage proportional to its importance in the number. In this way, a 1 in the second place of a number would give, say, 2 mV, and a 1 in the fourth place would give 8 mV, with a 1 in the seventh place giving 64 mV. Adding these voltages would then provide a voltage amplitude proportional to the complete digital number, 74 mV in this simple example. This, in essence, is the basis of all D–A converters.

Reverting to a 4-bit number, Figure 9.20 shows the basis of a practical method. Four resistors are used in a feedback circuit which makes the output of the buffer amplifier depend on the voltage division ratio. This in turn is determined by the resistor ratio, so that switching in the resistors will give changes in voltage that are proportional to resistor value. In this example, each resistor can be switched into circuit by using an analogue switch, and the switches are in turn controlled by the outputs from the bits of a parallel output register. If the highest order bit in the register is a 1, the resistor whose value is marked as R is switched in, if the next bit is 1, then resistor 2R is switched in and so on.

The result of this is that the voltage from the output of the buffer amplifier (an analogue adder circuit) will be proportional to the size of the digital number. The attraction of this method is its simplicity – and that is

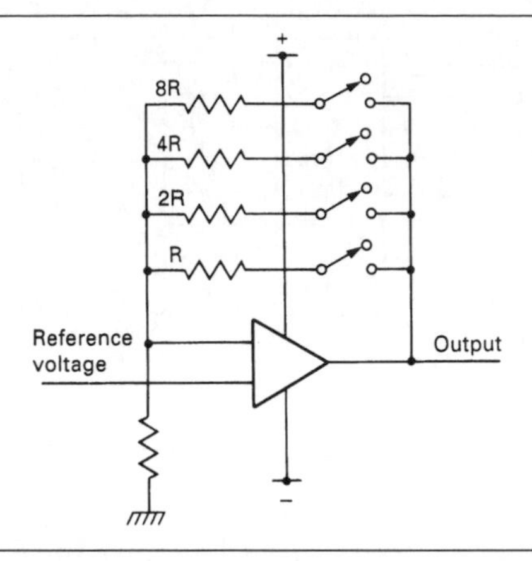

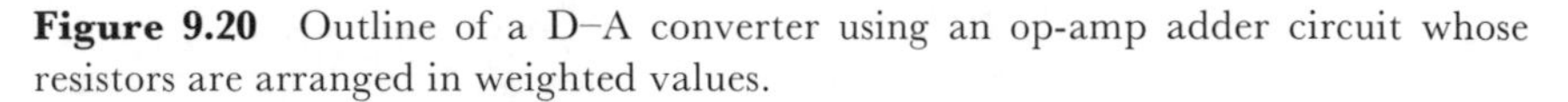

Figure 9.20 Outline of a D–A converter using an op-amp adder circuit whose resistors are arranged in weighted values.

also its main problem. The resistor switching type of converter is widely used and very effective, but only for a limited number of bits. The problems that arise are the range of resistance that is required, and the need for quite remarkably precise values for these resistors. Suppose, for example, that the circuit can use a minimum resistance value of around 2 kΩ. This is a reasonable value, assuming that the amplifier has an output resistance that is not negligible and that the resistance values must be large compared with the resistance of the analogue switch when it is on. Now each resistor in the arrangement will have values that rise in steps of two, so that these values are 4 kΩ, 8 kΩ, 16 kΩ ... all the way to 256 kΩ for an 8-bit converter.

Even this is quite a wide range, and the tolerance of the resistor values must also be tight. In an 8-bit system, it would be necessary to distinguish between levels that were one 256th of the full amplitude, so that the tolerance of resistance must be better than one in 256. This is not exactly easy to achieve, even if the resistors are precisely made and hand-adjusted, and it becomes very difficult if the resistors are to be made in IC form. The problem can be solved by making the resistors in a thick-film network, using computer-controlled equipment to adjust the values, but this, remember, is only for an 8-bit network.

When we consider a 16-bit system, the conversion looks quite impossible. Even if the minimum resistance value is reduced to around 1 kΩ, the maximum will then be of the order of 65 MΩ, and the tolerance becomes of the order of 0.006%. The speed of conversion, however, can be very high, and for some purposes a 16-bit converter of this type would be used, with the resistors in thick-film form and adjusted for precise value. For any mass-produced application this is not really a feasible method. The requirement for close tolerance can be reduced by carrying out the conversion in

4-bit units, because only the highest order bits require the maximum precision.

CURRENT ADDITION

The alternative to adding voltages is the addition of currents. Instead of converting a 16-bit number by adding 16 voltages of weighted values (each worth twice as much as the next lower in the chain), we could consider using 65 535 current sources, and making the switching operate from a decoded binary number.

Once again, it looks easier if we consider a small number, three digits this time, as in Figure 9.21. The digital number is *demultiplexed* in a circuit that has eight outputs. If the digital number is 000, then none of these outputs is high; if the digital number is 001, then 1 output is high; if the digital number is 010, then two outputs are high; and if the digital number is 011, then three outputs are high; and so on. If each output from this demultiplexer is used to switch an equal amount of current to a circuit, the total current will be proportional to the value of the binary number.

The attraction of this system is that the requirement for precise tolerance is much less, since the bits are not weighted. If one current is on the low side, another is just as likely to be on the high side, and the differences will cancel each other out, something that does not happen if, for example, the one that is low is multiplied by 2 and the one that is high is multiplied by 1024. The effort of constructing 65 535 current supplies, each switched on or off by a number in a register, is not quite so formidable as it would seem in the current state of IC construction.

The main compromise that is usually made is in the number of currents. Instead of switching 65 535 identical current sources in and out of circuit, a lower number is used, and the current values are made to depend on the place value of bits. Suppose, for example, that 16 383 currents are to be used, but that each current could be of 1, 2, 4 or 8 units. The steps of current could be achieved by using resistors whose precision need not be

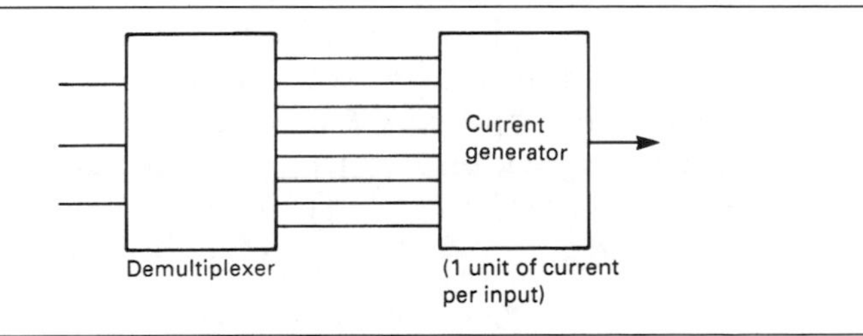

Figure 9.21 A block diagram for a current-adding D–A converter for a 3-bit system.

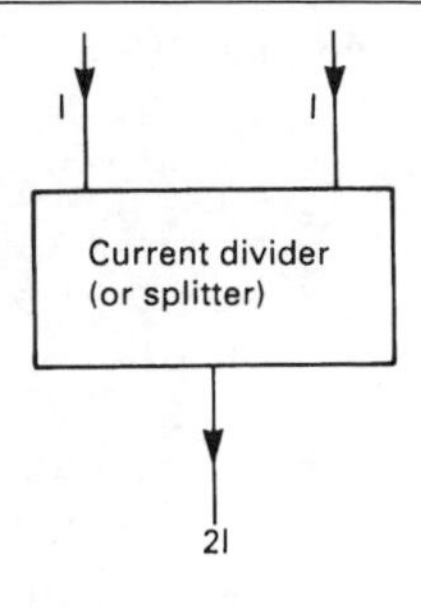

Figure 9.22 The basic current divider unit as used in D–A converters.

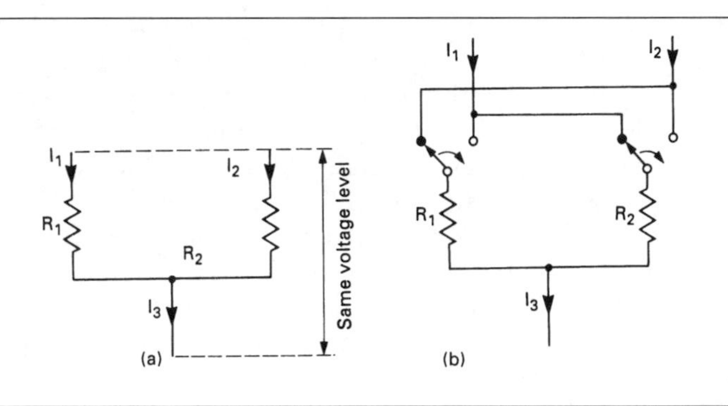

Figure 9.23 (a) The current divider which needs perfect matching of resistors, and (b) the switching principle which can use wider tolerance components.

too great to contemplate in IC form, and the total number of components has been drastically cut, even allowing for the more difficult conversion from the digital 16-bit number at the input to the switching of the currents at the output. This is the type of D–A converter that is most often employed in fast-working converters.

Another possibility is to use a current analogue of the resistor switching method used in the voltage step system, so that the currents are weighted in 2:1 steps, and only 16 switch stages are needed. This is feasible because current dividers can be manufactured in IC form. The principle of a current divider is shown in Figure 9.22, with currents shown as flowing into the terminals at the top rather than out. The distinction between a current adder and a current divider depends on the direction of current. To act as a precise divider, however, the input currents must be identical to a very close tolerance, and this cannot be achieved by a resistive network alone.

In this form of converter, first described by Plassche in 1976, the resistor network is used along with current switching. The principle is shown in Figure 9.23, and is easier to think of as a current adder rather than a

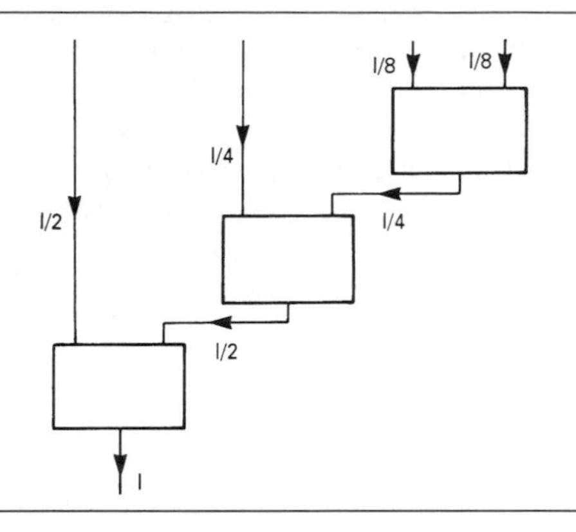

Figure 9.24 A set of current-division units connected so as to generate a binary sequence of currents.

divider. The two input currents I_1 and I_2 form the current I_3, and if the resistors R_1 and R_2 are equal, then $I_1 = I_2 = I_3/2$, giving the condition (I_1 or I_2) being exactly half of the other current I_3. The snag, of course, is that resistors R_1 and R_2 will not be identical, particularly when this circuit is constructed in IC form.

The ingenious remedy is to alter the circuit so that each current is chopped – switched so that it flows alternately in each resistor. For one part of the cycle, I_1 flows through R_1 and I_2 through R_2, then in the other part of the cycle, I_1 flows through R_2 and I_2 through R_1. If the switching is fast enough and some smoothing is carried out, the differences between the resistor values are averaged out, so that the condition for the input currents I_1 and I_2 to be exactly half of I_3 can be almost met even if the resistors are only to about 1% tolerance. Theory shows that the error is also proportional to the accuracy of the clock that controls the switching, and this can be made precise, to better than 0.01% with no great difficulty.

Now if a set of these stages is connected as in Figure 9.24, the ratio of currents into the stages follows a 2:1 step, and this can be achieved without a huge number of components and without the requirement for great precision in anything other than a clock signal to control switching. In addition, the action of the circuit can be very fast, making it suitable for use in the types of A–D converter that were discussed above. The only component that cannot be included into the IC is the smoothing capacitor that is needed to remove the slight current ripple that will be caused by the switching. This can be an external component connected to two pins of the IC.

The level of conversion is another point that needs to be considered. If we take a voltage conversion as an example, the amplitude of the signal steps will determine whether the peak output is 10 mV, 100 mV, 1 V or whatever. The higher the peak output, the more difficult is the conversion because switching voltages rapidly is, in circuit terms, much simpler when the voltage steps are small, since the stray capacitances have to be charged to lower levels. On the other hand, the lower the level of the conversion, the greater the effect of stray noise and the more amplification will be needed. A conversion level of about 1 V is ideal, and this is the level that is aimed at in most converters.

CONVERSION PROBLEMS

The problems that arise with the conventional type of D–A converter are closely connected with the nature of the digital signal and the ever-present problems of precise current generation. A 16-bit D–A converter requires the use of 16 current sources, and the ratio of a current to the next lower step of current must be exactly 2. For a conventional type of current generator system in which the current source provides current I, the most significant bit will be switching a current equal to I/2 and the least significant bit will be switching a current equal to I/65 536. With all switches open (all bits zero), the current will be zero; with all switches closed (all bits 1), the current will be equal to I – I/65 536. Note that the current can never be equal to I unless the number of bits is infinite; in this example, I/65 536 is the minimum step of current. The value of I/2 is taken as being the current corresponding to zero signal, so that zero is being represented by 1000000000000000 – current converters cannot deal with negative currents.

Maintaining the correct ratios between all of these currents is virtually impossible, although many ingenious techniques have been devised. In particular, this scheme of D–A conversion is very susceptible to a form of cross-over distortion, when the zero current level changes to the minimum negative current. In digit form, this is the change from the current I/2 (represented by the digital number 1000000000000000) to the value which is the sum of all the current from I/4 down to I/65 536, corresponding to the digital number 0111111111111111. This step of current ought to be small, equal to I/65 536, but if any of the more significant current values are incorrect even to the extent of only 0.05%, the effect on this step amount will be devastating – a rise of seven times the minimum current instead of a fall of the minimum step amount, for example.

This amount of error is virtually unavoidable, and it will cause a cross-over error to occur each time the signal passes through the zero level, corresponding to a current of I/2. In addition to this cross-over problem, all converters suffer from 'glitches', which are transient spikes that occur as the bits change. These are caused by small variations in the time when switches open and close, and they are also most serious when the signal level is at its minimum, because they cannot be masked by the signal level, and the number of bits that change is at its greatest when there is a change

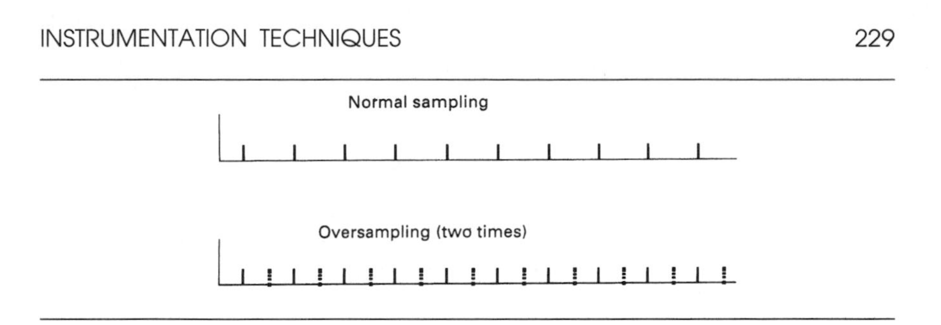

Figure 9.25 Normal sampling and oversampling in a D–A converter. The oversampling system adds pulses between the existing signal pulses.

from zero (I/2 current) to a small negative value. Both of these problems are answered by the adoption of bitstream methods.

OVERSAMPLING

The sampling of an analogue signal creates a set of pulses which are still amplitude modulated and which correspond to the analogue signal plus a set of sidebands around the sampling frequency and its harmonics. After the D–A converter has done its work, this is the signal that will exist at the output and it requires low-pass filtering to reject the higher frequencies.

This, however, calls for the use of filters with a very stringent specification, and such filters have an effect on the wanted part of the frequency range that is by no means pleasant. A simple way out of the problem would be to double the sampling rate. This is always feasible, since A–D converters are stretched as it is to cope with high-frequency signals. What can be done, though, is to add pulses between the output pulses from the A–D converter (Figure 9.25). This creates a pulse stream at double the original rate, and so makes the frequency spectrum quite different (Figure 9.26). This now makes the lowest sideband well above the limit of the wanted signal and easily filtered out. In addition, if the added pulses are midway in amplitude between the original pulses, interpolated values, the conversion from pulses to smoothed audio can be considerably smoother, just as if the real sampling rate had been doubled. Oversampling of four times or even more is now quite common.

BITSTREAM METHODS

The most significant development in digital conversion in recent years has been the use of bitstream technology. The output of a bitstream converter is not the voltage or current level that is the output of a conventional D–A converter but, as the name suggests, a stream of 0s and 1s in which the ratio of 1s to 0s represents the ratio of positive to negative in the analogue

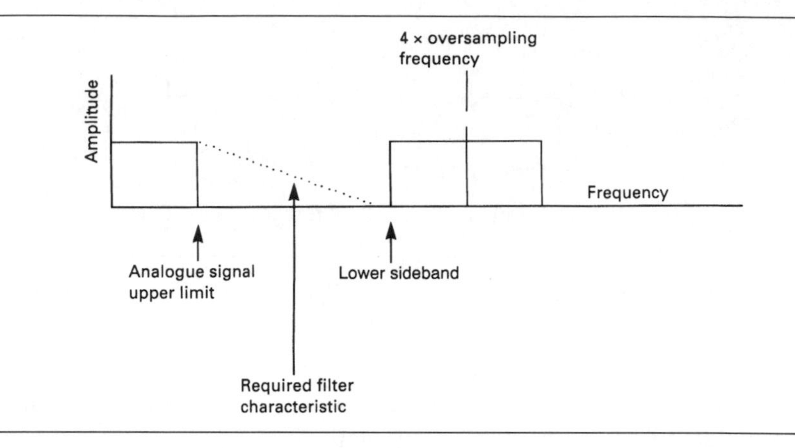

Figure 9.26 The frequency spectrum of a system using 4 × oversampling, showing that the filter characteristic is easier to produce.

signal. For example, a bitstream of all 1s would represent maximum positive voltage, one of all 0s would represent maximum negative voltage, and one of equal numbers of 1s and 0s alternated would represent zero voltage.

The enormous advantages of bitstream conversion are that only one current generator is required, there are no ladder networks needed to accomplish impossibly precise current division, and the output signal requires no more than a low-pass filter – the bitstream signal is at such a high frequency that a suitable filter is very simple to construct. The disadvantage is that the system must at this point handle high frequencies – of the order of 11–33 MHz for the bitstream converters now being used for audio-frequency signals. The essence of the bitstream converter is that instead of using a large number of levels at a relatively slow repetition rate, it uses very few levels (two) at a much higher repetition rate. The rate of information processing is the same, but it is accomplished in a very different way.

The principles are surprisingly old, based on a scheme called delta modulation which was developed for long-distance telephone links in the 1940s and 1950s. The normal pulse-code modulation system of sampling and converting the amplitude of the sampled signal into a binary-coded number has well-known drawbacks due to the noise that is generated because of the quantization process.

Delta modulation uses the difference between samples, which is converted into binary numbers. If the rate of sampling is very high, the difference between any consecutive pair of samples is very small, and it can be reduced to 1 bit only. A signal of this type can be converted back to analogue by using an integrator and a low-pass filter. Because of the very severe distortion that occurs when a delta modulator is overloaded, the sigma delta system was developed. This uses sampling of the signal

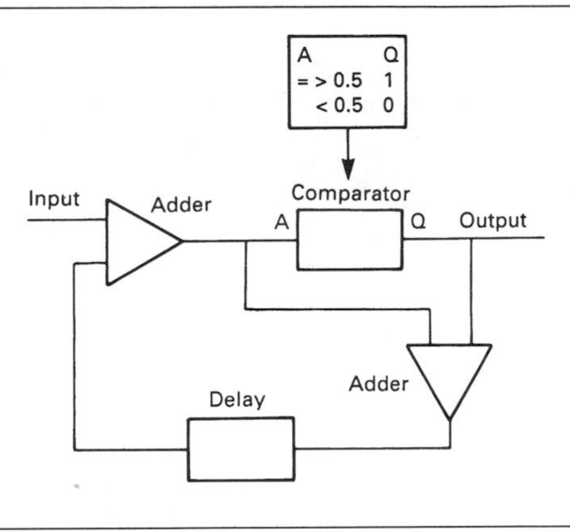

Figure 9.27 Outline of a simple noise-shaper circuit used for bitstream conversion.

amplitude rather than differences, and forms a stream of pulses by a feedback system, feeding the quantified signal back to be mixed with the incoming samples in an integrator. This results in a noise signal which is not white noise, spread evenly over the whole range of frequencies, but coloured noise, concentrated more at the higher frequencies. Because of this characteristic, this type of modulator is often termed a noise-shaper. This is the type of technology, used on other digital systems, which has led to bitstream and other forms of D–A converters in which the number of bits in the signal is reduced.

Current bitstream D–A converters consist basically of an oversampling stage followed by the circuit called the noise-shaper. The oversampling stage generates pulses at the correct repetition rate, and the noise-shaper uses these pulses to form a stream of 1s and 0s at the frequency determined by the oversampling rate. The important part of all this is the action of the noise-shaper. Figure 9.27 shows the principles of a noise-shaper, which consists of two adders, a converter, and a delay – the amount of the delay is the time between pulses. If we concentrate on the converter for the moment, this is a circuit which has a very fast response time and which will give an output of 0 for numbers corresponding to less than half of full output, and an output of 1 for numbers corresponding to more than half of the maximum level. The generation of the bitstream is done rather like the conversion of a denary number into binary, by successive comparisons giving a 0 or 1 and feeding the remainder of a comparison back to be subtracted from the next (identical) input signal. Note that 0 and 1 from the converter corresponds to 0 or I_{max} current values in a conventional D–A circuit rather than the digital 0 and 1 of a number.

Imagine a number at the input causing either a 0 or a 1 signal to be developed. The second adder works with one inverted signal, so that its output is the difference between the original signal amplitude number and

In	Out	Fed back
0.7	1	−0.3
0.7−0.3=0.4	0	+0.4
0.7+0.4=1.1	1	+0.1
0.7+0.1=0.8	1	−0.2
0.7−0.2=0.5	1	−0.5
0.7−0.5=0.2	0	+0.2
0.7+0.2=0.9	1	−0.1
0.7−0.1=0.6	1	−0.4
0.7−0.4=0.3	0	+0.3
0.7+0.3=1	1	0.0 end

Figure 9.28 An example of signal levels in a simple noise-shaper, showing how the bitstream is generated (bitstream contains seven 1s and three 0s; 70% full amplitude or 0.7 V).

the maximum signal (output 1) or minimum signal (output 0). This difference is delayed so that it appears at the input adder while the same oversampled copy of the input is still held there – the amount of delay will fix how many times this action can be carried out on each input. If the error signal is negative, it will be subtracted from the input, otherwise it is added, and the result is fed to the converter. This again will result in a maximum or minimum output signal and an error difference, and the error difference is once again fed to the input adder to be added to the new input and so to be converted. At each step, the output is a 1 or a 0 at many times the pulse repetition rate of the oversampled signal, and with the ratio of 0s and 1s faithfully following the amplitude corresponding to the input numbers.

Figure 9.28 illustrates the principle. The input number represents an amplitude of 0.7 and this input will cause the converter to issue an output of 1. The difference is −0.3, and when this is added to the next identical input it gives a net input of 0.4 which causes a 0 output with an error of +0.4. The diagram shows the successive steps which end when the net input to the converter is 1, so that there is no error signal from the output – this ends the conversion which, in this example, has taken 10 steps – implying that the output frequency will be 10 times the input frequency. The output is a stream of seven 1s and three 0s, corresponding to a wave that is 70% full amplitude or has an amplitude of 0.7 of the maximum.

Chapter 10

Switch principles

10.1 Principles

Switching is part of a sensor action for many devices, so that the topics relating to switch contacts are of considerable importance in specifying some types of sensors. In addition, switches, as an electromechanical component, are a vital part of any sensing or transducing equipment.

Switches are almost as old as present-day electricity applications, which is probably why their action is taken so much for granted. The function of a switch is to make or break current in a circuit, and in the early 19th century this would have meant only a DC circuit. Later, switching had to be applied to AC, later still to audio and radio frequencies and video signals, and most recently to digital signals. Early switches were simple mechanical devices, the knife switches that can still be seen in old factories. This century has seen the development of much more elaborate mechanical principles, and of switches with no mechanical moving parts at all. In this book, however, we are mainly concerned with mechanically-operated switches, and although some electronic types are dealt with, relays and contactors have been excluded.

The switch, then, is a device with a circuit path that can be made or broken. When the circuit path is made, the circuit resistance through the switch is low, of the order of mΩ for most switches. With the circuit path broken, the circuit path through the switch has a high resistance, several MΩ or higher. In addition, the resistance between the switch circuit path and the body of the switch, including the operating mechanism and the mountings, is generally required to be high. This resistance, and the maximum voltage that can be applied to this insulation, is often of major importance in mains voltage switches, and is a vital feature of legislation regarding switch suitability. For some applications to high-frequency signals the capacitance between the switch circuits and the switch body is of considerable importance, and will decide the design of the switch.

10.2 Contact resistance

The resistance of the (mechanical) switch circuit when the switch is on (made) is determined by the switch contacts, i.e., the moving metal parts in each part of the circuit which will touch when the switch is on. The amount of the contact resistance depends on the area of contact, the contact material, the amount of force that presses the contacts together, and also the way that this force has been applied. For example, if the contacts are scraped against each other (a *wiping* action) as they are forced together, then the contact resistance can often be much lower than can be achieved when the same force is used simply to push the contacts straight together. In general, large contact areas are used only for high-current operation, and the contact areas for low-current switches (as used for electronics circuits) will be small.

The true area of electrical connection will not be the same as the physical area of the contacts, because it is generally not possible to construct contacts that are precisely flat, or have surfaces that are perfectly parallel when the contacts come together. This problem will be familiar to anyone who has renewed the ignition points in a car in the days before electronic ignition circuits. The actual area of contact of the switch is revealed by the contact burning when the switch is used in inductive circuits, but is not necessarily obvious in other cases. Since the size of the switch to a large extent determines the amount of contact pressure that can be used, and the area of contact is rather indeterminate, the main factor that affects contact resistance is the material used to make the contacts themselves.

WIPING CONTACTS

One method of reducing contact resistance in difficult conditions is to use a wiping action. As the name suggests, the mechanical action of the switch is such that the contacts are rubbed against each other as they make connection. This will normally remove any thin films of oxide or other contaminants, and so ensure a much lower contact resistance than could be obtained by bringing the contacts together in a more straightforward manner. The disadvantage of a wiping action is that it can abrade the contacts, and if these consist of steel or nickel-alloy plated with gold, the gold plating can be removed by this abrasion. Wiping contacts are best suited to switches that are infrequently used and are located in contaminating environments, and those in which silver is used as the contact material.

The switch designer has the choice of making the whole of a contact from one material, or of using electroplating to deposit a more suitable contact material. By using electroplating, the bulk of the contact can be made from any material that is mechanically suitable, and the plated coating will provide the material whose resistivity and chemical action is more suitable to the electrical function of the switch. Plating also makes it possible to use materials such as gold and platinum, which would make the switch impossibly expensive if used as the bulk material for the contacts. It is normal, then, to find that contacts for switches are constructed from steel or from nickel alloys, with a coating of material that will supply the necessary electrical and chemical properties for the contact area.

The choice is never easy, because the materials whose contact resistance is lowest are, in general, those that are most vulnerable to chemical attack by the atmosphere. In addition, some materials that can provide an acceptably low contact resistance exhibit *sticking*, so that the contacts do not part readily and may weld shut.

The other main problems are burning and oxidation. The spark current that passes at the time when contacts are opening can induce melting of the contacts, or cause the metal to combine chemically with the oxygen in the atmosphere (oxidation or *burning*). Oxidation, rather than melting, is the more severe problem, because the oxides of most metals are non-conductors. This means that oxidation causes a large rise in contact resistance, even to the point of making the switch useless.

The usual choice of materials is illustrated in Table 10.1. From the point of view of resistance alone, silver is the preferred material, since silver has the lowest resistivity of all metals. Unfortunately, silver is very badly corroded by the atmosphere, particularly if any traces of sulphur dioxide exist (where coal or oil is burned, for example), and silver contacts will

Table 10.1 Comparison of commonly-used contact materials for switches and relays.

Material	Advantages	Disadvantages	Uses
Silver	Low resistance	Corrodes	High current, high contact pressure
Palladium-silver	Less easily contaminated	Higher resistance	General use
Silver-nickel	Resists burning and sticking	Higher resistance	General use
Tungsten	Hard, high melting point	Oxidizes easily	High-power switching
Platinum	Stable, resists chemical attack	High voltage, low current use only	Specialized
Gold	Resists corrosion, can be plated onto other metals	Low current only	Widespread in electronics

have a short life if there is any arcing at the contacts (see below). Silver contacts are desirable if the lowest contact resistance is needed for high-current use, but the action of the switch will have to be such that the contact pressure is high, and the contacts are wiped as they are brought together. Silver-coated contacts can be used with fewer problems when the switch is sealed in such a way as to exclude atmospheric contamination, or if the contacts are surrounded by inert gases, but these are not simple solutions to the corrosion problem.

Gold plating is a very common solution to the contact problem, particularly in switches for low-current electronics use. The contact resistance can be moderately low, and the gold film is soft, so that moderate pressure can result in a comparatively large area of contact. Of all metals, gold is about the most resistant to corrosion, although the combination of hydrochloric acid and electric current can cause gold to be attacked quite rapidly, making this material unsuitable if the atmosphere contains traces of chlorine or hydrochloric acid.

Since passing electric current through sea-water causes chlorine to be generated, gold-plated contacts can be severely corroded when the switches are used at sea. The switches are suitable only for the lower currents, but since most switch applications in electronics are for low currents this is no handicap. Gold plating of contacts is often mandatory in the specification of switches for military contracts.

Contact coatings based on the 'noble' metals are often employed. These metals are so named because, like gold, they exhibit a high resistance to chemical attack; typical metals in this group are platinum, palladium, iridium and rhodium. All of them have rather high contact resistance, but are very stable and resist chemical attack. Platinum is particularly useful for contacts that will be used for low currents and for high voltage levels. Rhodium and iridium platings provide a high level of resistance to corrosion along with stable contact resistance, and are suitable for medium-voltage, medium-current applications. Some advantages can be gained by using thick films of contact materials that are alloys of the noble metals with silver. For example, palladium–silver has a much better resistance to contamination than silver itself, although with a higher contact resistance than silver. It is one of the general contact materials that can be used in switches for various types of application.

The metals tungsten and molybdenum, although not of the platinum group, are also used as contact materials for special purposes. Tungsten in particular is very resistant to burning caused by contact arcing, and is used for high-power switching. Its disadvantage is that the surface will oxidize easily, causing contact-resistance problems. The most common general-purpose contact alloy material, however, is nickel silver. The cost of nickel silver is such that it can be used as a bulk material rather than as a coating, and although its contact resistance is higher than that of pure silver, it is much more resistant to chemical attack than the pure metal. It also resists burning, and the contacts do not tend to stick together.

10.3 AC and DC switching

The type of current that is being switched has a very considerable effect on the life of the switch contacts. If we confine ourselves to currents that are not signals, then the switch may have to handle either AC or DC. When contacts are made, the type of current is almost immaterial, because there is very little sparking across contacts that are rapidly coming together. Any *sparking* or *arcing* takes place almost exclusively when contacts are being separated. This is because conditions for sparking or arcing are ideal at the instant when two contacts are separating. At this instant, the gap between the contacts is very small, so that only a small voltage difference is needed to make current cross the gap. In this respect, sparking means an intermittent discharge between the contacts, and arcing is a continuous discharge.

Arcing is a much more damaging effect, because very high temperatures can be generated locally at the points where the arc originates. In addition, the air itself between the contacts will be heated to very high temperatures, and this has the effect of ionizing the air, making it conduct and so encouraging the arcing, which will then continue even when the contacts are separated by considerably more than the normal sparking distance.

A DC supply is much more likely to cause arcing than an AC supply, given the same conditions of contact opening speed. This is because the voltage across the opening contacts will remain fairly steady, whereas for an AC supply the voltage will drop to zero in each half-cycle. If the opening time of a switch used for AC is reasonably short, then the contacts will have opened appreciably while the voltage across them is, on average, low. In addition, the effect of metal transfer (see below) will be more serious when DC is switched. This difference is reflected in the very different ratings that apply for AC and for DC use of any given switch. One example is a miniature mains rocker switch rated at 250 V 4 A AC and 28 V 4 A DC, so that the effect of using DC is effectively to derate the switch operating voltage by a factor of nine times. This is not at all unusual, although the DC voltage rating might be increased if the current rating were decreased.

10.4 Inductive loads

Whether the supply that is being switched is AC or DC, the presence of an inductor of appreciable size in the switched circuit has a large effect on the switch ratings. Once again, the presence of an inductor (of the order of 0.05 H upwards) when a circuit is *made* presents no difficulties. The effect of an inductor on making a circuit is to cause the current in the circuit to rise to its normal value more slowly than for a resistive circuit.

The problems arise when an inductive circuit is broken. When current through an inductor is decreased, a voltage is induced across the inductor, and the size of this voltage is equal to inductance multiplied by rate of change of current. The size of the inductance will be expressed in Henrys and the rate of change of current in A/s.

Suppose, for example, that a circuit contains a 0.5 H inductor, and a switch would break a current of 2 A in a time of 10 ms (0.01 s). The rate of change of current would be $2/0.01 = 200$ A/s, so that the voltage across the inductor is $0.5 \times 200 = 100$ V. Now this is comparatively small, but if the circuit happened to be a 6 V DC circuit, with the switch contacts rated for 6 V DC, then the presence of 100 V across the switch contacts will cause a considerably greater amount of arcing than would be present in a resistive circuit. In practice, the effective breaking time is likely to be less than 10 ms, so that the voltage induced by breaking the circuit will be higher than this example suggests. If the circuit uses AC, then there is a reasonable chance that the current may not be at its peak at the time when the circuit is broken and, as before, the amount of arcing is likely to be less. Note, however, that the induced voltage across the inductor at the instant when the circuit is broken is not an AC voltage, and follows much the same pattern whether the current that flowed was AC or DC. The effect of breaking the inductive circuit is a pulse of voltage, and the peak of the pulse can be very large, so that arcing is almost certain when an inductive circuit is broken unless some form of arc suppression is used.

10.5 Arcing and switch life

Arcing is one of the most serious effects that reduce the life of a switch. During the time of an arc, as we have seen, very high temperatures can be reached, both in the air and on the metal of the contacts. The ionization of the air gives rise to a *plasma*, a quantity of completely ionized air that is a comparatively good conductor of electric current. The temperatures that are reached in this plasma can cause the metal of the contacts to vaporize and be carried from one contact to the other. This effect is very much more serious when the contacts carry DC, because the metal vapour will also be ionized, and the charged particles will always be carried in one direction. If, for example, the metal forms positive ions, then it will always be carried from the positive contact of the switch to the negative contact. This will result in the familiar effect of the positive contact developing a crater and the negative contact developing a mound. Since the surfaces of these parts of the contacts are rough, this greatly reduces the contact area, lessening the ability of the contacts to take their rated current. As usual, AC causes less trouble, because the arc is usually quenched within a half-cycle, and since the current direction will not always be the same at the time when the arc exists, the transfer of metal is not always in one direction.

Arcing is also a means by which contacts become contaminated, because in the plasma that exists during arcing any contaminant elements in the atmosphere become ionized, and are transferred to one of the contacts. By this means, contaminants that were present in the atmosphere become permanently embedded in the contacts, with detrimental effects on the contact resistance. Arcing can also cause severe oxidation of the metal of the contacts, and this will also raise contact resistance. Fortunately, the transfer of metal in arcing inhibits oxidation to some extent, because it causes a fresh set of metal surfaces to appear at each switching-off action.

Arcing, therefore, will greatly reduce the life of switch contacts, and it is an effect that cannot be completely eliminated. Arcing is almost imperceptible if the circuits that are being switched run at low voltage and contain no inductors, because a comparatively high voltage is needed to start an arc. For this reason, then, arcing is not a significant problem for switches that control low voltage, such as the 5 V or 9 V DC that is used as a supply for solid-state circuitry, with no appreciable inductance in the circuit. Even low-voltage circuits, however, will present arcing problems if they contain inductive components, and these include relays and electric motors as well as chokes. Circuits in which voltages above about 50 V are switched are the most susceptible to arcing problems, particularly if inductive components are present, and some consideration should be given to selecting suitably rated switches, and to arc suppression, if appropriate.

10.6 Arc suppression

The total elimination of arcing is never possible, although arcing becomes insignificant if the current carried by a switch is low (of the order of a mA), and with a non-inductive load. If a mechanical switch is to have a very long life, consideration should be given to using it in the input side of a solid-state switching arrangement, using transistors (bipolar or MOS), thyristors or Triacs. This solution may not be appropriate to every type of switching application, but for the switching voltages and currents that are used in electronics it can often be a way out of a reliability problem. One snag is that the use of a mechanical switch along with an electronic switch requires a power supply to the semiconductor part of the circuit, which will need to be applied before the mechanical switch can be operated. This may mean that two switches are needed: the second applies power to the circuitry that will then switch the main load. If the switches are then used in the wrong order, the arcing problem will return. For very large switchgear, solutions such as oil-quenching and air-blast arc suppression are used, but in this book we are confining our attention to the smaller size of switch, and hence to purely electrical and electronic methods.

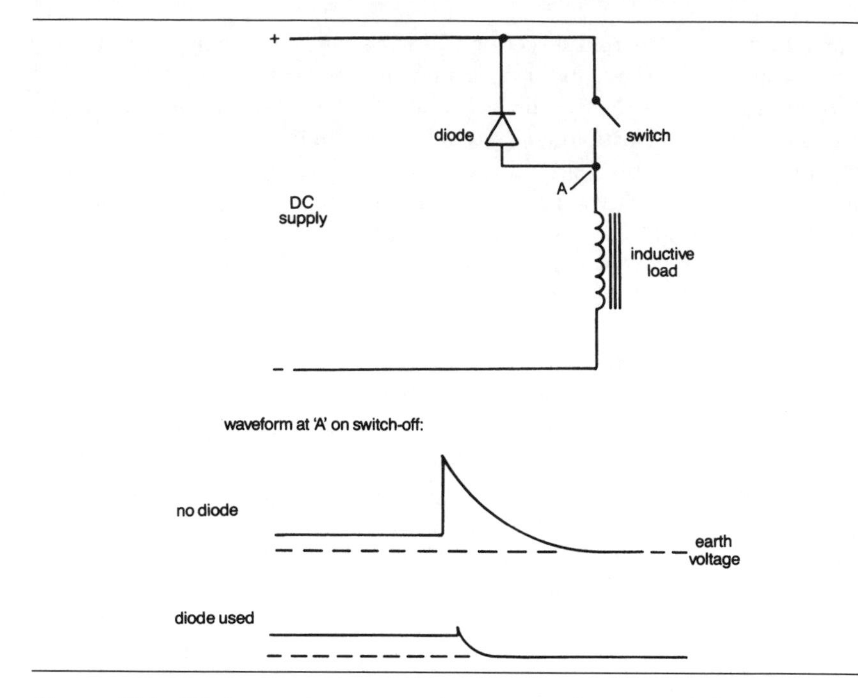

Figure 10.1 Diode transient suppression connected across switch contacts in an inductive circuit.

Most arc-suppression circuits hinge on delaying the change of voltage across the switch contacts, and the techniques that are used depend on the type of supply (AC or DC) and the nature of the load (inductive or non-inductive). Since DC supplies and inductive loads present the main problem, we shall deal with these first. The most effective arc suppression for circuits of this type is diode suppression, as illustrated in Figure 10.1. The principle is that when the circuit is broken, the sudden drop of current through the inductive load will induce a voltage that is always in the reverse direction to the applied voltage across the switch. By using a diode connected so as to conduct for this polarity of voltage, the surge can be forced to pass a pulse of current back to the power supply, or into a capacitor wired between the switch live terminal and ground. This form of protection is particularly effective when the supply voltage is low, because it protects the switch against the much higher voltages which are caused by breaking the inductive load, and which are high enough to cause arcing.

Diode protection is much less effective when the circuit operates at a higher voltage, e.g. 100 V or more, because the voltage across the contacts cannot be less than the supply voltage, and this is large enough to promote arcing. The arcing will, however, be considerably less severe than it would be if the diode were not present. Whatever voltage level is used in the circuit, the diode must be adequately rated both for the voltage and for the

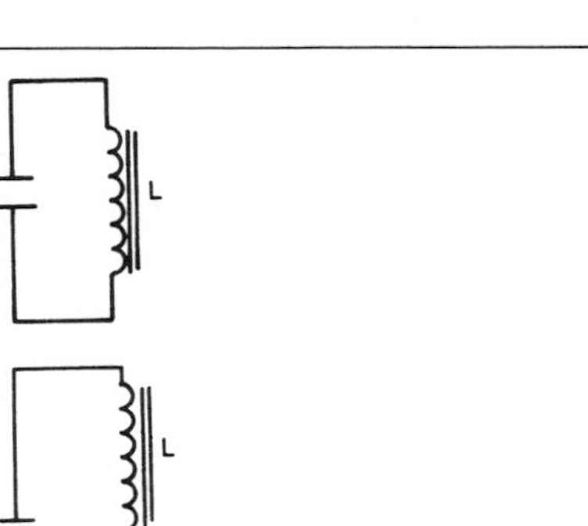

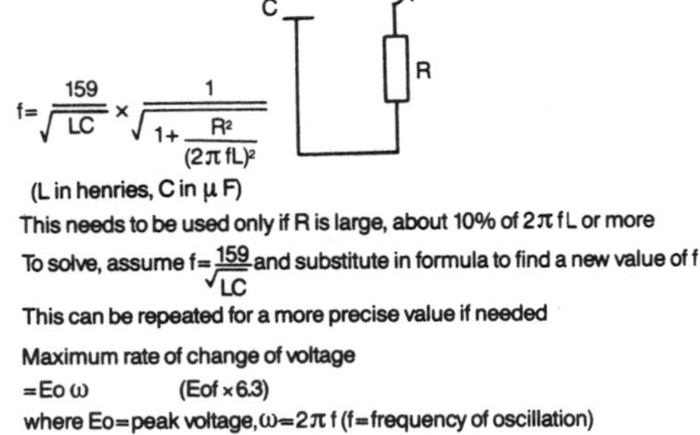

Figure 10.2 Resonance in capacitor–inductor circuits. The resistance in a switch suppression circuit is usually large, but the precise frequency of resonance is not important in any case.

current that can flow. The reverse voltage rating of the diode should be adequate for the voltage applied in the circuit, and preferably higher so as to give a margin of safety.

Note that in such a circuit, if the diode were to break down to a short-circuit the circuit would be switched on, and this could affect safety. No form of arc-suppression diode or other circuit should be applied across any form of safety switch. For such switches, arc suppression is unimportant because the switches should not normally be opened with power applied, and the risk of failure of a suppression component might have disastrous results.

The other main electronic method of arc suppression in DC circuits involves the use of capacitors, often along with resistors and/or inductors. The use of a capacitor alone relies on the rate of charge of the capacitor being lower than the rate of change of voltage across the switch when it is opened. This will certainly be true for a non-inductive circuit, and will be true for an inductive circuit if the resonant frequency of the capacitor with the inductive load is low. Figure 10.2 gives a guide to resonant frequency and the relationship between rate of change of voltage and frequency.

For many purposes, a resonant frequency of 50 Hz is a good choice, because the AC rating of the switch is then applicable to its use in the DC arc-suppressed circuit. The capacitor that is used must be rated for the

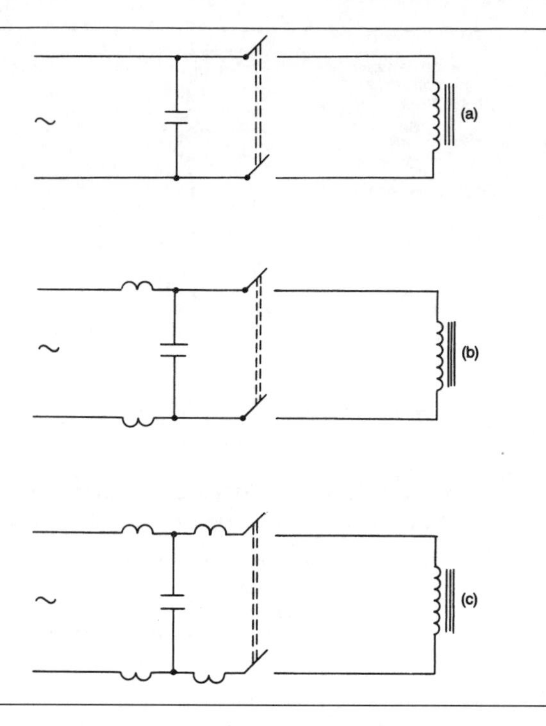

Figure 10.3 Circuits used on AC supply lines for the suppression of transients caused by switching. This can include the interference caused by the use of thyristors and Triacs. The capacitors must be rated for at least 250 V AC working.

peak voltage that can be expected, and must be of the paper or plastic dielectric type. Never specify an electrolytic capacitor for this type of action, although ceramic or mica types can be used additionally as suppressors for high-frequency oscillations if needed.

Note also that this capacitor method of arc suppression, using comparatively large values of capacitors, applies to DC circuits. For an AC supply, the presence of a capacitor will provide a leakage path when the switch is open-circuit, and this can cause problems, not least with safety. Any suppression components in an AC supply should use capacitors only across the lines (Figure 10.3), never across the switch contacts themselves, although RF suppression capacitors of a few pF are permissible in some circuits.

The use of capacitors with resistors and inductors leads to more elaborate suppression circuits. A series capacitor–inductor suppression circuit will have a series resonance at which its impedance is a minimum, and a rate of change of voltage corresponding to this resonant frequency will pass easily through the suppression circuit in preference to the (open) switch contacts. This type of suppression is particularly useful if diodes cannot be used. By adding a resistor, either to the capacitor–inductor circuit or to the capacitor alone, the rate of change of current through the arc-suppression

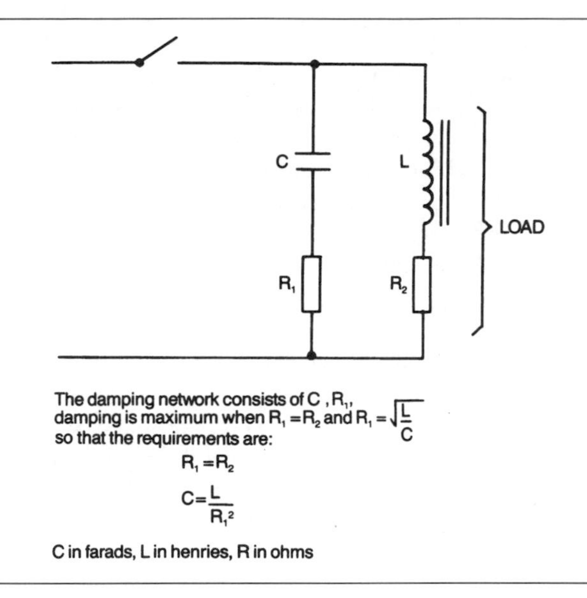

Figure 10.4 Using a capacitor–resistor damping network connected across switch contacts.

circuit can be reduced. This in itself does not assist in arc-reduction – quite the reverse – but it can greatly reduce radio-frequency interference (RFI) caused by the arc-suppression circuit itself. This in turn should be rather less than the RFI that would be caused by the arcing. Figure 10.4 shows circuits and formulae relating to resistor–capacitor–inductor suppression circuits.

For AC supplies, the nature of the supply voltage itself ensures that any arcing should be quenched in a fraction of a cycle, so that the AC current rating for switches is generally higher than the DC rating. The use of diode arc suppression is ruled out, and in general, capacitors and inductors are used only in order to suppress RFI, using mica capacitor and ferrite-cored inductors. Another method of suppression is the provision of non-linear resistors across the switch contacts. The type of non-linear resistor that can be used has a voltage–current characteristic that shows a decrease of resistance as the voltage across the resistor increases. In many applications, however, the polarity of any excessive voltage across a switch can be predicted, so that a semiconductor diode can be used for absorbing any surges.

10.7 Environmental effects

The effect of temperature on a switch will be to make changes in its rated characteristics. The normal temperature range for switches is typically

−200°C to +800°C, with some rated at −50°C to +100°C. This range is greater than is allowed for most other electronic components, and reflects the fact that switches usually have to withstand considerably harsher environmental conditions than other components. The effect of very low temperatures is due to the effect on the materials of the switch. If the mechanical action of a switch requires any form of lubricant, then that lubricant is likely to freeze at very low temperatures. Since lubrication is not usually an essential part of switch maintenance, the effect of low temperature is more likely to be an alteration of the physical form of materials such as low-friction plastics and even contact metals.

Plastics that are used as low-friction bearings can become sticky at very low temperatures, causing irregular mechanical action. The metals that are used for contacts can alter their crystal shape, causing a large increase in contact resistance. This change of crystal shape is very sharp, and will occur at a clearly defined temperature. The best-known change of this type is the change of tin from its metallic form to the form of a dull-grey powder at a temperature of about −20°C, but other metals will exhibit less dramatic changes at various temperatures. At the high-temperature end of the scale, the most serious threat is to insulation resistance, which can decrease very considerably at the higher temperatures. Contact resistance can also be increased because of the growth of layers of oxide on contact surfaces, and if a switch is to be used at constant high temperatures, near its rated limit, a wiping action is desirable (see above). For electronics purposes, exposure of a switch to high temperatures is more likely than exposure to very low temperatures.

10.8 Sealing and flameproofing

The flameproofing of a switch does not refer to protection of a switch against fire, but to the prevention of fire caused by the switch. A flameproof switch would be specified wherever flammable gas can exist in the environment (for instance in mines, chemical stores, and processing plants that make use of flammable solvents). In such an atmosphere sparking or arcing at switch contacts could ignite flammable gas in the atmosphere, with potentially devastating results. A flameproof switch is one which is sealed in such a way that sparking at the contacts can have no effect on the atmosphere outside the switch. This implies that no vapour from the atmosphere can penetrate the switch in any way, either through cable entry points, or where the switch actuating mechanism enters the switch. This makes the preferred type of mechanism the push-on, push-off type, since the push button can have a small movement and can be completely encased along with the rest of the switch.

Flameproofing cannot necessarily be satisfactorily achieved simply by sealing the switch, because a sealed switch can often have a very short life because of the accumulation of acids caused by the effect of sparking and arcing on the confined atmosphere. A flameproof switch requires materials that are selected so as to minimize attack by such acids, and a sealing material that will not contribute unwanted vapours into the switch. For preference, it should be filled with an inert gas such as nitrogen. The materials used for sealing nowadays are usually synthetic silicone rubbers, because these are themselves very inert, and can also withstand high temperatures without deterioration. Since flameproof switches are intended for situations where safety is of paramount importance, all such switches must conform to the appropriate British Standard BS 6458.

The sealing of a switch may also be intended to provide waterproofing, and this is covered by the BS 2011 set of tests, Parts 2.1 Q and R. A major problem of waterproofing is that the switch becomes non-vented, so that any arcing will result in a corrosive atmosphere unless the switch is surrounded by inert gas before sealing. Any switch that is part of a circuit likely to be immersed for long periods should preferably handle low voltage and current with a non-inductive load. This is not difficult to arrange if the switch is part of a system for which the main circuits are in air.

10.9 Connections to switches

Connections to switches may be made by soldering, welding, crimping, screw fastening or by various connectors or other plug-in fittings. The use of soldering is now comparatively rare, because unless the switch is mounted on a PCB which can be dip-soldered this will require manual assembly at this point. Welded connections are used where robot welders are employed for other connection work, or where military assembly standards insist on the greater reliability of welding. By far the most common connection method for panel switches (as distinct from PCB-mounted switches) is crimping, because this is very much better adapted for production use. Where printed circuit boards are prepared with leads for fitting into various housings, the leads will often be fitted with bullet or blade crimped-on connectors so that switch connections can be made. This assumes the use of switches with suitable terminations, and the most common switch termination of this type is the flat blade, so the connecting cable should carry a matching receptacle. Piggyback receptacles can be used when wiring has to be carried from one switch to another.

10.10 Non-contact switching

Low-voltage switching can be carried out by semiconductor devices, notably FETs, with no mechanically moving contacts. The use of devices such as FETs and thyristors is more correctly classed as a type of relay

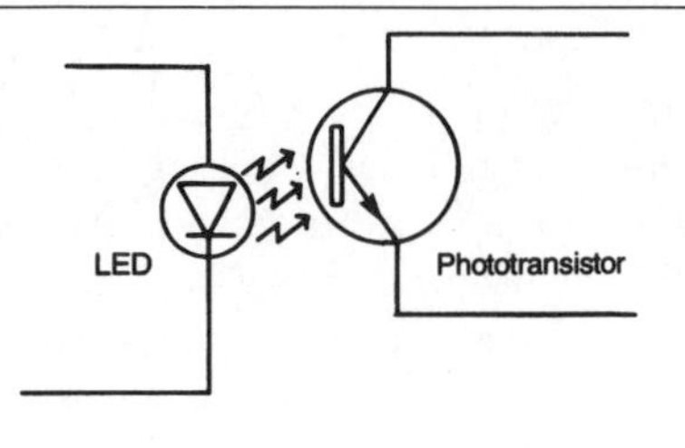

Figure 10.5 Principle of the opto-isolator, consisting of an LED and photransistor.

action, however, since the switching is in response to an electrical signal. The use of the Hall effect is a true contactless switching effect, because a non-electrical input (the movement of a magnet or a change in a magnetic field) causes a current path to be made or broken. Switching of this type is now very common, notably in the contact-breaker mechanisms of car ignition circuits that use semiconductor switching methods.

Contactless switching has a number of advantages, particularly in hazardous environments. The switching action does not depend on any form of mechanical action, so that problems of rusting, jamming, wear and fatigue no longer exist. Since the 'contacts' are internal to the semiconductor's material they are not liable to corrosion, particularly if the whole switch is encapsulated. Low voltages and currents can be used, avoiding spark risks even if wires are damaged. The switching action can be very fast and in many cases can have variable sensitivity.

The main drawback of contactless switches is that the action generally requires low-voltage DC at some point – I am excluding the action of thyristors and Triacs here. The control action of the switch may, however, need to make or break circuits that operate at higher voltages, and this requires some form of relay action. The conventional contact form of relay can always be used, and in some applications may be required by regulations, but the alternative is a contactless form of relay, the opto-isolator. The opto-isolator is shown in outline in Figure 10.5. It consists of an LED whose light can be focused onto a photocell, with no electrical contact between the LED and the photocell. When the LED is illuminated, the light path to the photocell ensures that there will be an output from the photocell, and this can be amplified and used for switching purposes. The devices are encapsulated to ensure that all the light from the LED will reach the photocell, and no ambient light can reach the photocell. Table 10.2 shows some typical opto-isolator characteristics.

Since the LED can be part of a low-voltage circuit, and the photocell can be part of a circuit that is connected to mains or other high-voltage supplies, the opto-isolator can be a very useful linking device. One particular advantage is that the opto-isolator action can be fast, allowing switching at audio and even video frequencies. It is therefore a unique way

Table 10.2 Characteristics of an opto-isolator intended for high-voltage isolation work.

The RS OPL 1264A is a high-voltage opto-isolator which is BAS EEFA approved. It consists of a gallium–arsenide–phosphorus photoemitter coupled to an NPN phototransistor, and provides isolation to 10 kV.

Minimum current transfer ratio (at $I_f = 10\,mA$, $V_{ce} = 5\,V$) is 25%
Rise time is 4 μs (typical)
Fall time is 3 μs (typical)

Diode portion:
V_f max. at 20 mA and 25°C is 1.5 V
V_r min. at 25°C is 3.0 V
I_f typical is 10 mA
Max. power dissipation at 25°C is 50 mW

---

of transmitting a signal which may contain both high frequencies and DC levels between two very different voltage levels. The most obvious use is to link a low-voltage circuit, such as a touchpad, to a high-voltage circuit, such as a thyristor controller, but this use is often prohibited by wiring regulations, so that some care has to be taken when specifying the use of opto-isolators. Since thyristors can be triggered by pulses applied via transformers, this alternative is more commonly used.

Chapter 11

Switch mechanisms

11.1 Non-signal switches requirements

The category of non-signal switches includes all varieties of switch that are placed in supply voltage lines, whether DC or AC, or which make and break contacts between DC levels that are not of critical amplitudes. Switches that are used in power supply leads will often have a mechanical action that provides for a rapid switchover, the snap-over type of mechanism that is illustrated in Figure 11.1. This provides a form of mechanical hysteresis, so that the position of the switch operating mechanism for making connection is not the same as the position for breaking connection. This snap-over action is not provided on all types of mechanical switches, however, and may be undesirable if a slow opening rate is required. In this chapter we shall confine our attention to manually operated switches, because any discussion of relays and contactors would extend the scope of the book too far into more specialized topics.

The simplest switch requirement to fulfil is for high-voltage (in the region of a few hundred volts), low-current, non-inductive load, AC or DC. For such a requirement, the type of switch and the contact materials are not particularly critical and the only stipulation is that the switch should be adequately rated for the voltage and current required. The value of contact resistance, for example, will make only a small difference in the voltage across the closed contacts, and this difference will be negligible compared to the voltage that the switch is handling. The non-reactive load implies that there will be no current surge on making connection, nor any voltage surge on breaking connection. Because of this, no special precautions have to be taken against arcing, and a long life can be expected. Only if special environmental conditions have to be allowed for will there be any difficulty in selecting a suitable switch from any supplier.

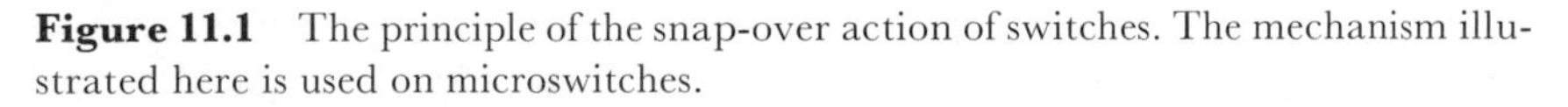

Figure 11.1 The principle of the snap-over action of switches. The mechanism illustrated here is used on microswitches.

Rather more care needs to be taken if the switching has to cope with high currents at high voltage, and considerably more care is needed if surges exist at switch-on or at switch-off. Taking these one at a time, the handling of high currents requires large contact areas, and contact materials that are of low contact resistance. The dissipation of power at the contact is equal to I^2R, where I is the current (steady or surge) and R is the value of contact resistance. For example, for a current of 10 A and a contact resistance of 5 mΩ the power dissipated at the contact is $0.005 \times 100 = 0.5$ W. This may seem insignificant, but a lot depends on how large the contact area is, and how well the contacts can dissipate the resulting rise of temperature. Just to put these figures into perspective, the 5 inch, 1 kW hotplate on an electric cooker will typically have a total surface area of 12 667 mm^2, giving a dissipation of 0.07 W/mm^2. To achieve the same dissipation per unit area for switch contacts dissipating 0.5 W would require an area of contact of about 7 mm^2, corresponding to a good fit of contacts of diameter 3 mm. From all this, you can see that a contact resistance of only 5 mΩ for a 10 A switch is by no means negligibly low, and that switch contacts, if thermally isolated, would run very hot.

Even if the voltage drop across the contact resistance is negligible, the effect of heat dissipation may not be, so that the size of the contact resistance and the area of the contacts are important considerations in any switch that handles large currents in the range of 1 A upwards. These factors will, of course, have been allowed for in the rating of the switch, but not all manufacturers will calculate their ratings making the same assumptions about thermal dissipation and the ratio of true contact area to measured area. If a large-current switch is intended for a sensitive application, in which failure would entail considerable consequent damage (for example, shutting down a complete installation), then it would be wise either to use more generously rated switches, or to measure the temperature of a few samples after an 8-hour period at full rated current in the expected maximum ambient temperature. The contact resistance should also be measured before and after such a test, and any significant increase will point to local overheating, which will not necessarily show up in the temperature measurements.

Switching for low-voltage supplies brings another set of problems, depending on whether the current is large or small. Most low-voltage switching is likely to involve large currents, so that a low contact resistance is an important requirement from both the dissipation and the voltage-drop considerations. In addition, low-voltage DC supplies are likely to use large capacitors on each side of the switch, and the capacitor on the load side will charge at the instant of switch-on, causing a surge current that can be very large (at least 10 times the average supply current). If this might cause difficulties, non-linear surge-suppressing resistors can be included in the supply line, but the voltage drop across such a device is usually prohibitive for low-voltage supplies. If possible, circuits should be arranged so that no large capacitors are connected on the load side of the switch between supply voltage lines. Unfortunately, the requirement for the minimum value of contact resistance in a switch for low-voltage, high-current applications often conflicts with the equally important requirement of long life. As always, there is a trade-off between contact resistance and contact durability.

The easiest requirement to fulfil for low-voltage supplies is for low current (in the range of a few mA to a few hundred mA). At these currents, contact resistance is not the main consideration, small contact areas can be used, and only the possibility of surge current or voltage can make for any difficulty in switch selection. The mechanical action of the switch is seldom important. There is no need for a snap action, so the full range of actions can be considered, and the physical size of the switch is unimportant from the electrical point of view. Switch size should, however, be considered from the ergonomic point of view. Many very small switches are available, but they are decidedly difficult to use simply because they are so small. Long lines of such switches should be avoided, no matter how satisfying they may look, because it is highly likely that at some stage a user will unintentionally switch over two at once. When switch operating levers are so close that a human fingertip can cover more than one, trouble can be expected eventually.

11.2 Ratings and specifications

The published ratings for a switch will invariably start with the voltage and current rating of the contacts, showing both DC and AC ratings. These ratings should be taken as absolute maxima if the highest standard of reliability is required. For example, if a switch is rated at 30 V, 4 A DC and 250 V 2 A AC, then it is reasonable to use the switch with 30 V, 4 A either

AC or DC, and up to 250 V 2 A AC, but not to exceed the absolute limits of 250 V AC or 30 V DC, and the absolute limits of 4 A at low voltage and 2 A at high voltage. You should not, in other words, work on the basis that you can trade voltage for current and assume that if the voltage is less than 30 V, then the current can be more than 4 A. As has been indicated earlier, the voltage rating and the current rating for a switch are imposed through different constraints, and a change in one quantity does not necessarily affect the other, unlike power dissipation in a load.

The other important quantity, which is not always quoted, is contact resistance. When this is quoted, it will be in units of mΩ, typically in the region of 5–20 mΩ. The figure that is quoted is usually the initial contact resistance, meaning the contact of the switch as supplied, since contact resistance tends to increase with the age of a switch when the switch is put into service. The rise in contact resistance will be greater if the switch does not use wiping contacts, if the environment is corrosive, and if arcing is frequent. Contact resistance is seldom of the greatest importance in non-signal switches, and for that reason is seldom quoted for these switches.

The reliability of a switch will be quoted in terms of the average number of operations to failure under accelerated test conditions. These test conditions may be unusually severe, so that for 'average' applications it is normal to find that the actual life of a switch is greater than the quoted amount. In many cases this may be quite irrelevant, because the equipment in which the switch is installed will have failed or been scrapped long before the switch fails. The expected life is usually quoted as two figures, the mechanical life and the electrical life. The mechanical life expresses the average number of operations before a mechanical failure, such as a broken spring or contact leaf. This is usually longer than the electrical life, which is the average number of operations up to the point of electrical failure through unacceptable contact resistance, contact welding, or failure to make contact. A typical figure for this quantity is 50 000 operations. To put this into perspective, if a piece of equipment is switched on and off once a day in a 5-day week cycle, the switch failure can be expected at about 96 years. In other words, a switch in a reasonable environment is not likely to be the component that determines the effective life of a piece of equipment.

The problem is that a switch, of all the components in an electronic circuit, is least likely to be located in a favourable environment. This applies particularly to front panel switches that will be exposed to shop-floor atmosphere and a good deal of heavy mechanical use. Whereas the other components are encased, possibly encapsulated, and do not, in the main, have moving contacts, a panel switch is located where the surrounding atmosphere can reach the contacts.

It is also subject to large forces on its actuating levers. Manual operation of a switch can vary from gentle pressure to violent prodding of a push button or fast flicking of a toggle lever. The actuating mechanism must be strong enough to withstand the extremes of treatment that are likely, but must not, as far as possible, allow such treatment to affect the contacts. The provision of strong and freely moving actuating parts makes it difficult to seal the interior of a switch in any way, and because of the effects of arcing, sealing is seldom desirable in any case. Expected life is therefore a quantity that can be very misleading unless it has been measured under conditions that approximate to the working conditions of the switch.

11.3 Snap-over action

The old-style knife switch had an action for which switching speed depended entirely on the operator, since the moving contact(s) would be mounted on the switch handle that the operator held. Modern switches for supply use do not rely entirely on the movement of the operator to determine opening or closing speeds, so that the switch consists mechanically of two parts, the contact part and the actuating part. Of these, the contact part consists of a spring-loaded toggle (Figure 11.2) which has two stable positions (the positions in which contact is made). In any other position of the toggle, apart from a precariously balanced central position, the spring action will ensure that the moving contact is in one stable position or the other. This spring also determines the force with which the contacts are held together.

The actuating portion of the switch will, in its motion, displace part of the spring toggle until the spring action flips over the switch. In this way the shape of the toggle and the strength of the spring determine the minimum switching time. No matter how slowly the switch is actuated, the toggle action should ensure that the moving contact snaps over, and no normal style of actuation should allow the moving contact to rest in any position

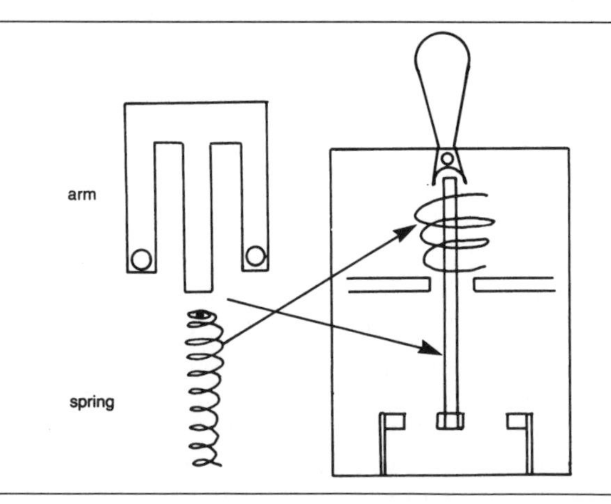

Figure 11.2 The toggling action of a lever switch.

except its contact (or on/off) states. The action of the operator can, however, affect the switching time in the sense that fast operation can speed up switching. Note that the action of a biased switch (see below) is very much more directly affected by the operator.

The actuator of the switch does not affect the static contact pressure, but very fast actuation can greatly increase the dynamic pressure, i.e., the momentary pressure due to the momentum of the moving contact assembly. This can in turn lead to excessive switch bounce, in which the contacts close and then bounce open again before closing finally. Switch bounce is more of a hazard for signal-carrying switches, but it can cause increased severity of arcing in non-signal switches, and the excessive mechanical force that accompanies it can also shorten the life of a switch.

11.4 Contact configurations and actions

Switch contact configurations are primarily described in terms of the number of poles and number of throws or ways. A switch pole is a moving contact, and the throws or ways are the fixed contacts against which the moving pole can rest. The term *throw* is usually reserved for mains switches, mostly single- or double-throw, and *way* for signal-carrying switches. A single-pole, single-throw switch (SPST) will provide on/off action for a single line, and is also described as single-pole on/off. Such switches are seldom used for AC mains nowadays and are more likely to be encountered on DC supply lines. For AC use, safety requirements call for both live and neutral lines to be broken by a switch, so that double-pole single-throw (DPST) switches will be specified for this type of use. Double-throw switches are not so common for mains switches in electronics use – their domestic use is in two-way switching systems, and they have some limited applicability for this type of action in electronics circuits. A double-throw switch will also be described as a change-over type. The word *throw*, used to describe the number of fixed contacts, is a reminder of the snap-over action that is required of this type of switch.

Where a single or multiple contact is required to make connection to more than two fixed contacts per pole, then the fixed contacts are referred to as poles. This type of configuration is much more common in signal switches, and switches of up to 12 ways per pole are available in the traditional wafer form, as noted in Chapter 12.

The configuration of a double-throw switch also takes account of the relative timing of contact. The normal requirement is for one contact to break before the other contact is made, and this type of break-before-make action is standard. The alternative is make-before-break (MBB), in which the moving pole is momentarily in contact with both fixed contacts during the change-over period. Such an action is permissible only if the voltages at the fixed contacts are approximately equal, or the resistance levels are such that very little current can flow between the fixed contacts. Once again, this type of switching action is more likely to be applicable in signal-carrying circuits. There may be a choice of fast or slow contact make or break for some switches.

A switch may be *biased*, meaning that one position is stable, and the other, off or on, is attained only for as long as the operator maintains pressure on the switch actuator. These switches are used where a supply is required to be on only momentarily – often in association with the use of a hold-on relay – or can be interrupted momentarily. A biased switch can be of the off-on-off type, in which the stable condition is off, or the on-off-on type, in which the stable condition is on. The off-on-off type is by far the more common. The ordinary type of push-button switch is by its nature off-on-off biased, although this is usually described for such switches as momentary action (see, however, alternate action below), and biased switches are also available in toggle form.

11.5 Toggle switches

The toggle switch accounts for the majority of switch sales, and is available in the largest range of current and voltage ratings for electronics use. In addition to the large range of electrical characteristics, toggle switches are also available in a very wide variety of mechanical styles of mounting and actuation. The most common mounting method is the traditional threaded bush, generally of 0.468 inch diameter (11.9 mm) and 10–11 mm length. The bush uses two clamping nuts so that the body of the switch need not be placed hard against the panel on which it is mounted (Figure 11.3). Both clamping nuts should always be used, because it is bad practice to butt the switch body against a panel with a single retaining nut on the other side. The use of bush mounting allows the switch to be mounted in

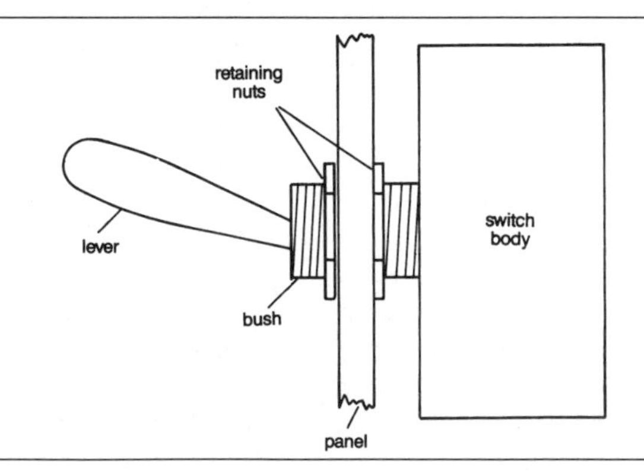

Figure 11.3 The use of clamping nuts to ensure that switch fastening places no stress on the switch body.

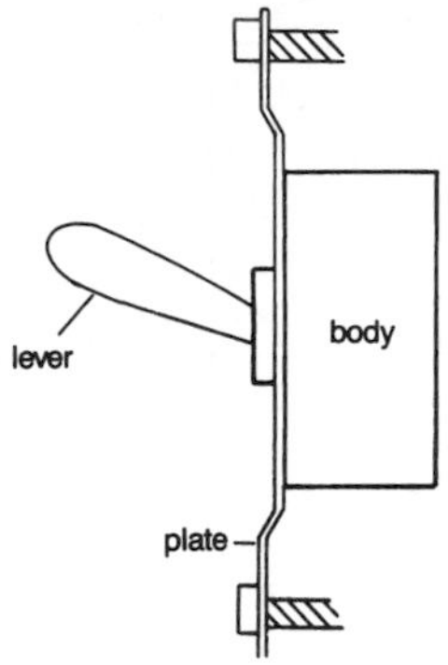

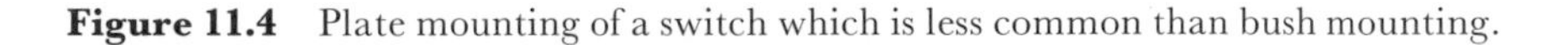

Figure 11.4 Plate mounting of a switch which is less common than bush mounting.

any position on a panel, but the conventional position allows the toggle to be operated in a vertical plane. Alternative mountings include plate mounting, in which the body of the switch at the toggle end is extended into a plate (Figure 11.4) which is drilled for fitting to a panel, and PCB mounting, in which the switch terminals are pins that will fit on a 0.1 inch (2.5 mm) grid. Since toggle switches for PCB mounting cannot be rotated relative to the printed board, such switches are available in vertical or horizontal toggle movement styles. Toggle switches for PCB mounting can also be obtained with plain (non-threaded) bushes for push-fitting through panels.

The actuating toggle itself can be obtained in a variety of lengths, and in rounded or paddle (spade) shapes. A long paddle-shaped toggle can often be an advantage for operating a switch that has a high spring pressure, and can also be easier to operate in an emergency than a short, round toggle. Coloured covers can be obtained for some round-toggle types, allowing for faster identification of switches. Some types can also be obtained with locking lever action to prevent casual or accidental switching. To operate such a switch, the toggle must be pulled out slightly against the tension of a spring before the switch-over movement can be made. Waterproof sealing glands are available for some sizes, and PVC splash covers are also available for a lesser degree of protection. Connections are made by all the available methods of solder tags or buckets, screw fittings, PCB pins, and slide-on receptacles.

The contact materials for toggle switches will normally be silver-plated nickel, although the smaller switch sizes may allow a choice of solid silver or gold-plated contacts.

11.6 Rocker switches

Rocker switches are a more recent development in switch technology. The

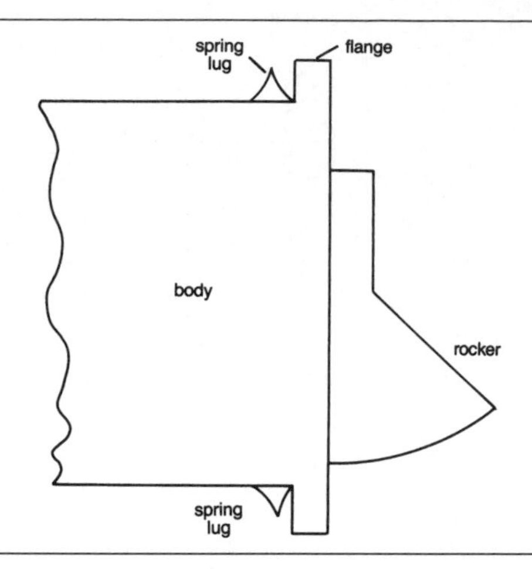

Figure 11.5 The use of spring lugs to retain a switch in a rectangular hole of precise dimensions.

aim is to make the switching action easier from the ergonomic point of view: a rocker switch can be operated by a push action as well as by the conventional flicking action, or by any combination of these movements. The rocking actuator of the switch is much more closely coupled to the internal toggling arms, and will snap over at the same rate. For small rocker switches the external rocker is part of the assembly that bears the moving contact(s). The actuating rocker is much wider than the lever of a toggle switch, even a paddle-shaped lever, and results in lower pressure on the operating finger. This extra width requires a broader switch body, so that rocker switches often allow space for signal lights. The considerable difference in the shape of the actuator means that bush mounting is not used, and rocker switches are for the most part fixed by lugs that have a snap action against the mounting panels (Figure 11.5), although a few types use mounting plates that can be bolted into place.

The mechanism of some types of rocker switch can make it possible for the operator to hold the switch in a 'neutral' position. If this would be undesirable such switches should not be used, although this does not preclude the use of other designs of rocker switch. The range of electrical and mechanical parameters is much less than for toggle switches, but rocker switches do have particular advantages: they are available in colours to match panel schemes, and can be obtained illuminated. Illumination in non-signal switches generally makes use of a miniature neon with a series resistor of a size suitable for 240 V AC operation (but not suitable for low-voltage use).

11.7 Slide switches

The mechanism of a slide switch does not permit any form of snap action:

the speed of making or breaking depends purely on the operator, and there is nothing to stop a switch being placed and left in a mid-way position. Such switches are therefore specified more for use in signal circuits, and in supply circuits that use low-voltage AC or DC. The main advantages of using a slide switch are that the contacts usually have a wiping action that ensures clean surfaces, and that the contact pressure can be quite high without making the switch action difficult. Fastening can be achieved using bolts through lugs, PCB mounting or clips. Single slide switches are becoming a rare sight in electronics equipment; they are being replaced by DIL pattern switches (which are of the slide pattern) for signal voltages, and by miniature rocker switches for supply voltage use.

11.8 Push-button switches

Push-button switches challenge the range of toggle switches for variety, although many push-button types are intended purely for signal uses. The two basic types of push-button switch are the momentary-action and alternate-action types. The internal construction of a momentary-action switch (Figure 11.6) demonstrates the principles. The moving contact is held on the shaft that is actuated by the button, and the contact is made (in this example) for as long as the button is pushed. Releasing the button allows the main spring to separate the contacts, subject to the range of travel of any internal spring that is used to buffer the moving contact from the shaft. This type of momentary-use switch can be of push-to-make or push-to-break varieties.

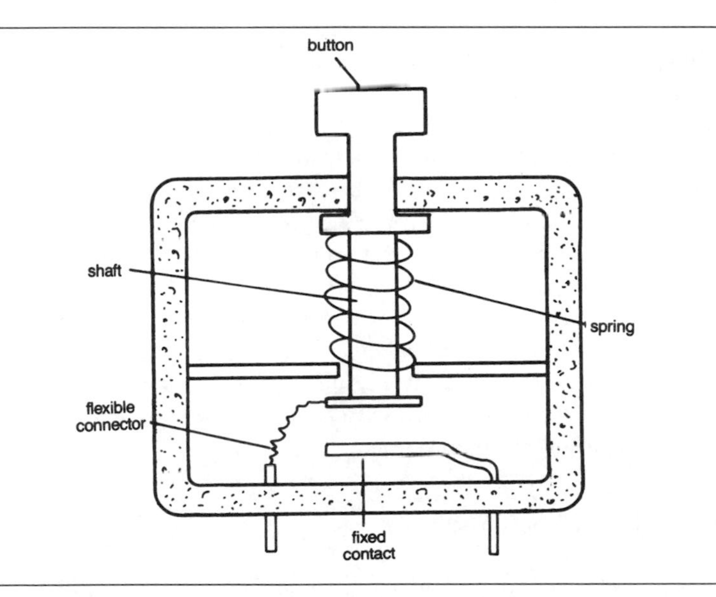

Figure 11.6 Operating principle of the momentary-contact push-for-on switch.

The other basic type is the alternate-action type. In this version, pushing the button will reverse the effect of the previous push, switching either on or off. The mechanism is usually rotary and permits only fairly small contact pressures, so these switches are not intended for high current uses. For the mains switching versions, 250 V and 3 A is a typical rating, although higher current ratings are available. The term *latching* is often used of the alternate mechanism, but latching has a specialized meaning when applied to a push-button switch. A latching push-button switch will allow the button position to indicate whether the switch is on or off, for example by having the button fully out when the switch is off, and partly in when the switch is on. The mechanism of a latching switch is basically that of the alternate type, but with the push button connected to the internal mechanism rather than simply pushing.

Push-button switches of either basic type are available with solder lugs or pins, or PCB mountings, but seldom with snap-on receptacles. Like rocker switches, many types of push-button switches are available with illuminated buttons in a variety of shapes and colours. The illumination is usually by miniature filament bulbs that will require a low-voltage supply, rather than by the neon which is used for rocker switch types.

11.9 Rotary and key action

The most common type of rotary switch is the wafer type (this is dealt with in the following chapter along with other signal circuit switches), but for more specialized purposes rotary mains switches are available. A rotary action switch is often found incorporated into a potentiometer, providing the combined on/off/volume control action that was very common at one time for mains radios and TV receivers, and is still found on some audio equipment. Separate rotary action mains switches are still available, but mainly for replacement purposes. One modern type is intended to discourage casual use by having a shaft with a screwdriver slot, but the main use for rotary mains switches in supply circuits nowadays is to permit key action for security purposes. The variations in key mechanism include removable or trapped keys, and common or random key patterns. The use of a trapped key allows a circuit to be switched on by a keyholder, but since the key must then remain in the switch the equipment can be switched off by anyone, as safety regulations will usually stipulate. If the key can be removed, some other switch-off mechanism in the form of an emergency button should be provided.

11.10 Microswitches

The microswitch is a component with a comparatively short history but a rapid development in use. The ‘micro’ part of the title is slightly misleading

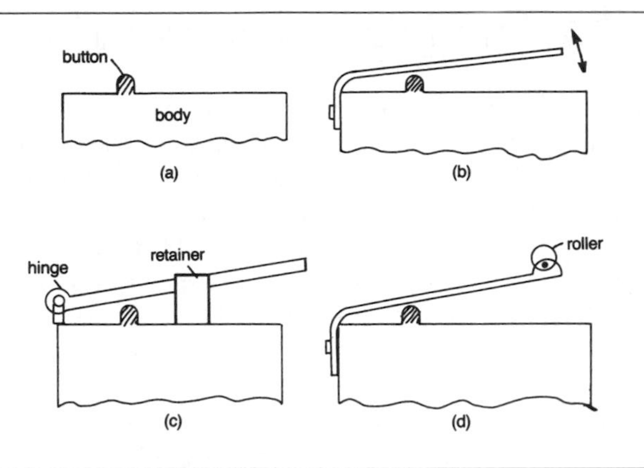

Figure 11.7 Microswitch actuation: (a) plain button, (b) leaf, (c) lever, (d) roller leaf.

– these switches need not be particularly small physically (although some are) and often have contact areas as large as conventional toggle switches. The distinguishing feature of a microswitch is its toggling action, which calls for a very small amount of movement of the actuating button or lever. For a basic microswitch operated directly, the movement will typically be 0.4 mm before contacts are changed over, with a differential of 0.05 mm. The differential is the minimum change in operating-pin position between on and off, and it is this quantity in particular that is responsible for the name of the switch. The mechanism also permits the operating pin to move beyond the position required for switching over, typically 0.15–1.5 mm depending on type. These small actuating movements are achieved along with voltage and current ratings of mains standard – typically 480 V, 15 A.

The microswitch itself is operated by pushing a pin of small diameter (typically 1.5 mm), so environmental sealing is much easier than it is with other types. In addition, a variety of external mechanisms can be used to actuate the pin. The operating forces call for quite high pressures on the pin of a conventional microswitch, so that the external addition of buttons, plain levers, and roller-arm levers (Figure 11.7) all make operation easier. The plain pin type is used for door interlock switching, and in any other application in which other moving parts can apply pressure directly. By using push buttons, operation by finger pressure is made easier; by using levers, lighter action at the expense of larger amounts of movement can be achieved.

Types of low-torque microswitch can be obtained, allowing mains switching at up to 5 A, but with such low torque operation that the action can be initiated by a small vane blown in an airstream, the weight of a coin, and so on. For use with signals or with low voltage supplies, subminiature microswitches can be obtained that combine very small size with the usual short operating travel, and in this case with small amounts of force as well.

The main applications for microswitches when they originally appeared were as door safety switches and alarm systems. The action is always of the momentary variety, so that microswitches are employed as main unit switches only in conjunction with hold-on relays, since there is no latching action in the microswitch. This has not deterred the rapid spread in the use of microswitches, some with mechanical latching mechanisms attached externally. One important factor must be the expected operating life of a microswitch, which is very long: about 10 million operations. This greatly exceeds the expected operating life of most toggle switches. Add to that the better sealing from external atmospheres and the high contact ratings that are available, and it is not difficult to see why microswitches are so often specified for actions that have nothing to do with door switching or security.

11.11 Switch mounting

The majority of switches are panel-mounted, and the methods (such as bush, snap-fit and flange mounting) have already been mentioned. When a large bank of switches is to be available on a panel, there may be a problem of removing the panel for servicing, and a good solution is a sub-frame. The switches are mounted on a sub-panel, with only the operating toggles or other mechanical parts protruding through holes in the main panel, and the sub-panel is attached in only a few places. In this way, only a few bolts need to be removed to allow the main panel to be detached, avoiding the need to remove each switch individually. A few types of switch require specialized mountings – the rotary edge-wheel type, for instance, is very difficult to accommodate. Microswitches in particular do not lend themselves well to panel mounting, and usually require the use of angle brackets. In general, any application involving the use of micro-switches will need some thought as to fixing methods, since the built-in fixings consist only of two holes in the casing of the microswitch that do not permit direct connection to any surface not parallel to the movement of the actuating pin.

11.12 Relays

The classical form of relay is illustrated in Figure 11.8. The coil, usually a solenoid, is wound on a former around a core, and a moving armature forms part of the magnetic circuit. When the coil is energized, the armature moves against the opposition of a spring (or the elasticity of a set of leaves carrying the electrical contacts) so as to complete the magnetic circuit. This movement is transmitted to the switch contacts through a non-conducting bar or rod, so that the contacts close, open or change-over

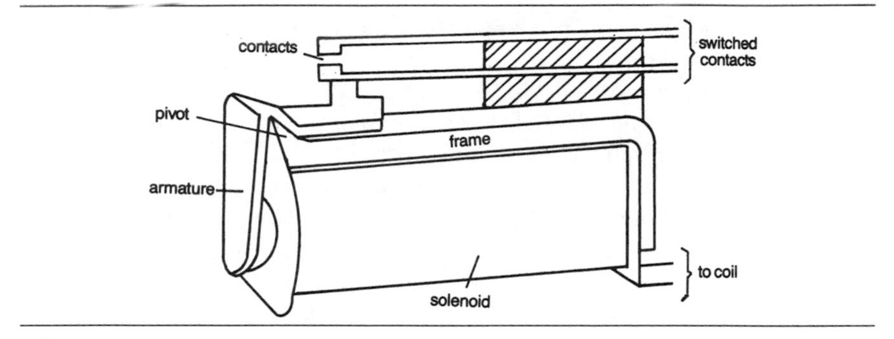

Figure 11.8 The classic type 8000 relay construction.

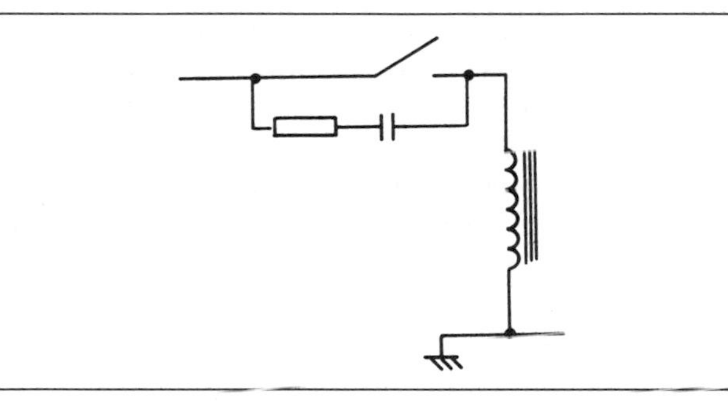

Figure 11.9 Using CR arcing suppression on switch contacts that operate a relay coil.

depending on the design of the relay. The contacts are subject to the same limits as those of mechanical switches, but the magnetic operation imposes its own problems of make and break time, contact force and power dissipation. Like so many other electronics components, relays for electronics use have been manufactured in decreasing sizes, and it is even possible to buy relays that are packaged inside a standard TO-5 transistor can.

Since the contact requirements are as for other types of switches, we shall look mainly at the magnetic operation of a relay. The operating coil constitutes an inductor as far as its circuit is concerned, and if this coil is being driven by a transistor then a diode should be used to prevent damage due to the back-EMF when the coil current is switched off. This point is noted in relation to reed relays in Chapter 12. If the relay is operated by a mechanical switch the need for protection is less stringent, but a CR network (Figure 11.9) across the switch contacts is often used to suppress sparking, not least because of the need to reduce electromagnetic impulse interference. If the relay is used in equipment that incorporates amplification of low-amplitude RF signals, much more thorough radio interference reduction will be needed both across the switch contacts and across the relay contacts.

The specification for a relay will include the permitted range of current or, more commonly, voltage that can be applied to the coil. At the minimum level and below, correct operation cannot be ensured. The relay may hold in at a voltage or current considerably lower than the specified minimum, but cannot pull in its armature if the relay has been off. The maximum rating is due to dissipation, so that exceeding the maximum voltage or current rating will cause overheating, and will also cause more severe contact bounce. The normal rated values that are quoted are usually close to the lower limits, so that a typical percentage tolerance is −8% to +60%. Failure to maintain the correct voltage on a line used to energize a relay can therefore lead to incorrect operation if the voltage falls by a comparatively small amount. For many purposes, using a large reservoir capacitor on such a supply eases problems of maintaining voltage, because if the capacitor can maintain the voltage level for long enough (typically 0.5–10 ms for small relays) then a drop in voltage of more than 8% will be harmless, since this will still be enough to hold the relay.

The use of a relay as a remote electrically controlled switch is just one aspect of relay use. Relays are used where the use of switches alone would require impossible mechanical interconnections or awkward electrical cabling. They can be placed where a mechanical type of switch could not be reached, in environments where sealed contacts must be used, or where the contacts have to handle voltages or currents that are beyond the range of normal switches. An important feature of almost any type of relay is its electrical isolation between the operating coil and the contacts. For some uses, safety considerations demand the use of relays when a mains (or higher) voltage supply has to be switched by low-voltage equipment with no ground connection. Relays are also used when a sequence of switching has to be carried out, initiated by making contacts on one single switch, or where a safety cut-out is required that will provide isolation from high voltages until reset.

Like switches, relays also exist in special forms. In the latching relay, a mechanical tooth and ratchet will maintain the contacts in a set position. The normal form of relay operates its contacts for as long as current flows in the coil, but a latching relay can be used to hold contacts either made or broken, reversing the action each time current is applied to the coil. A similar action can be obtained without any mechanical complications using a remanence relay, in which the coil current is applied for a controlled time (typically more than 10 ms and less than 1 min). This magnetizes a core made from an alloy that retains its magnetism after the coil current is switched off. The retained magnetism then holds the armature in place until a reverse current is passed through the coil. The reverse current is usually smaller than the forward current, and a circuit arrangement such as that in Figure 11.10 is used with a resistor to control the reverse current.

A third method of achieving a latching action is to incorporate a permanent magnet into the core path. If the strength of this magnet is

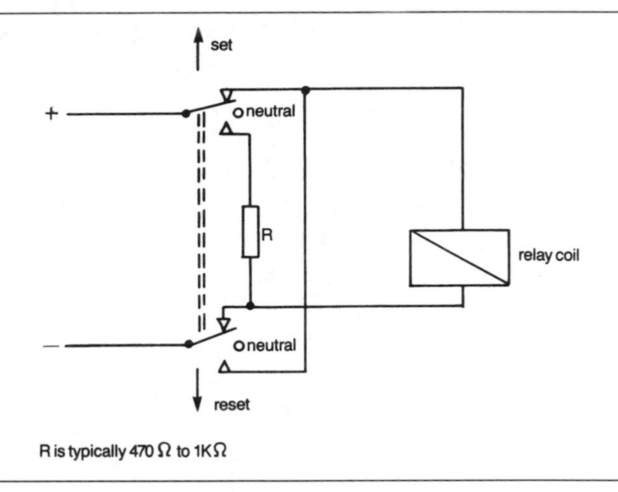

Figure 11.10 Switching a remanence relay. The relay will remain in its on or off state with no power applied.

midway between the holding and pull-in levels, then such a magnet cannot operate the relay unless current flows in the correct polarity in the coil. Once the armature has been pulled in, completing the magnetic path, the permanent magnet can maintain this state. To release the relay a lower current has to be applied in the reverse direction.

High sensitivity relays can also be bought off the shelf. These incorporate a transistor and protection diode so as to allow the relay to be operated with a very low current swing (although the voltage swing will be of the order of 0.6–1.2 V). The use of a Darlington pair circuit (Figure 11.11) allows very low current operation, typically of the order of 100 CIA, to be used. The use of transistor drivers can ensure that the operating voltage and current can be at safe, non-sparking levels, with currents low enough to be no hazard even in biomedical applications.

11.13 Semiconductor switching

In this section we shall not be concerned with the switching effect of a single bipolar or field-effect transistor but with the Schmitt trigger and the four-layer semiconductor switching devices such as the thyristor. All of these are electrically-operated switches, and so are not the primary concern of this book, but since their use is so widespread, and so often occurs in conjunction with conventional switches, we must deal with them even if only briefly.

The Schmitt trigger is an old-established switching circuit that has been implemented in turn with thermionic valves, bipolar transistors and field effect transistors. It is still the basis of many switching devices in IC form. A simple version, using bipolar transistors, is illustrated in Figure 11.12.

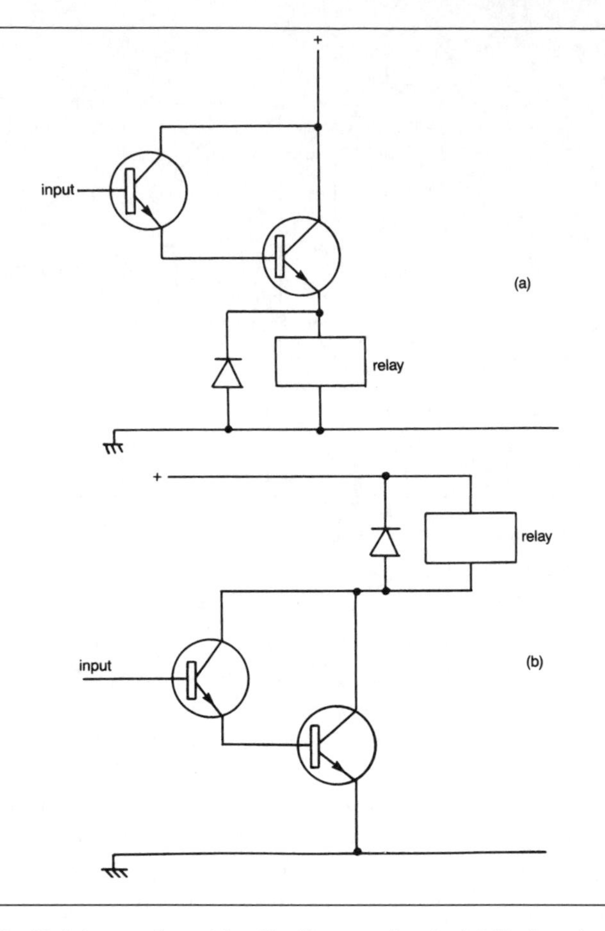

Figure 11.11 Driving a relay with a Darlington circuit. (a) Emitter loading, which provides large current sensitivity but requires the input signal to change voltage by rather more than would normally be needed to operate the relay. (b) Collector loading, providing both current and voltage sensitivity. The change of voltage at the input need only be enough to switch the transistor on and off.

The normal circuit bias is arranged so that T_{r2} is normally conducting, and the current that is passed through the common emitter resistor R_3 is sufficient to keep T_{r1} biased off. In many applications the DC voltage of the base of T_{r1} will be ground, but the sensitivity of the circuit can be increased by biasing this base slightly positive. A positive-going signal at the base of T_{r1} will eventually cause this transistor to conduct, and when this happens the fall in voltage at the base of T_{r2}, along with the drop in current through R_3, will make the switch-over very rapid. Within a microsecond or so, T_{r1} will be fully conducting and T_{r2} shut off. This persists only for as long as the base voltage of T_{r1} causes conduction, however, and when this base voltage reaches a lower level, the rise in collector voltage of

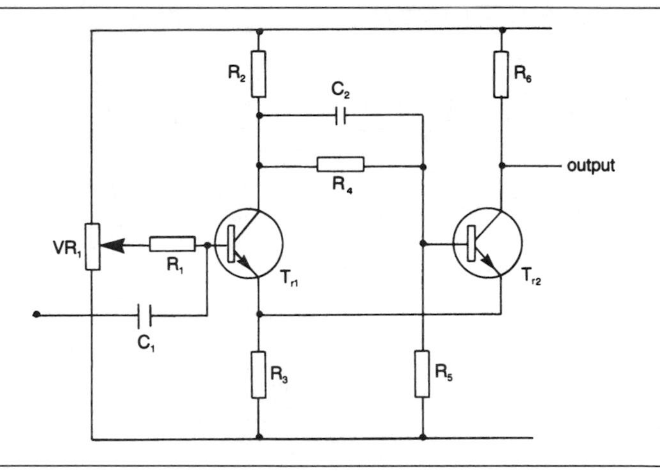

Figure 11.12 The discrete-transistor form of a Schmitt trigger. The circuit is nowadays more often used in its IC form.

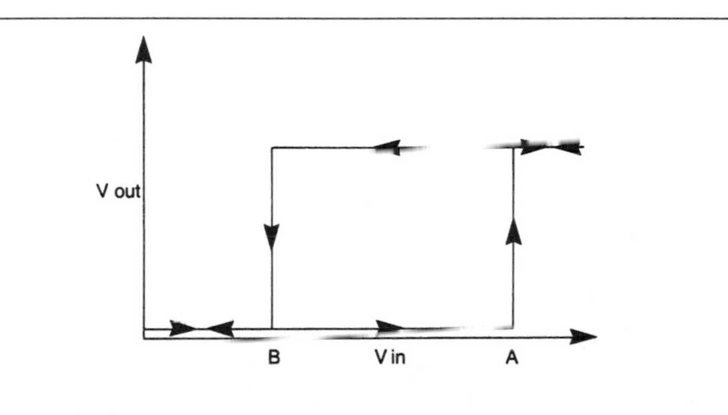

Figure 11.13 The input–output characteristics of a Schmitt trigger. A rising input voltage has no effect until its level reaches point A, when the output switches to a high level. When the input is lowered again, there is no effect until the voltage level reaches B. The switch-over can be very rapid, typically 20 ns or less for IC versions.

T_{r1} will allow T_{r2} to conduct, causing the circuit to snap back to its original state. Once again the change is very rapid, of the order of a μs in the version illustrated, and is very much faster in TTL IC circuits.

The important feature of the Schmitt trigger, apart from its very fast contactless switching, is the difference in voltage levels between switch-on and switch-off. This difference is called voltage hysteresis, and is illustrated in Figure 11.13, which shows the higher voltage at which the device switches over when an input voltage is rising, and the lower level at which it switches back. The difference between these levels serves to make the

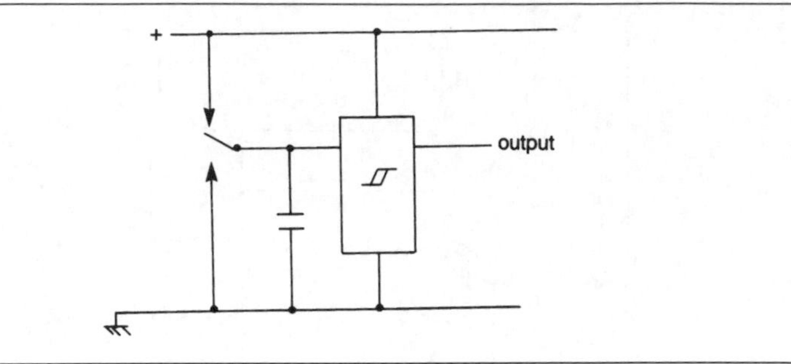

Figure 11.14 Using a Schmitt IC for debouncing a switch. This is often a more satisfactory method than the use of the RS flip-flop.

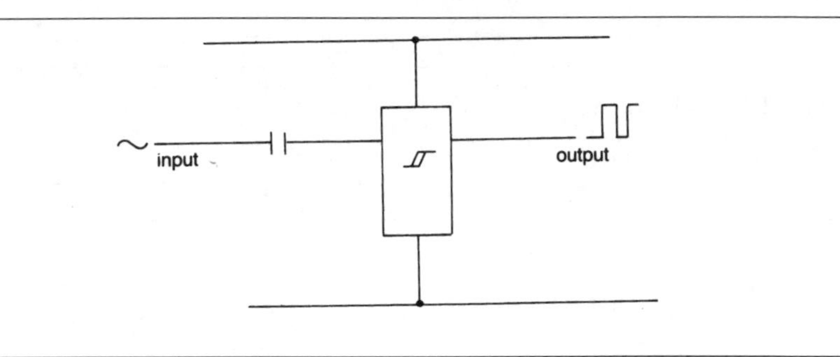

Figure 11.15 Using a Schmitt IC to convert from a sine wave to a square wave with very sharp rise and fall.

device bounceless, and one common use of the IC form of the Schmitt trigger is to provide debouncing for a mechanical switch, as illustrated in Figure 11.14. The Schmitt trigger can be used to advantage wherever a slowly changing or analogue voltage has to be used to generate a switch action, usually as part of a digital device. One example is in deriving trigger pulses from sine waves, using a circuit of the type outlined in Figure 11.15, but a more common application in a true switching application is to generate a switching output from the slowly changing inputs of sensors such as thermistors, Hall-effect devices, liquid level detectors, photocells and similar analogue devices.

An entirely different class of semiconductor device makes use of a single semiconductor crystal with four separate layers, as shown in Figure 11.16. The best known device of this class is the thyristor, whose action is very similar to that of the gas-filled thyratron device of earlier years. The thyristor uses connections to only three of the four semiconductor layers, referred to as anode, cathode and gate, respectively (Figure 11.17). The voltage to be switched is applied between the anode and cathode, with the anode positive with respect to the cathode. With the gate at cathode potential, no current will flow between anode and cathode, but when the

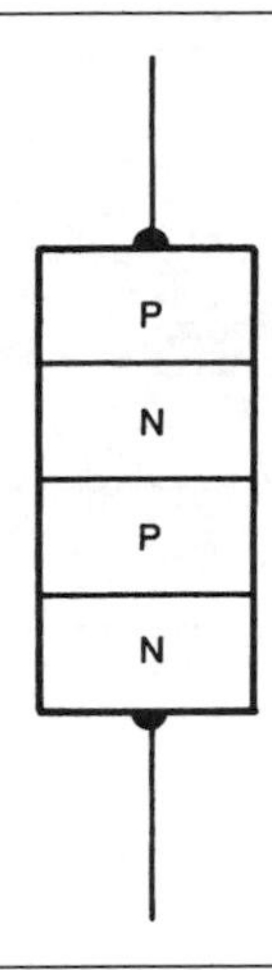

Figure 11.16 A four-layer semiconductor. The characteristics of a device formed in this way depend on what connections are made to the intermediate layers.

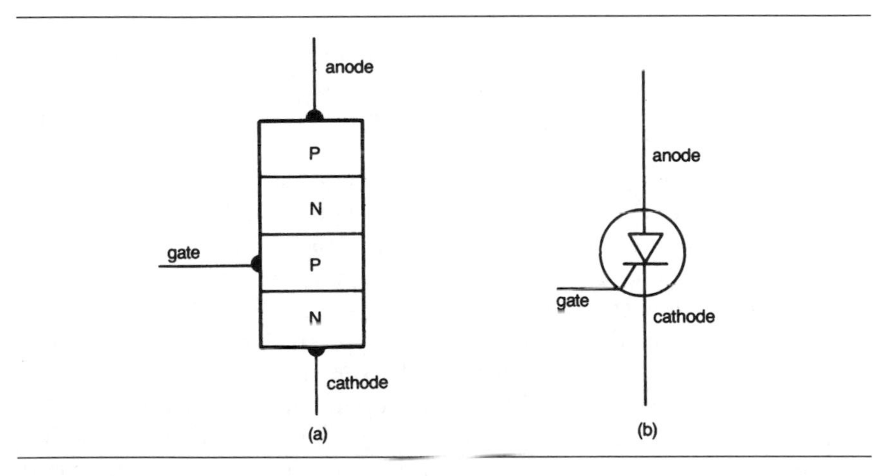

Figure 11.17 The thyristor, showing (a) layer connections and (b) the conventional symbol. Some older texts still use the name 'silicon controlled rectifier' and abbreviation SCR.

gate voltage rises to a given level (usually above 0.6 V) then conduction between anode and cathode is triggered. From then on it will continue until the voltage between anode and cathode falls to a very low value, around 0.2 V.

The gate can be triggered by a pulse, provided that the gate voltage and current are held at triggering level for a sufficient time, around a μs. The switch-on action is similar to that of a latching type of relay, but there are important differences. First, the power required for switching at the gate is

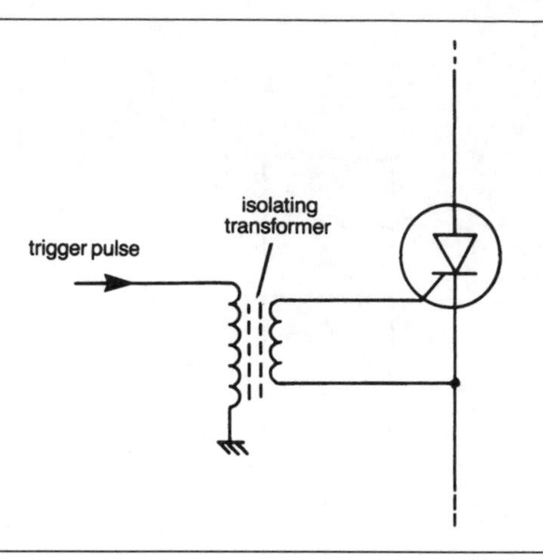

Figure 11.18 Operating a thyristor through a pulse transformer to isolate the triggering circuit from the switched circuit. Since only a brief pulse is required, the transformer can be a miniature type, often wound on a ferrite ring.

very low. Second, time is very short compared to that required for a mechanical device (although long compared to other semiconductor switches). Third, the device is unidirectional, i.e., current will not flow unless the anode is positive relative to the cathode. Finally, there is no method of turning off conduction other than reducing the voltage and current between anode and cathode to less than the critical holding level (but see gate turn-off devices, below). Note that the current requirements at the gate call for low-impedance circuits, since gate currents of 100 mA or so may be needed for a brief period.

The thyristor is used in relay-type circuits, particularly where the voltage to be switched is AC. This automatically provides for turn-off when the voltage falls to zero in the middle of each cycle, but in DC circuits turn-off can be arranged by shorting the supply briefly with a second thyristor. The thyristor can pass large currents and has the advantage of being contactless, unlike electromagnetic relays, but it has the disadvantage of having no isolation between the gate circuit and the circuit that is being switched. This arises because the gate voltage must be applied between gate and cathode. In many circuits, then, a pulse is fed to the gate-cathode circuit by means of a transformer, which can have very high resistance and breakdown voltage ratings between primary and secondary (Figure 11.18). Another triggering option is the opto-isolator (see Chapter 10). Some equipment specifications will not, however, accept the use of thyristors in place of electromagnetic relays, particularly in domestic appliances that are subject to high temperatures (such as central heating controllers).

For controlling full wave AC, circuits that use two thyristors connected anode to cathode (Figure 11.19) can be used, but a more common solution

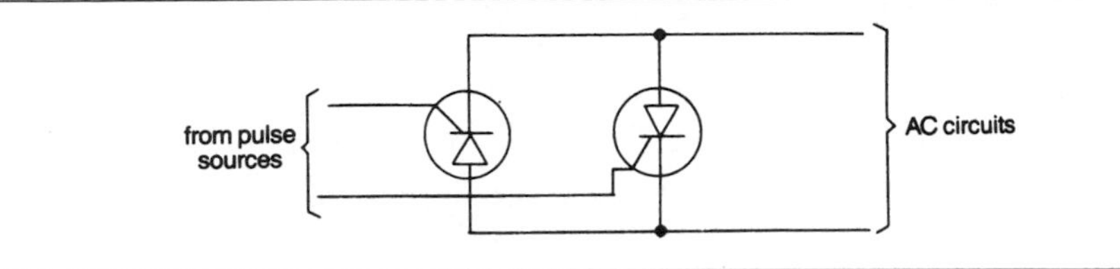

Figure 11.19 Full-wave AC switching using two thyristors back-to-back.

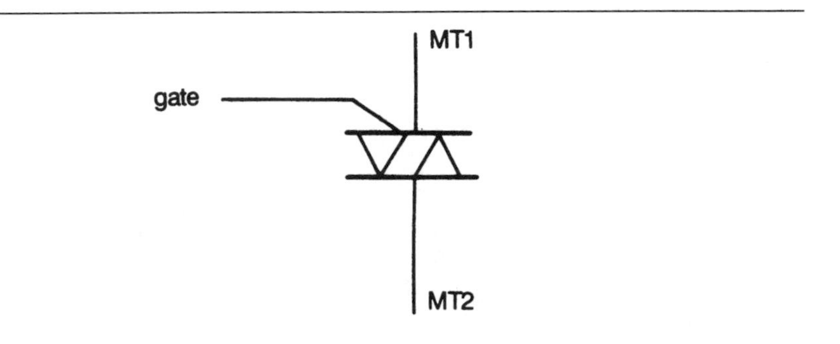

Figure 11.20 The Triac symbol. This device conducts between the MT1 and MT2 terminals when triggered by a pulse at the gate. The pulse can be of either polarity, but the triggering amplitude for a negative pulse is not the same as that for a positive pulse.

involves the use of a bi-directional device known as a Triac (the name is a trade mark), as shown in Figure 11.20. The electrodes of the Triac are labelled as MT1, MT2 and gate, with the voltage to be switched applied between MT1 and MT2 in either polarity. Triggering is also bi-directional for a swing of about 2 V between the gate and MT1, although the triggering current required depends on the polarity of the triggering signal. Like the thyristor, the Triac is switched off when the voltage and current between MTI and MT2 are at a very low level.

There is one thyristor-like device that can be turned off electrically, known as the gate turn-off (GTO) thyristor or switch. Typically, a device of this type will permit conduction between anode and cathode for a positive pulse of specified voltage, current and duration, and will turn off with a negative pulse of larger voltage but shorter duration. Other devices of the four-layer type include the silicon-controlled switch (SCS), a device in which a lead is taken to each of the four semiconductor layers. This allows the SCS to be used in two modes, either as a form of uni-junction transistor or as a thyristor.

Chapter 12

Signal-carrying switches

12.1 Special requirements

A switch intended to carry signals, as distinct from DC or AC supplies, must inevitably be subject to several requirements in addition to the normal needs already described. The type and frequency range of the signals that the switch is required to carry determine to a very great extent the special needs that will prevail: a signal-carrying switch for low-frequency AC in the range 5–240 V and with currents that are not less than 50 mA or so will be identical, to all intents and purposes, to a switch for AC supply use. The important differences in design appear when a switch is intended to carry signals of high frequency, or signals in which rapid changes of voltage are important, such as digital signals. A feature of signal switching that is of no importance in supply switching is that the switch can itself add noise to the signal or even generate spurious signals. Capacitance between contacts (and from contacts to ground) is of considerable importance, and the maintenance of low contact resistance becomes more difficult just as it becomes more important.

Contact resistance can be particularly important if signals of low-voltage level are being switched. At low signal voltage levels, there is no sparking or arcing which might expose fresh surfaces now and again, and unless a wiping action can be obtained, as is often the case, the contact resistance can become quite high. This in itself is not necessarily a hazard if the signal current levels are low, but if the switch is used for low-voltage levels at which current levels are not low, then the voltage drop across the switch contacts will be appreciable.

Since the greatest problem in low-level circuits is usually the maintenance of an adequate signal-to-noise ratio, any impedance in series with the signal current is very undesirable. For low-level circuit switching, gold-plated contacts are usually recommended. Although the resistance is never as low as can be obtained with silver, gold has the great advantage of being chemically inert, so that contact resistance is very stable. The softness of the metal also tends to ensure a greater area of contact, which compensates for the higher resistivity.

Other important contact considerations are capacitance and non-linearity. If the contact resistance is high, then a thin non-conducting film probably exists on the contacts, and this film will form a capacitor coupling between the contacts. For signals at very high frequencies, the reactance of this capacitive coupling may be lower than the resistance of the contacts, but like any other reactance will introduce a phase shift. Because of the low reactance of the layer, an increase in contact resistance may be unnoticeable in terms of signal strength, but the phase changes can cause problems that will not necessarily be attributed to the switch. For digital signals at lower frequencies that have short rise and fall times, the capacitive coupling will cause some differentiation of the signals, which may have no noticeable effect until it causes loss of amplitude.

A rectifying contact is considerably more troublesome. This can arise where there is metal to oxide contact on a switch, and in some cases it can be the result of severe arcing which has left a sharp whisker of metal making contact against an oxidized surface on the opposite contact. The result is a diode whose forward resistance and back resistance values are not very greatly different, but which differ nonetheless. If appreciable signal currents flow across this part-rectifying contact then the diode action will have the usual effects on the signal. One effect will be distortion of the signal, such as differences between positive-going and negative-going portions of the signal.

The other effect, which can be very serious, is DC bias. If the switch feeds a DC-sensitive input, such as a directly-coupled input to an operational amplifier, the amount of DC bias produced by the rectifying action of the switch contacts may be enough to bias the amplifier off or into non-linear action. For digital circuits, the effect can be to prevent any signal input from being effective. Modern contact materials make this condition unlikely, but there is always a possibility of a rectifying action arising through contamination, particularly in humid atmospheres, and its effects are seldom immediately blamed on the switch contacts.

For the rather special case of RF switching at high power, the problem of arcing is much more severe than it is for supply current switches. This is because the time of opening the switch contacts will inevitably be greater than the time of a cycle, so that the waveform will be at its peak voltage at least once while the contacts are opening. Since the contacts have capacitive reactance in any case, a considerable current can flow, and this will cause ionization of the air that, because of the short time between peaks of a high-frequency signal, cannot easily be damped out. Switches for high-power, high-frequency applications are therefore a very specialized branch of switching, and one for which a one-off switch may have to be designed to suit the particular application.

12.2 Speed of action

A mechanical switch operating with mains AC will normally have an opening or closing time that is short compared with the time of a cycle. This does not apply to switches that are used for higher frequencies, so that the rating of switch contacts for such signals is closer to the DC rating than to the AC (mains) rating. If switching has to be carried out in very short times, then semiconductor switching devices need to be used. These are dealt with later in this chapter, although more as adjuncts to mechanical switches than as a subject in their own right. Since switching speed is inevitably slow in comparison to the time of one cycle of wave, it is less relevant to the specification of switches for signal frequencies, and for low-power signals the time of switching is usually dependent on the action of the operator. A wafer switch, for example, will change connections in a time that is determined completely by how fast the operator turns the switch knob. For audio work, switches with a snap-over action are often used, but for all sorts of work where conditions are critical in terms of speed and noise-free operation the combination of a mechanical switch with a semiconductor switch is now more common.

12.3 Capacitance

Two figures for capacitance are significant in a switch intended for use with signals: the stray capacitance to ground and the capacitance across the open contacts. Taking the latter first, the open-circuit capacitance (Figure 12.1) determines to a large extent the efficacy of the switch in terms of open and closed circuit impedance. The figure, unfortunately, is seldom quoted but will be in the range of 10–20 pF for the common types of signal switches. This will cause a significant amount of signal to appear on 'open'

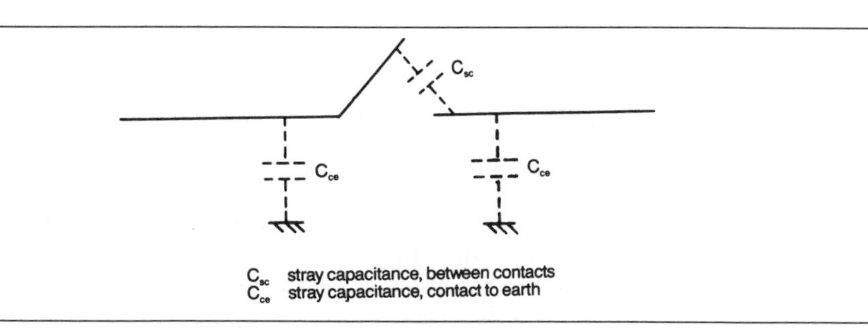

Figure 12.1 The stray capacitances that exist across switch contacts and from switch contacts to ground.

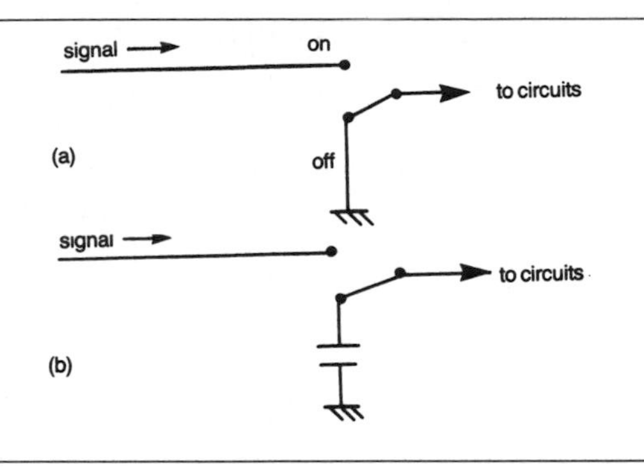

Figure 12.2 Grounding the 'cold' side of a switch to reduce the effect of signal coupling through stray capacitance. (a) No DC level present at switch; (b) use of a capacitor when DC is present.

contacts, particularly for high-frequency or fast-rising signals, unless the switches are operated in circuits of low impedance. For this reason, a simple open-and-close type of switching arrangement is seldom suitable for high-frequency signal switching, and the switch should ensure that no unwanted signal is coupled across by grounding the 'cold' side of the switch when the switch is off. Obviously if a DC bias exists on the cold side of the switch contacts, the grounding will be done through a capacitor. Figure 12.2 illustrates this principle applied to a single-way switch.

The type of switch mechanism very greatly affects the amount of capacitance that exists between the contacts. The contacts of a wafer switch lie in the same plane, with only a very tiny area of contact conductor parallel to any other conductor, so they have very low capacitance across the contacts. This is a major reason for the continued popularity of this switch type, which has remained almost unchanged (apart from improvements in materials) for some 60 years. The type of switch in which open contacts lie parallel to each other is the worst, from the point of view of open-contact capacitance, although if the open contacts are reasonably far apart, the amount of capacitance between them can be as low as can be achieved in a wafer design. Since speed of action is seldom an important factor for the mechanical switch, a comparatively large distance between open contacts can be specified in push-button and keyboard switches.

Another important aspect of capacitance in signal-carrying switches is the stray capacitance between contacts that carry different signals. If a switch is used to carry several signal channels, then capacitance between contacts can lead to signal crosstalk. If impedance levels and signal levels are both low, the amount of such crosstalk may be insignificant, but in the worst-case example of a high-level signal on one channel and a low-level signal on another, with comparatively high impedance, the amount of crosstalk would become intolerable. Multi-channel switching of this type is best handled by ganged assemblies in which the channels can be kept physically well separated. Once again, the familiar wafer switch copes best with this requirement if one wafer is used for each channel, with grounded screens placed between the wafers.

Stray capacitance to ground is, by contrast, much less of a hazard unless the switch is being used in a high-impedance circuit. By their nature, most switches are mounted on conducting panels, so that stray capacitance to ground is always significant, but since this stray capacitance has little or no effect on switching performance it can be treated like any other stray capacitance to ground in the circuit. Stray capacitance of this type becomes a problem only if the local ground happens to be the hand of the operator, which can occur when a signal-carrying switch is mounted on a non-conducting panel. This will have the effect of making the stray capacitance to ground much larger when the switch is being operated than at other times, and is the same effect as the *hand-capacity* problem that haunted amateur radio builders many years ago. This manifested itself as a change of tuned frequency as the user's hand approached or was taken away from the tuning control.

Once again, the problem may be negligible: it is only likely to cause difficulties if the switch happens to be part of a tuned circuit whose frequency can be changed by the alteration in capacitance to ground. This problem can also be solved by the use of a wafer switch with a grounding screen. In general, switches should not be mounted on insulating panels for any application in which variation of capacitance is undesirable.

All these remarks on stray capacitance underline the importance of good circuit design, in which switches are never placed in high-impedance circuits. Except for the most demanding of waveforms, with very fast rise and fall times, switch capacitance is never a problem when impedance levels are low. At high impedance levels, the switch capacitance to ground is just one of the stray capacitances to ground, and presents no special problems unless the capacitance varies. The main problem for higher impedance circuits is capacitance between switch contacts, and this can be reduced to very small amounts, as we have seen. The usual and least avoidable type of high-impedance circuit is the tuned circuit in which a switch is used to change inductors, capacitors or the complete LC tuned circuit for an oscillator. In this case, the stray capacitance across contacts has an effect that is constant, and which can be dealt with by any form of fine-tuning adjustment.

12.4 Noise

All signal noise is caused by random motion of electrons or other current-carriers in a circuit. Even in a piece of wire consisting of the same metal throughout there will be some noise caused by the irregular movement of the electrons at room temperature. This type of noise can be greatly reduced by lowering the temperature, but is usually negligible compared to the other noise sources in a circuit. In general, noise is generated mainly when a small number of electrons are moving comparatively rapidly. The main sources of noise are therefore semiconductor junctions and bad contacts, and it is this latter source that is the main concern for the switch user. In addition, however, noise can be generated at junctions that are not semiconductor junctions. Wherever any two different metals meet there will be a build-up of electrons that causes a measurable voltage difference, called the *contact potential*. This contact potential is mainly DC, but because of irregularities in the movement of the electrons around the contacts it will have an AC component, which is noise.

Any type of switching circuit will therefore introduce noise into a signal that is switched. The noise arises to some extent from variations in contact resistance, but can also be caused by the fluctuating contact potentials if the contacts are not clean and of identical chemical composition. A switch is not a major noise-producing component in circuits unless it has been used in a circuit that operates at very low signal levels. Switching at such low levels is unusual, because it is normally possible to amplify a signal before switching it, but if such switching is unavoidable, it should make use of switches with gold-plated contacts. In addition, if switching at low signal levels is completely unavoidable, no DC should pass through the switch, because DC flow is by far the most likely cause of noise in a switch (as it is in a potentiometer).

The type of noise that is caused by the switch contacts while they are closed is the usual broad-band noise, but there can be an additional problem of impulse noise when switch contacts close, caused by the charge and discharge of stray capacitance by the DC levels in the circuit. A familiar effect of this at one time was the noise pulse that accompanied audio circuit switching, for example when changing an amplifier input from disc to tape.

The root cause of this problem is the presence of DC on one of the switch contacts, or the use of different DC levels between contacts. The easy way out of this is, of course, to design the circuit so that switching is carried out at zero DC level at all times, but this may require the use of more coupling capacitors in the circuit path than is thought desirable, and in any case if the switch contacts are isolated to ground there will be a build-up of DC through the leakage resistance of any capacitor. The usual cure, if a purely mechanical switch is to be used, is to equalize the DC levels at the switch contacts, usually by the use of high-value resistors (Figure 12.3). This will lead to some signal breakthrough unless the signal circuits are at low impedance.

12.5 Wafer switches

The familiar wafer switch still fulfils most of the requirements for signal

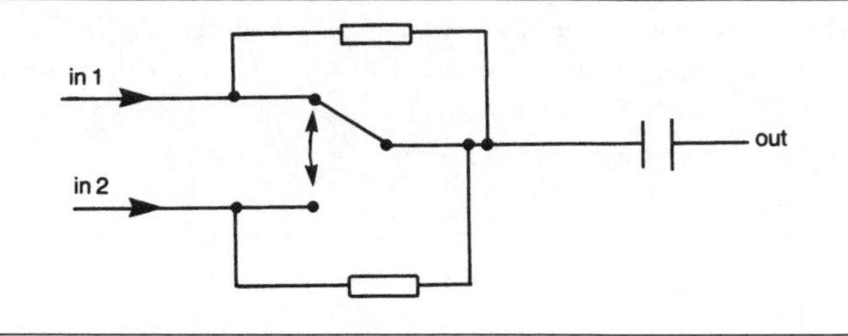

Figure 12.3 Leakage resistors are used to equalize DC levels when a switch is connected to a capacitive circuit.

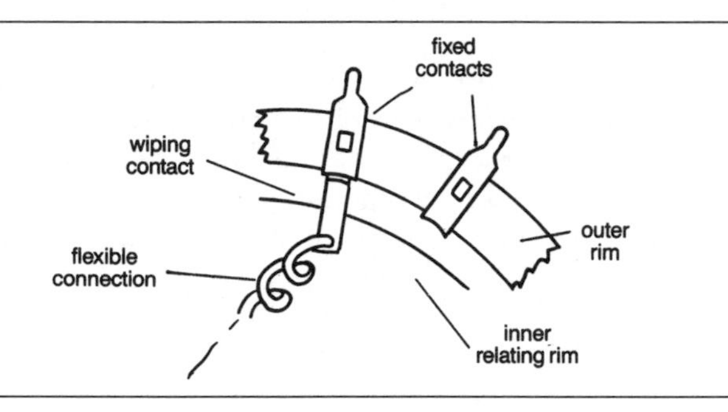

Figure 12.4 The basic construction of a wafer switch. The use of contacts in one plane ensures that stray capacitances are very low, and the construction also ensures high resistance values.

switching, particularly in low-power circuits. The layout of a wafer (Figure 12.4) ensures minimum capacitance between contacts, and also means that the contacts are cleaned by the wiping action each time the switch is used. Because of the wiping action, silver-plated contacts can be used to provide low contact resistance (and low noise) with less risk of contamination than arises when the contacts are non-wiping. The main virtue of the switch, however, is that it can be constructed to order, for example allowing wafers to be widely spaced and separated by grounded screens if this is desirable. This is possible because the rotation of the inner part of a wafer (the rotor) is controlled by a shaft that is located by a flat side, and this shaft can be cut to whatever length is needed or joined to other shafts.

The mechanical control of the switching shaft is carried out in the section that is mounted on the panel. The switching positions are usually at 30° intervals, and the movement of the shaft is regulated by a detent mechanism (Figure 12.5) which helps to ensure that the rotating contacts are not left between fixed contacts. The detent can also assist in speeding switching times, although this is not an important factor. Careless use of a wafer switch can leave contacts isolated between detent positions, but modern mechanisms are surprisingly efficient – surprisingly, that is, because the mechanisms are comparatively simple.

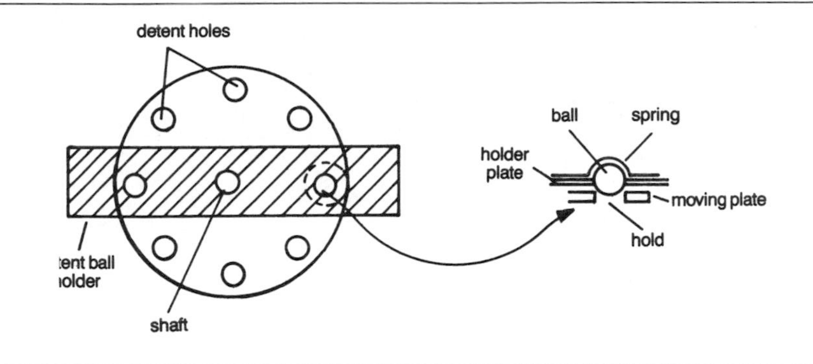

Figure 12.5 The principle of the spring-loaded ball detent mechanism for wafer switches.

The main points that distinguish the wafer switches of today from those of some 60 years ago are construction and materials. The old type of wafer switch came as a ready-made item, using wafers of laminate plastic (usually of the Bakelite family). The changes in wafer switches over the last twenty years have included the provision of switch kits that allow the user to specify the type and positioning of wafers, and of materials such as diallyl phthlate, usually glass-fibre filled, and acetals for exceptionally high insulation resistance coupled with considerable mechanical strength. The traditional form of wafer switch is still used to a very great extent, but modified wafer shapes for PCB mounting are now available. These, however, inevitably have higher capacitances between contacts than the traditional type of wafer because of the length and close spacing of the tracks to the contact pins.

Wafers allow mainly break-before-make operation, and the standard switchings are 1-pole 12-way, 2-pole 6-way, 3-pole 4-way, 4-pole 3-way, and 6-pole 2-way. These are the standards for the traditional type of wafer switch, and not all of these switchings may be available in PCB wafers. A few make-before-break wafers are available, usually as 1-pole 12-way or 2-pole 6-way switchings. The usual length of shaft allows for the use of 6–10 wafers depending on the size and type. In addition, insulating spacers and grounding screens can be fitted between wafers, and some types allow for a dummy wafer that consists of a stator only, with contacts to allow components to be wired between wafers without recourse to the main PCB. Another useful feature of most of these wafer kits is that a miniature mains switch can be fitted at the end of a shaft to allow for mains switching in addition to signal switching. This is not desirable if the wafers handle low-level or high-impedance signals. At the switch mechanism end, a key lock can be fitted to allow the switch to be locked in positions at 60° angles. Note that this allows only alternate contacts to be used, since the conventional wafer contact positions are at 30°.

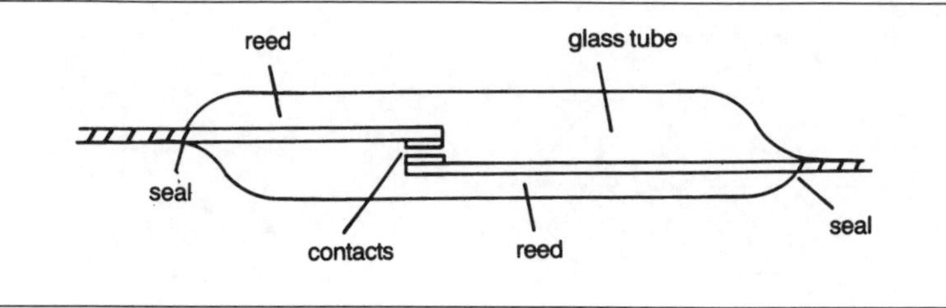

Figure 12.6 A reed switch (or relay). The reed arms are usually made from a nickel alloy that are sealed into glass.

12.6 Reed switches

Reed switches are widely used for signal switching, particularly at the lower signal frequencies. The main attraction of the reed type of switch is that the contacts are enclosed, and cannot therefore be damaged in adverse environments. In addition, the switch is magnetically operated either by a solenoid coil or by a permanent magnet, so that remote control is possible and variable capacitance to the hand of an operator is not a problem. In this respect, the reed switch is normally used as a secondary switch or relay in the sense that the operating current to its coil will normally be controlled by another switch. Reed switches are generally classed as relays for the purposes of cataloguing electronic components. The reed switch can therefore be mounted directly on a PCB deep inside the circuit, possibly in an inaccessible place, and operated from any conventional type of low-voltage switch on a panel. This avoids the problems of bringing a signal cable out to a panel-mounted switch and back again.

The basic reed principle is illustrated in Figure 12.6. The thin metal reeds are sealed into a glass tube, and bent so as to make contact because of the effect of a magnetic field. The usual arrangements of contacts are normally open or changeover, and the reed portion can often be specified separate from the actuating coil or magnet. The use of the solenoid coil allows a relay type of operation, but the use of a permanent magnet allows mechanical operation (by moving the magnet to or from the reed tube) which can be custom-made to whatever pattern is needed. The typical magnet distances are of the order of 7–11 mm to operate the reed and 13–16 mm to release, so that a movement of around 5–20 mm would normally be used to ensure reliable opening and closing of the reed circuit. The predominant use of reeds, however, is with the solenoid, using up to 100 ampere-turns to operate the reed.

One important feature of reed-switch use is the switching time which is fast by the standards of mechanisms. This will be of the order of 1–2 ms to make, and can be as low as 0.2 ms to break, with reed resonant frequencies in the order of 800–2500 Hz. Fast switching speed is not necessarily an important feature of signal switches, but if signals have to be sampled, or if there has to be switching from one signal source to another at frequent intervals, then the reed type of switch has considerable applications.

An important further consideration for high-frequency or fast rise-time signals is the very low capacitance between open contacts, which can be less than 1 pF for the normally-open type of reed. The use of silver or gold-plated contacts allows low contact resistance values of around 100 mN, and for applications that demand very low and stable contact resistance, mercury-wetted reeds can be obtained. The mercury-wetted type also has the advantage of providing very low-bounce operation, a point that will be taken up later.

The other features of reed switches are high insulation resistance of the order of 100 000 MΩ (because of the use of glass encapsulation), and high breakdown voltage of at least 200 V. Note that the use of nylon as an insulator can affect these figures adversely, because the insulating properties of nylon deteriorate as the temperature rises. The operating currents are also quite surprisingly high for a device with small contacts and limited contact force, of around 0.25–2 A. The mechanical life, depending on type of use and reed construction, ranges between one million and one hundred million operations.

The normal operation of the reed relay by its solenoid coil can raise one problem which is common to any device that uses an inductor. When the current through the coil is broken, the usual back-EMF is generated, and this can cause sparking between switch contacts (of the switch used to operate the solenoid) or breakdown of a transistor if this is used as the coil driver. When a reed switch and solenoid are constructed as one unit, the coil will usually incorporate a protection diode to avoid the effects of the back-EMF, and this implies that the polarity of connections to the coil will have to be observed. This in turn means that such a coil should not be used if you are operating the coil from an AC supply (as a 100 Hz sampling circuit, for example). If you are using a separate coil and reed, you will need to incorporate the diode for yourself (Figure 12.7) if you are using a DC supply to the coil.

12.7 Solid-state switches

The subject of solid-state switching could fill several books, and in this chapter we are concerned more with selection and use than with theory. In addition, the more specialized types of electronic switches were dealt with in the previous chapter. The fundamental principles of solid-state switching are that a transistor (bipolar or FET) has a low-impedance state and a high-impedance state, and can be switched from one state to another with a small change of voltage. The bipolar transistor (Figure 12.8) has a high impedance between collector and emitter when the base terminal is at the emitter voltage or up to 0.5 V above the emitter voltage ('above' in

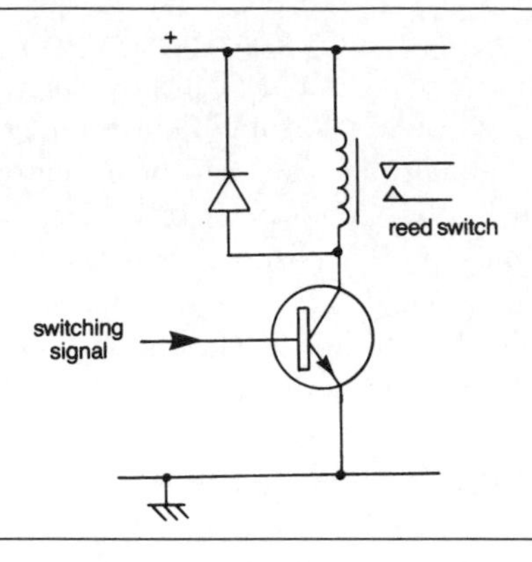

Figure 12.7 Using a diode to protect a transistor from the effect of a reed-solenoid load.

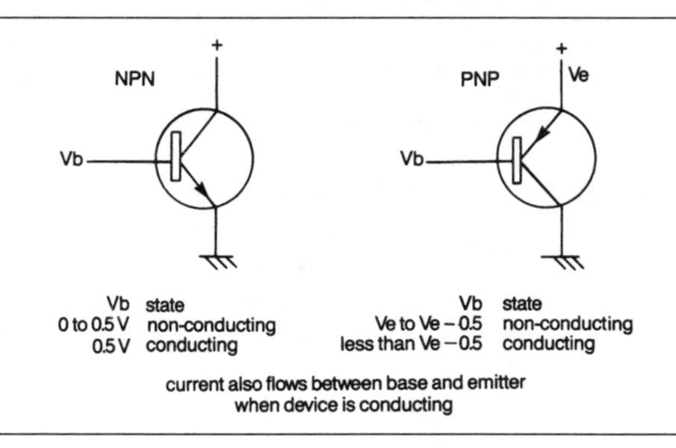

Figure 12.8 Bipolar transistors used as switches. The disadvantage of using a bipolar transistor is that the operating input (to the base) is not isolated from the switching terminals (emitter and collector) and requires some current.

this sense meaning closer to collector voltage). When the base voltage is (for a silicon transistor) above about 0.55 V then the collector–emitter path becomes conducting. In this respect the transistor acts like a form of relay. There are, however, two snags. One is that the conductivity is one way from collector to emitter only; the other is that the base to emitter circuit conducts at the same time as the collector to emitter circuit conducts. This makes it completely impossible to isolate the switched circuit from the controlling circuit.

The use of bipolar transistors in signal-switching circuits is therefore a matter of careful design, making use of bridge circuits. Since the base voltage can be DC and can be decoupled, the bipolar transistor bridge can be used to switch signal circuits by means of a low-voltage DC supply to the bases. In general, though, bipolar transistors are used more for DC switching than for signal purposes. This is because the minimum collector

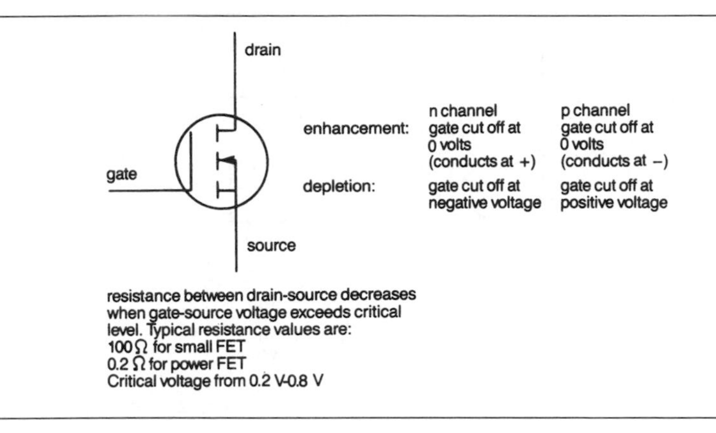

Figure 12.9 The N-channel FET used as a switch. No current has to be supplied to the operating terminal (gate), and isolation of input from output can be very high, particularly in a bridge arrangement.

to emitter voltage is around 0.2 V, so that signal amplitudes of less than that amount do not pass through the switch. For low-level signals, then, this causes an unacceptable amount of signal waveform distortion.

The use of FETs, particularly the MOSFET type (Figure 12.9), for signal switching is much simpler and more frequently encountered. The resistance between the source and the drain is governed by the voltage at the gate, and the gate is almost completely isolated electrically from the other two electrodes. In addition, current will flow between source and drain even when the voltage difference is very low, and the geometry of the device can be such that current will continue to flow even if the polarity is reversed (provided that the gate voltage is maintained).

Once again, bridge or two-transistor circuits are normally used, and the complete arrangement is available in IC form, usually as dual or quad single-pole single-throw switches – Figure 12.10 shows some typical characteristics. Switches of this type can be used with fast-changing waveforms such as video signals, and have only negligible switching transients. They are also widely used in audio preamplifiers because they permit switching that is free of the usual transients, and are extremely useful for switching data signals.

12.8 Chopper switches

At one time, vibrating mechanical switches were used for DC to AC conversion, particularly for car radios in which their use posed a filtering problem that was never completely overcome. With the coming of transistors to permit low-voltage operation of radios, the main applications for the technique of chopping were in DC amplifiers, using the principle illustrated

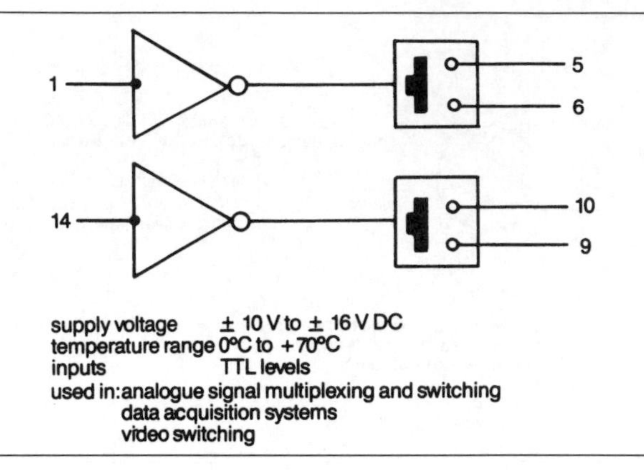

Figure 12.10 The H1200 FET switch IC which features low 'on' resistance that is almost constant throughout the frequency range, and high 'off' resistance. Switch transients are very small. (Courtesy of RS Components.)

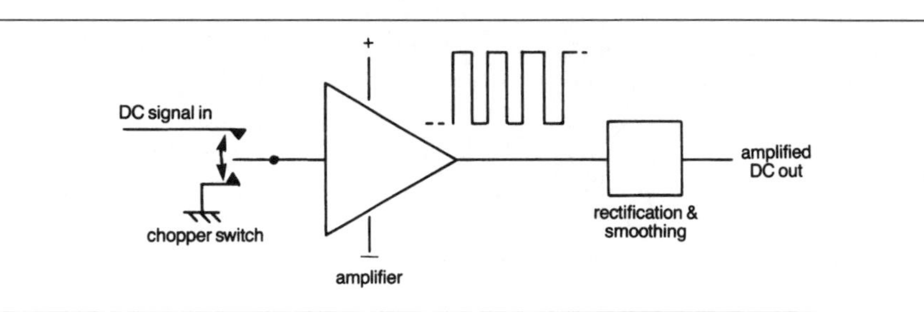

Figure 12.11 Principle of a chopper amplifier for amplification of DC and very low-frequency signals.

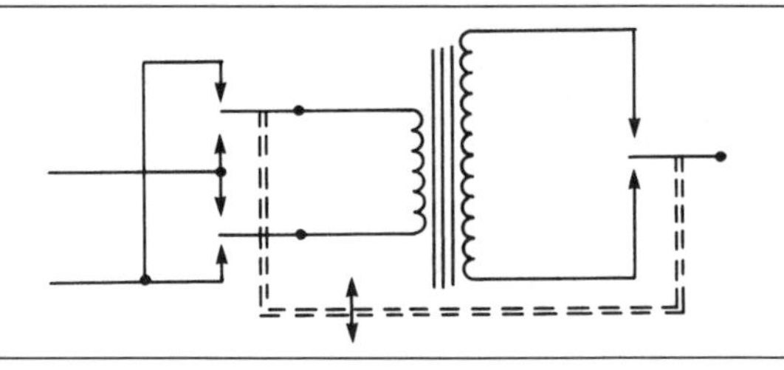

Figure 12.12 The synchronous switch principle, formerly used for DC–DC converters but now replaced by semiconductor devices.

in Figure 12.11. The input of an amplifier is connected alternately to the small DC signal and to ground, so that the input to the amplifier is a square wave at the frequency of operation of the switch.

The early mechanical vibrating switches were often used synchronously (Figure 12.12) so that no rectification was needed, but more modern chopper amplifiers have used diode bridges as rectifying units. The important point, however, is the type of switching used. Reed switches were a useful intermediate stage, but only the use of modern FET amplifiers permitted the use of reasonably high frequencies for chopper amplifiers. An offshoot of this type of circuit has been the 'DC–DC' transformer, which consists of a DC to AC chopper, amplifier and rectifier in one single package. Such circuits are used extensively in digital circuitry where the main supply is +5 V but a few units need a low-current 12 V supply, or some require a negative supply.

12.9 Digital switching

The switching of digital signals is a more specialized topic than analogue switching because of the range and nature of digital signals. Excluding the switching of supply voltages, switching actions for digital circuits can be of three separate types. First of all, a switch may be used to provide a signal into the circuit. An example of this is a push button used to allow a digital circuit to operate step by step. Another use is to select parameters for the digital circuit, such as the use of DIL switches in devices, such as computer printers, to set up the way that the machine operates. The third method is the switching of digital signals on lines from one unit to another, like the 'T-switches' used to allow one computer to control one of two different printers. This last type of switching can make use of the same types of FET switches as are used for analogue switches, but the way that the switches are used may be different. It is usually undesirable to switch over a digital circuit while the voltages are active, so this type of switch is likely to make use of an inhibit pin. While any line is active, a voltage can be generated to keep the inhibit pin at its inhibit level so that the switch will not operate until all the switched lines are at zero level.

Control switches for digital work are generally simple DIL types, usually in banks of eight. Their switching action is on DC voltages only, although the DC voltage does not necessarily control the digital circuit directly. Instead, the circuit interrogates the switch settings at the time when the operating voltage is first applied, and the settings that exist at that instant are used from then on. Changing a switch position in such a digital circuit will not therefore cause any effect until the circuit has been switched off and then on again. This avoids the effects, possibly catastrophic, of unintentional operation of these setting switches while the circuit is active.

The most critical application is for switches that cause a signal input, usually in the form of a push-button switch that delivers a single pulse when pressed. The problem here is that no simple mechanical switch delivers a single voltage change, because the elastic nature of the contacts means that contacts will make, then bounce apart and remake at least once, delivering a multiple pulse. Contact bounce in switches that handle

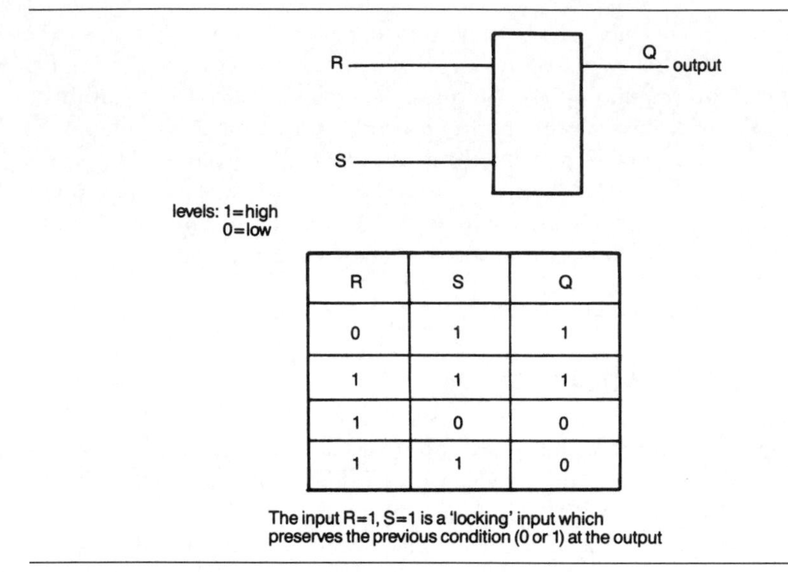

R	S	Q
0	1	1
1	1	1
1	0	0
1	1	0

Figure 12.13 One form of RS flip-flop and its truth table. The important feature for switch debouncing is that the $R=1$, $S=1$ state maintains the output that was caused by the previous input. Some RS flip-flops use $R=0$, $S=0$ for this *locking* or *latching* state.

analogue signals will cause noise, mainly impulse noise, but in digital circuits it can cause total failure of the switch action. If the switch is meant to be used to show the digital action step by step, it is hardly very useful if the steps are sometimes single, sometimes double, sometimes triple. Since the bounce is almost unavoidable (although mercury-wetted reed switches are almost bounce-free) the methods used to eliminate the bounce must be electronic.

The main method is the use of an RS-type flip-flop. The characteristics of such a flip-flop are shown in Figure 12.13, and the important point is that the output does not change while the inputs are at the same voltage. In the circuit shown in Figure 12.14, then, the bounce of the switch will only serve to hold the output voltage at its changed level, rather than allowing the voltage to change back. It is unusual to find switches sold together with debouncing circuitry, and the problem of debouncing is usually left to the designer.

In some cases (when the switch bounce time is very short) a simple resistor–capacitor circuit of the type illustrated in Figure 12.15 can be used. This relies on the bounce time being short compared to the RC time constant, so that the voltage level cannot change by much during the time when the switch contacts bounce open. If a switch that uses such a debounce circuit has to be replaced, then it must be replaced by a switch with identical or very similar bounce characteristics. This is seldom easy,

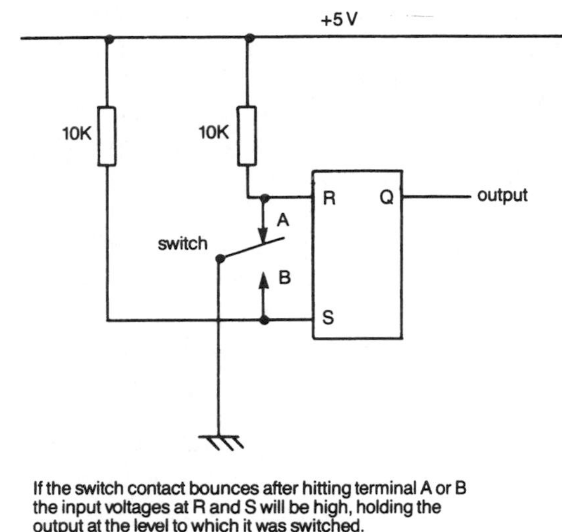

Figure 12.14 Using an IC form of RS flip-flop to provide a debounced output from a changeover switch.

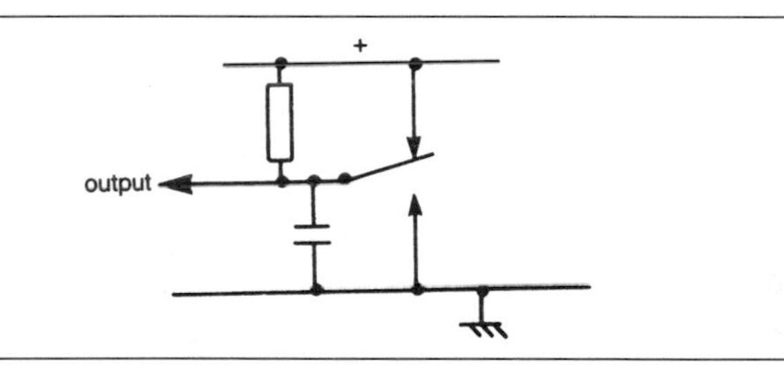

Figure 12.15 A simple form of debouncing that can sometimes be useful – the capacitor maintains the voltage level with little change during the bounce period. This is useful only when the voltage levels are low because the presence of the capacitor can cause more contact arcing.

because the bounce characteristics of switches are seldom documented, and specialized equipment (such as a storage or digital oscilloscope) is needed to measure these characteristics.

Digital equipment that makes use of signal-input switches will not necessarily incorporate any visible switch debouncing, however, in the form of ICs or other circuits. This is because the debouncing of a switch can be carried out by the operating software of a microprocessor circuit. The opening of the switch can be detected in two different ways, described as polling or interrupt. In the polling system, the state of the switch (open or closed) is checked at intervals, and if this state changes it is rechecked to find if the change is permanent. In this way, the closing of a switch causes no bounce problems because the switch closure has no effect until the polling check reports that the switch has been found closed on two, possibly more, consecutive readings.

The other method uses the closing of the switch to send a signal to the microprocessor, which then interrupts whatever it is processing at that instant. The switch state is then checked, and checked again after an interval longer than the bounce time, and the latter state taken as its final state. In each case the time delay, which is the important feature of any debouncing system, is obtained by a part of the microprocessor software. If problems arise due to insufficient debouncing (such as double letters from a keyboard when a letter key has been struck sharply once), then the remedy is usually a modification to the software, and this may have to be in the form of a patch, a piece of added software, rather than a permanent change. The main problems of this type arise with computer keyboards, of which the old TRS-80 Mark 1 was a notorious example, although many older machines of the PC class were prone to give bounce effects, mainly on the 'n' and 'm' keys.

12.10 Dual-in-line (DIL) switches

The miniature DIL or DIP type of switch is designed specifically for PCB mounting, and is used as a pre-set mainly in digital equipment, although it is equally well suited for pre-set applications in analogue circuitry. The main feature of the switch is its very small size. This makes operation with the finger very difficult, so the usual method is to use a screwdriver blade. These switches should preferably be mounted where a slip with the screwdriver blade will cause the least possible damage, electrical or mechanical, to the rest of the circuit. The usual formats are 4-pole or 8-pole on/off, and the action is a slide or a lever toggle. A changeover action can be obtained on suitable types (in which one lever operates two switches) by connecting a pair of switch pins together. DIL switches are not intended for continual use, and have a comparatively low life expectancy of around 20 000 operations. To put this in perspective, you might expect to alter the settings once or twice a year, at the most once a week.

One of the main problems attending DIL switches is mechanical damage. Designers have an unfortunate knack of placing these switches where they are difficult to see and even more difficult to operate. One printer mechanism, for example, requires the casing of the printer to be removed simply in order to set the line-feed switch, and many users of this printer cut a hole in the casing above the switch so it can be operated without dismantling the machine. This, however, means that the switch is being operated by a screwdriver stuck through a hole in the casing, which is the main source of damage to the switch and to surrounding circuitry. It is not usually possible to make amends for inconsiderate design, but in some instances the switch can be replaced by a miniature socket and the switch relocated at the end of a cable.

12.11 Keypads

The keypad type of switch, used either as the small numerical keypad or as the full-blown alphanumeric computer keyboard, is usually mechanically simple and requires debouncing circuitry or software. The switch is usually arranged so that contact is made about halfway along the travel of the key, and a spring is used to impart some 'feel' to the keyboard and to prevent the contacts from being pushed together with excessive force by the sort of typing that is more thump than touch in nature. Just as the mechanical quality of the switches can vary from the simple phosphor-bronze leaves to the Hall-effect magnetic, so the electronics of the keyboard can range from simple matrix to the completely debounced and decoded. The basic keyboard, however, is always connected using on/off switches in a matrix pattern.

The basis of a matrix keyboard is illustrated in Figure 12.16. The keys are arranged in lines and columns, and each switch will have one connection made to a line and the other to a column. This greatly reduces the cabling to the keyboard, because an array of 64 switches in an 8×8 matrix needs only 16 connections. The matrix arrangement, incidentally, applies only to the electrical connections – the physical arrangement of the keys on the keyboard can be anything that is required by the designer, although very

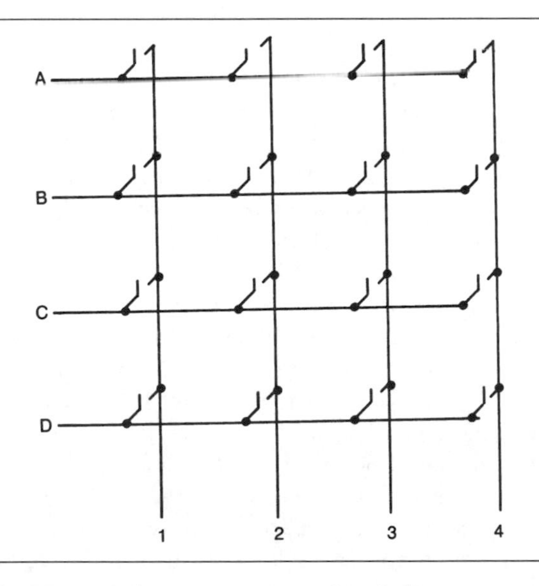

Figure 12.16 The matrix arrangement of switches on a typical keypad or keyboard.

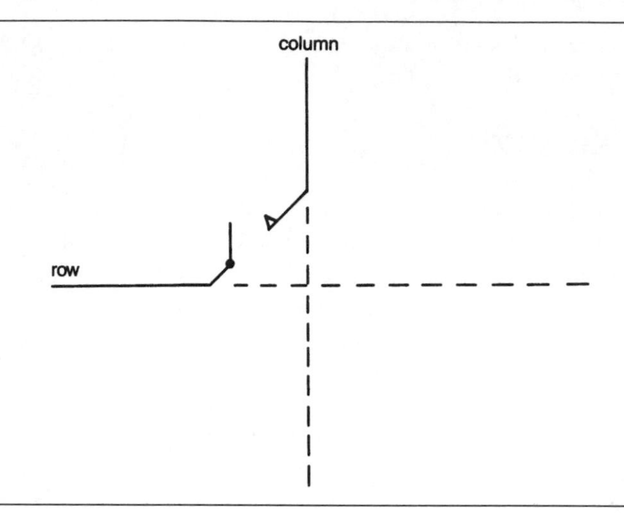

Figure 12.17 Each switch makes a contact between one column and one row. The matrix of rows and columns must be connected to a decoder to convert each keypress into a signal on a separate line.

few keyboards depart from the calculator or telephone pad style for numeric, or the typewriter arrangement for alphanumeric use.

Because of the matrix connection, the closure of a key does not have the effect of a conventional switch in causing connection between two points that are unique to that switch. In a switch matrix, each row or column wire will be used by more than one switch, and the closure of one switch will cause a connection between one row and one column (see Figure 12.17). This particular combination will often be unique to that switch, although on some arrangements the simultaneous pressing of two keys can produce a false output: for example, on an old PC keyboard, careless pressing of OM can produce OM7. A matrix keyboard or keypad is therefore of little use in its natural state and needs to be decoded, in the sense that combinations of connections between rows and columns have to be translated into signals that are related to the keys.

The decoding is usually carried out in the keyboard itself, using a matrix IC, and when this is done the keyboard is sometimes described as decoded, intelligent or smart. Using decoding, or at least partial decoding, at the keyboard allows the connection to the computer to be made using a simple serial link. This is why it is so easy to convert a keyboard termination between the usual choice of Serial or PS/2 by means of a simple adapter, since both use the same signal methods. Now that keyboards use at least 102 keys, this serial connection is essential to avoid bulky cables between the keyboard and the computer. A later form of serial connection, USB (Universal Serial Bus) is now becoming common for keyboard and other connections.

- When the keyboard or keypad is part of a computer system, the decoding will invariably be carried out by software, and this software will include the key debouncing circuits and provision for making the key action repeat if the key is held down.

Where a keypad is used in equipment that does not include a microprocessor (now rapidly becoming an endangered species) a decoder IC can be used. The use of such ICs provides much more than simple key detection, and it would be difficult to justify the use of any other method of decoding. Just to give one example, even a low-cost decoder could be expected to provide effective debouncing and also key *roll-over*. This latter term means that if two keys are struck in rapid succession, the output of the decoder will be the signal for the first key struck, followed at an interval by the signal for the second; simple decoders will usually give a false signal for such an event. In addition, the decoder will usually incorporate a latch circuit that stores the result of a key closure even after the key has been released. This allows the main system to poll the decoder at intervals that are not synchronized with the pressing of the keys.

12.12 VHF and UHF switching

The switching of VHF and UHF signals presents problems that do not arise in the consideration of other switches. The frequency of the signals implies that even very small amounts of stray capacitance can be significant, and the inductance of the switch components can lead to the switch having a resonant frequency which may be in the VHF or UHF range. In addition, the switch will normally be located in a line whose impedance will be of the order of 50 Ω, and the switch should also have this impedance value in order to avoid mismatching on the line. Very simple designs exist and are used, for example, for switching the input to a TV receiver between an aerial source and a TV game source. These switches are not intended for anything other than casual domestic use, however, and have large losses. The better quality switches for this frequency range are of coaxial construction so that matching can be maintained, and the operation is more likely to be by solenoid. This permits remote control, so avoiding hand capacitance effects.

For switching signals at low level that are not on a line, mechanical switching is seldom useful, and has been replaced by the use of low-capacitance diodes, usually of the Schottky type. The bias on these diodes is taken through a well-decoupled connection to a remote switch, so that the main switching action is simply a low-voltage DC one. The design of switching for UHF/VHF signals at high-power levels is a very specialized matter and usually calls for a one-off solution rather than any form of off-the-shelf switchgear.

Appendix A

Suppliers of sensors and transducers

Name	Address	Telephone (Int'l)
Alcyon Electronique	Beynes, France	+33 1 34947700
AMS OptoTech GmbH	Martinsried, Germany	+49 89 895770
Assemtech Europe Ltd.	Thorpe-le-Soken, Essex, UK	+44 1255 862236
Balluff Inc.	Florence, KY, USA	+1 606 7272200
Bokam Engineering Inc.	Santa Ana, CA, USA	+1 714 5132200
Bourns Electronics Ltd.	Camberley, Surrey, UK	+44 1276 692392
Cal Switch	Carson, CA, USA	+1 310 6324300
Carel Components	London, UK	+44 181 9469882
Cherry Mikroschalter GmbH	Auerbach/Opf., Germany	+49 9643 180
City Technology Ltd.	Portsmouth, Hants, UK	+44 1705 325511
Clarostat	Richardson, TX, USA	+1 972 4791122
Compact Air Products	Westminster, SC, USA	+1 864 6479521
Contelec AG	Biel, Switzerland	+41 32 3665600
Control Transducers Ltd.	Bedford, UK	+44 1234 217704
Crouzet Automatismes S.A.	Valence, France	+33 475 448123
CTL Components plc	London, UK	+44 181 5430911
CUI Stack	Beaverton, OR, USA	+1 503 6434899
Dawn Electronics	Carson City, NV, USA	+1 702 8827721
Dräger Medizintechnik GmbH	Lübeck, Germany	+49 451 8824455
Dynamics Research Corp., Metrigraphics Div.	Wilmington, MA, USA	+1 978 6586100
E-T-A Circuit Breakers Ltd.	Aylesbury, Bucks, UK	+44 1296 420336
Elmwood Sensors Ltd.	North Shields, Tyne & Wear, UK	+44 191 2582821

Eltrotec Elektro-GmbH	Adelberg, Germany	+49 7166 50300
ENM Co.	Chicago, IL, USA	+1 773 7758400
Etalon Inc.	Lebanon, IN, USA	+1 765 4832550
Ferroperm a/s, Piezoceramics Div.	Kuistgaard, Denmark	+45 49 127100
Furness Controls Ltd.	Bexhill, E Sussex, UK	+44 1424 730316
Glas Platz	Reichshof-Allenbach, Germany	+49 2261 55557
Hartmann & Braun	Sensycon, Alzenau, Germany	+49 6023 923364
Herga Schaltsysteme GmbH	Eltville, Germany	+49 6123 4068
Honeywell, Sensing and Control Div.	Newhouse, Strathclyde, UK	+44 1698 481297
Imasonic	Besançon, France	+33 381 403130
In Vivo Metric	Healdsburg, CA, USA	+1 707 4334819
Industrial Control Solutions Ltd.	Worksop, Notts, UK	+44 1909 501188
Interface	Scottsdale, AZ, USA	+1 602 9485555
International Electronics & Engineering (IEE)	Findel, Luxembourg	+352 42 4737318
Introtek International	Edgewood, NY, USA	+1 516 2425425
Kistler Instrumente AG	Winterthur, Switzerland	+41 52 2241111
Lascar Electronics Ltd.	Salisbury, Wilts, UK	+44 1794 884567
LEM S.A.	Geneva, Switzerland	+41 22 7061111
Lightwave Medical Industries Ltd.	Vancouver, BC, Canada	+1 604 8754529
Ludlow Technical Products	Godalming, Surrey, UK	+44 1483 426986
Mecmesin Ltd.	Broadbridge Heath, W Sussex, UK	+44 1403 256276
Medical Measurement Systems	Enschede, Netherlands	+31 53 4308803
MESA Systemtechnik GmbH	Konstanz, Germany	+49 7531 93710
Metrigraphics Div. Dynamics Research	Wilmington, MA, USA	+1 978 6586100
Minco EC AG	Niederuzwil, Switzerland	+41 71 9527989
Minco Products	Minneapolis, MN, USA	+1 612 5713121
Minnesota Wire & Cable Co.	St. Paul, MN, USA	+1 651 6421800
Monitran Ltd.	Penn, Bucks, UK	+44 1494 816569
Motorola Semiconducteurs S.A.	Toulouse, France	+33 561 199070
Mountz Inc.	San Jose, CA, USA	+1 408 2922214
Neuricam S.p.A.	Trento, Italy	+39 0461 260552
Oesch Sensor Technology	Sargans, Switzerland	+41 81 7230422
Optel-Thevon	Montreuil, France	+33 1 48575833
Optolink Ltd.	Washington, Tyne Wear, UK	+44 191 4166546

Penny + Giles Controls Ltd.	Christchurch, Dorset, UK	+44 1202 409409
Phelps Dodge High Performance Conductors	Puurs, Antwerp, Belgium	+32 3 8609191
Photek Ltd.	St. Leonards on Sea, UK	+44 1424 850555
PI Ceramic GmbH	Lederhose, Germany	+49 3660 48820
Poly-Flex Circuits Ltd.	Newport, Isle of Wight, UK	+44 1983 526535
Roithner Lasertechnik	Vienna, Austria	+43 1 5865243
Sachs (Hugo Sachs Elektronik)	March-Hugstetten, Germany	+49 7665 92000
Select Engineering Inc.	Fitchburg, MA, USA	+1 978 3454400
SensLab GmbH	Leipzig, Germany	+49 341 2354001
SensoNor A/S	Horten, Norway	+47 33035141
Sensor Devices GmbH	Dortmund, Germany	+49 231 9144740
Sensor Products Inc.	East Hanover, NJ, USA	+1 973 8841755
Sensor Scientific Inc.	Fairfield, NJ, USA	+1 973 2277790
Sensors Testsysteme GmbH	Ratingen, Germany	+49 2102 856800
Sensortechnics GmbH	Puchheim, Germany	+49 89 800830
SenSym Foxboro ICT	Milpitas, CA, USA	+1 408 9546700
Sentronic GmbH	Dresden, Germany	+49 351 8718654
Servomex International Ltd.	Crowborough, E Sussex, UK	+44 1892 652277
Spectrol Electronics Ltd.	Swindon, Wilts, UK	+44 1793 521351
SRT Resistor Technology GmbH	Cadolzburg, Germany	+49 9103 79520
Steiger Laser Consult	Gernlinden, Germany	+49 8142 12055
Tedea-Huntleigh USA	Canoga Park, CA, USA	+1 818 8846860
Terwin Instruments Ltd.	Grantham, Lincs, UK	+44 1476 565797
Texas Instruments	Attleboro, MA, USA	+1 508 2363287
Top Sensor Systems	Eerbeek, Netherlands	+31 313 670160
Vaisala Oy	Helsinki, Finland	+358 9 89491
Vector Technology Ltd.	Abertillery, Gwent, UK	+44 1495 320222
Victory Engineering Corp.	Springfield, NJ, USA	+1 973 3795900
Warner Electric	South Beloit, IL, USA	+1 815 3893771
Watlow Electric Manufacturing Co.	St. Louis, MO, USA	+1 314 8784600
Wenglor Sensoric GmbH	Tettnang, Germany	+49 7542 5990
ZEVEX Inc.	Salt Lake City, UT, USA	+1 801 2641001

You can also use any Internet search engine to look for websites relating to these companies or to specialized equipment.

Appendix B

Glossary of terms

Absolute pressure transducer A transducer that compares applied pressure to the pressure in an internal cavity, held close to zero pressure. The output of the transducer is a voltage that increases with increasing applied pressure.

Accuracy The error figure that takes account of the factors of hysteresis, non-linearity, and repeatability. This is expressed as a percent of the full-scale output.

Axial load Any load whose line of force is along the main axis, or parallel to that axis.

Bandwidth See *Frequency response*.

Bridge A circuit consisting of four impedance elements in diamond formation, such as the Wheatstone Bridge for resistance measurement.

Bridge resistance The value of the individual resistors or impedances that make up a complete Bridge circuit.

Calibration The comparison of outputs from a sensor or transducer with the outputs of a reference standard.

Damping The reduction of oscillation, typically at a resonant frequency. Damping can be mechanical, using a medium such as oil, or electrical, using a resistor to dissipate oscillatory energy.

Dead volume The volume of any space whose characteristics are not intended to be measured, such as the internal volume of a transducer used for pressure measurement.

Deflection The movement of a pointer or diaphragm from its rest position to or from its loaded condition.

Diaphragm A membrane which is deflected from its rest position when pressure is applied.

Electrical excitation Any voltage or current that needs to be connected to the input terminals of a transducer.

Frequency response The range of frequencies for which the output of a transducer will follow the varying input. Also known as the *Bandwidth.*

Full scale The maximum response that a sensor or transducer can produce for the maximum permitted input. Also called *Rated capacity.*

Gauge pressure The pressure relative to atmospheric pressure. Absolute pressure is gauge pressure plus atmospheric pressure.

Hysteresis The difference between readings of the same nominal quantity when one reading is made with the quantity increasing and the other with the quantity decreasing.

Input impedance The impedance (resistance and reactance) measured across the terminals of a sensor or transducer at room temperature. Conventionally this is done with no load applied, and with the output terminals open-circuited.

Insulation resistance The DC resistance, usually in MΩ, measured between any electrical connecting pin and the transducer body.

Linearity The maximum difference between the calibration curve for a transducer and a straight line drawn between zero and full scale. This is usually expressed as a percent of full-scale output.

Line pressure The maximum pressure encountered in a pipe or pressure vessel.

Load The force or torque applied to a sensor or transducer.

Load button The point of application of a force on a load cell.

Measured medium Any physical quantity that is measured.

Mounted resonant frequency The resonant frequency of a transducer when it is connected and loaded.

Output The electrical signal at the output terminals of a sensor or transducer when an input is applied.

Output impedance The impedance (resistance plus reactance) measured at the output terminals of a sensor or transducer at room temperature. This measurement is usually made with no input and with the input terminals open-circuited.

Phase shift The phase angle between the output signal and the input signal of a sensor or transducer.

Primary axis The main load axis of a sensor or transducer. This is normally its geometric centreline.

Psi Abbreviation of pounds per square inch, old unit of pressure.

Psia Abbreviation of pounds per square inch absolute.

Psid Abbreviation of pounds per square inch differential.

Psig Abbreviation of pounds per square inch gauge.

Pull plate An attachment to a load cell designed to connect the cell so that forces are applied along the centreline.

Range The lower and upper limits of values that a transducer is intended to measure.

Rated capacity See *Full scale*.

Repeatability The ability of a sensor or transducer to provide the same output for the same input, irrespective of increasing or decreasing input or other conditions.

Resolution The smallest measurable change in input.

Safe overrange The maximum input that can be applied to the sensor or transducer without causing damage.

Sensing element The part of a sensor or transducer that is affected by an input.

Sensitivity For any sensor or transducer, the ratio of change in output to a change in the value of the input.

Strain gauge A measuring device whose input is force and whose output is an electrical signal.

Temperature compensation The use of circuits or components that reduce measuring errors due to changes of temperature.

Transducer Any device that converts energy from one form to another.

Transverse sensitivity Signal output as a result of an input directed perpendicular to the normal axis.

Vibration error The maximum change in the output of a sensor or transducer due to mechanical vibration.

Wetted parts The portions of a sensor or transducer that come into contact with the gas or liquid whose properties are being measured.

Zero adjustment Any device used to adjust the output signal of a sensor or transducer to zero when zero input is applied.

Index